2nd International Symposium on the Modern Physics of Compact Stars and Relativistic Gravity 2013

Journal of Physics: Conference Series Volume 496

Yerevan, Armenia
18-21 September 2013

ISBN: 978-1-63266-721-2
ISSN: 1742-6588

Printed from e-media with permission by:

Curran Associates, Inc.
57 Morehouse Lane
Red Hook, NY 12571

Some format issues inherent in the e-media version may also appear in this print version.

Printed by Curran Associates, Inc. (2014)

Institute of Physics
Dirac House, Temple Back
Bristol BS1 6BE UK

Phone: 44 1 17 929 7481
Fax: 44 1 17 920 0979

techtracking@iop.org

Additional copies of this publication are available from:

Curran Associates, Inc.
57 Morehouse Lane
Red Hook, NY 12571 USA
Phone: 845-758-0400
Fax: 845-758-2634
Email: curran@proceedings.com
Web: www.proceedings.com

TABLE OF CONTENTS

Preface

The international conference "The Modern Physics of Compact Stars and Relativistic Gravity" took place in Yerevan, Armenia, from 18–21 September 2013. This was the second in a series of conferences which aim to bring together people working in astrophysics of compact stars, physics of dense matter, gravitation and cosmology, observations of pulsars and binary neutron stars and related fields. The conference was held on the occasion of 100th birthday of the founder of the Theoretical Physics Chair at the Department of Physics of Yerevan State University and prominent Armenian scientist Academician Gurgen S Sahakyan.

The field of compact stars has seen extraordinary development since the discovery of pulsars in 1967. Even before this discovery, pioneering work of a number of theoretical groups had laid the foundation for this development. A pioneer of this effort was Professor G S Sahakyan who, together with Professor Victor Ambartsumyan and a group of young scientists, started in the early sixties their fundamental work on the properties of superdense matter and on the relativistic structure of compact stellar objects. This conference explored the vast diversity of the manifestations of compact stars, including the modern aspects of the equation of state of superdense matter, its magnetic and thermal properties, rotational dynamics, superfluidity and superconductivity, phase transition from hadronic to quark matter, etc. The articles on these subjects collected in this volume are evidence of liveliness of the field and of the continuous feedback between theory and the experiment. A part of this volume is devoted to the cosmology and the theories of gravity — the subfields of astrophysics that are of fundamental importance to our understanding of the universe. The reader will find here articles touching on the most diverse aspects of these fields such as modern problems in Einstein's classical theory of gravity and its alternatives, string theory motivated cosmological models, theories of cosmic microwave background, quantum field theory in curved background, Casimir effect, etc. Thus, it is fair to say that the present volume covers a large number of actively pursued subjects of modern relativistic astrophysics.

We would like to thank all those individuals and organizations that helped us in organizing a successful conference. These include the members of the international advisory committee as well as the local organizing committee (whose names are listed separately above) and the sponsors of the conference — the Volkswagen Stiftung (Hannover, Germany), the HIC for FAIR Institute (Frankfurt am Main, Germany) and the State Committee for Sciences of Armenia.

Edvard Chubaryan, Aram Saharian and Armen Sedrakian
Yerevan / Frankfurt-Main, 1 February, 2014

List of participants

Ahmad Raji, Amid	Islamic Azad University of Tehran
Alaverdyan, Grigor	Yerevan State University
Alvarez-Castillo, David E	JINR, Dubna
Baiko, Denis	A F Ioffe Physical Technical Institute
Barsukov, Dmitriy	A F Ioffe Physical Technical Institute
Blaschke, David	University of Wroclaw
Brillante, Alessandro	Goethe University, Frankfurt
Campos, Miguel	Roraima Federal University
Colucci, Giuseppe	Goethe University, Frankfurt
Darehmoradi, Mohammad	TSML, Teheran
Dieperink, Lex	KVI, Groningen
Evnin, Oleg	Chulalongkorn University
Evslin, Jarah Markar	IHEP, Chinese Academy of Sciences
Gevorkyan, Ashot	Institute for Informatics and Automation Problems NAS of Armenia
Goglichidze, Oleg	A F Ioffe Physical Technical Institute
Grigorian, Hovik	Yerevan State University
Grigoryan, Hovhannes	New York University
Grigoryan, Levon	Institute of Applied Problems in Physics
Hajyan, Gevorg	Yerevan State University
Hambaryan, Valeri	AIU, University Jena
Harutyunyan, Gohar	Yerevan State University
Hayrapetyan, Mekhak	Yerevan State University
Katanaev, Mikhail	Steklov Mathematical Institute
Khudaverdian, Hovhannes	University of Manchester
Kluzniak, Wlodek	Copernicus Astronomical Center
Kotanjyan, Anna	Yerevan State University
Margaryan, Amur	Alikhanyan National Science Laboratory
Mekhitarian, Vahram	Institute for Physical Research, National Academy of Science of Armenia
Mkrtchyan, Karapet	Scuola Normale Superiore, Pisa
Pienkos, Monika	University of Silesia
Piloyan, Arpine	Yerevan State University
Pogosian, Levon	Simon Fraser University
Rezzolla, Luciano	Goethe University, Frankfurt
Rischke, Dirk	Goethe University, Frankfurt
Roepke, Gerd	Institut fuer Physik, Uni Rostock
Saharian, Aram	Yerevan State University
Sargsian, Misak	Florida International University
Sedrakian, Armen	Goethe University, Frankfurt
Sedrakian, David	Yerevan State University
Shahabasyan, Karen	Yerevan State University
Sinha, Monika	Goethe University, Frankfurt
Stein, Martin	Goethe University, Frankfurt
Van Eysden, Anthony	NORDITA
Vartanyan, Yuri	Yerevan State University

Committees and supporting organizations

International advisory board
M Alford (St Louis)
E R Bezerra de Mello (Paraiba)
G S Bisnovaty-Kogan (Moscow)
F Burgio (Catania)
K Kokkotas (Tbingen)
D Lai (Cornell)
B Link (Montana)
J Margueron (Orsay)
C Pethick (Copenhagen)
M Oertel (Meudon)
L Rezzolla (Potsdam)
D Rischke (Frankfurt Main)
M Sargsian (Miami)
F Weber (San Diego)
D N Yakovlev (St Petersburg)

Organizing committee
R Avagyan
E Chubaryan
A Saharian
A Sedrakian
A Yeranyan
G Colucci (secretary)

VolkswagenStiftung

HIC for FAIR
Helmholtz International Center

STATE COMMITTEE
OF SCIENCE OF RA

Journal of Physics: Conference Series **496** (2014) 011003 doi:10.1088/1742-6596/496/1/011003

The founder and head of the Chair of Theoretical Physics of the Yerevan State University

L Sh Grigoryan[1]

Institute of Applied Problems in Physics, 25 Hr. Nersessian str., 0014 Yerevan, Armenia

E-mail: levonshg@mail.ru

Abstract. The paper is dedicated to the Centenary of an Academician of NAS RA, Professor G.S. Sahakyan's birth, the Man that founded and headed the Chair of Theoretical Physics (CTP) of the Yerevan State University for almost half a century. The reference to school days of G.S. Sahakyan is made, information about his 7 years long service in the forces in the fields, about the establishment and administration by him of the Chair of Theoretical Physics in the Yerevan State University, about his collaboration with academician V.A. Ambartsumian, about the research associates of the G.S. Sahakyan's Chair, the students of CTP and the advancement of theoretical physics in Armenia is given. The personality characteristics of G.S. Sahakyan as a principal investigator and leader of CTP are analyzed.

1. The chronological sequence and facts of activities

Sahakyan Gurgen Serob (born on 10.09.1913, in Sarnaghbyur Village of Armenian SSR – deceased on 26.03.2000, in Yerevan, the Republic of Armenia), was an Armenian - Soviet theoretician physicist, the founder of gravitation theory and relativistic astrophysics schools in Armenia, that made considerable contribution to the development of theoretical physics and training of young specialists in physics as well as the development of scientific brainpower in Armenia, defended the Thesis of the Candidate of Physical-Mathematical Sciences (1950, Moscow), of the Doctor of Physical-Mathematical Sciences (1963, Moscow), assigned the Professorship (1964, Yerevan), Honored Man of Science of Armenian SSR (1970), a correspondent Member of AS Arm. SSR (1965), an Academician of AS of Arm. SSR (1982), the founder and Head of Theoretical Physic Chair of the Yerevan State University (1951-1985), the Dean of the Physics Faculty of YSU (1967-1972), Manager of the theoretical laboratory associated with the Chair of Theoretical Physics and leading scientist (1989-2000). Mr. Sahakyan participated in the Finnish and the Great Patriotic Wars (1939-1945), was awarded with Orders of Red Star and of the Friendship of Peoples.

At the beginning of 1950 Mr. Sahakyan was the first to introduce the concept of nucleon form factor to the science literature and proposed to determine the structure of this particle by means of investigation of the features of large-angle elastic scattering of ultrarelativistic electrons on the nucleons.

Mr. Sahakyan made weighty contribution to the development of the physics of neutron stars. Along with his followers (under his guidance) he had elaborated the theory of degenerated stellar matter consisting of elementary particles, developed the models of neutron stars consisting of

[1] To whom any correspondence should be addressed.

degenerated matter. These investigations were based on pioneering works of Mr. Sahakyan conducted in collaboration with Academician V.A. Ambartsumian at the beginning of the sixties of the Twentieth Century [2-4]. In [2-4] the existence of superdense stars consisting of a gas of elementary particles (nucleons, hyperons, resonances, mesons and leptons) was substantiated for the first time, a new effect of the General Relativity (the anomalous defect of mass) was discovered.

Photo 1.

The group of scientists headed by Mr. Sahakyan substantiated the possibility of forming a sufficiently dense disk-shaped plasma magnetosphere around the rotating neutron stars and showed that the external boundary of the magnetosphere may emit short-wave cosmic radiation.

Mr. Sahakyan has developed the theory of degenerated plasma at densities of the order, as well as those higher and lower than the nuclear density, in which the contribution of negatively charged pionic component of the nuclear matter is consistently taken into account. He made a substantial contribution to investigations of external manifestations of neutron stars (the pulsars, bursters, clusters of neutron stars and of white dwarfs in compact galactic nuclei) and, in particular, to the solution of the problem of radiation from pulsars. Mr. Sahakyan made essential contribution to the development of relativistic theory with variable gravitational constant.

Mr. Sahakyan is the author of a number of monographs and textbooks [1,5-10], issued in Armenian [7], Russian [1,5-10], English [6], German [5] and French [5] languages. He managed the works of the III (1972) and the VII (1988) Soviet Conferences on the theory of Relativity and Gravitation and of All-Union Symposium (1985) in Yerevan-City, was the member of Scientific-and-Technical Council of the Ministry of Higher and Secondary Education of the USSR. Under his research supervision numerous scientists from Armenia, Ukraine and Czechoslovakia had defended their Candidate and Doctor theses.

The bronze bust of Mr. G.S. Sahakyan set up in the foyer of the central building of YSU symbolizes the outstanding services of the scientist for the development of Science and Education in the leading institute of higher education of the Republic of Armenia.

2nd International Symposium on the Modern Physics of Compact Stars and Relativistic Gravity IOP Publishing
Journal of Physics: Conference Series **496** (2014) 011003 doi:10.1088/1742-6596/496/1/011003

2. An exceptional destiny

If you can fill the unforgiving minute
With sixty minutes worth of distance run,
Yours is the Earth and everything that's in it.
R. Kipling

In the month of September, 1913, a boy was born in the village named "Cold Spring", in the family of Serob Sahakyan that named him Gurgen. Many years since then in the mark of respect in the Russian manner he was referred to as Gurgen Serobovitch (G.S.). At first he studied in the Secondary school and then he attended a Technical Training School to be a builder, and at last also the physical-mathematical faculty of YSU (1934-1939). The destiny of Gurgen was unusual. He served in the forces in the field during 7 years. First he was in action during the Finnish War (1939–1941), then in action all through the World War II (1941–1945). He served in the Army in Belarus, Poland and Austria. The diaries of Mr. Sahakyan with records made in the forces in the field were retained. The records were open up in Leningrad on December 13, 1939 (figure 2) and accomplished in German on December 3, 1945 (figure 3).

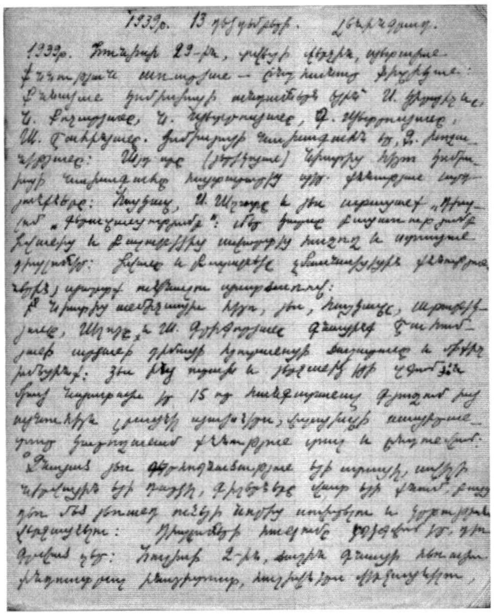

Figure 2. The first record in the war diary of Sahakyan G.S. in Leningrad on December 13, 1939.

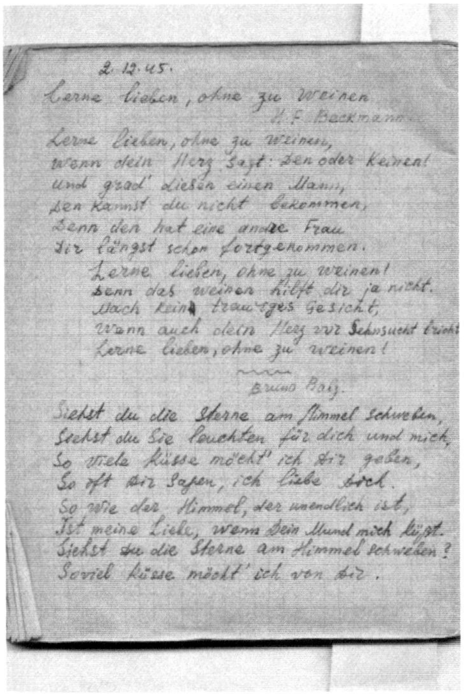
2.12.45.
Lerne lieben, ohne zu weinen
H. F. Beckmann
Lerne lieben, ohne zu weinen,
wenn dein Herz sagt: den oder keinen!
und grad' lieben einen Mann,
den kannst du nicht bekommen,
denn den hat eine andre Frau
dir längst schon fortgenommen.
Lerne lieben, ohne zu weinen!
denn das weinen hilft dir ja nicht.
Mach kein traurges Gesicht,
wenn auch dein Herz vor Sehnsucht bricht
Lerne lieben, ohne zu weinen!
Bruno Balz.
Siehst du die Sterne am Himmel schweben,
Siehst du sie leuchten für dich und mich,
So viele Küsse möcht' ich dir geben,
So oft dir sagen, ich liebe dich.
So wie der Himmel, der unendlich ist,
Ist meine Liebe, wenn dein Mund mich küßt.
Siehst du die Sterne am Himmel schweben?
Soviel Küsse möcht' ich von dir.

Figure 3. The last record in the war diary of Sahakyan G.S. made in German on December 3, 1945.

The pages of Mr. G.S. Sahakyan's diary with notes on physics have been retained. In figure 4 one of these pages is shown with notes on the Special Theory of Relativity. The records were made on March 26, 1943. At that time sub-lieutenant Sahakyan G.S. was a technician of high-performance equipment. The document issued to him after 50 days (figure 5) witnesses that.

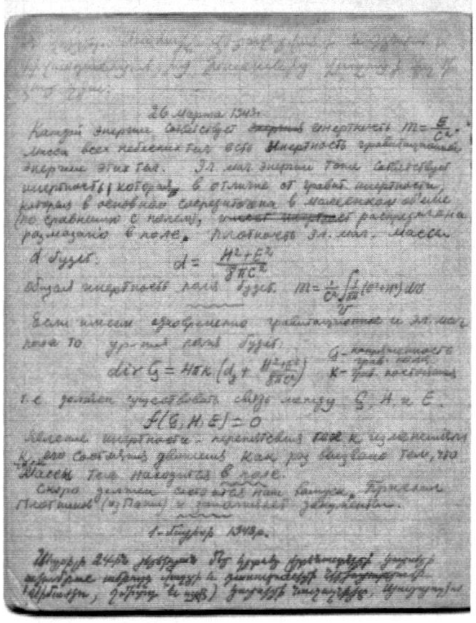

Figure 4. A synopsis on the Special Relativity made by Sahakyan G.S. on the front. The date of record was March 26, 1943.

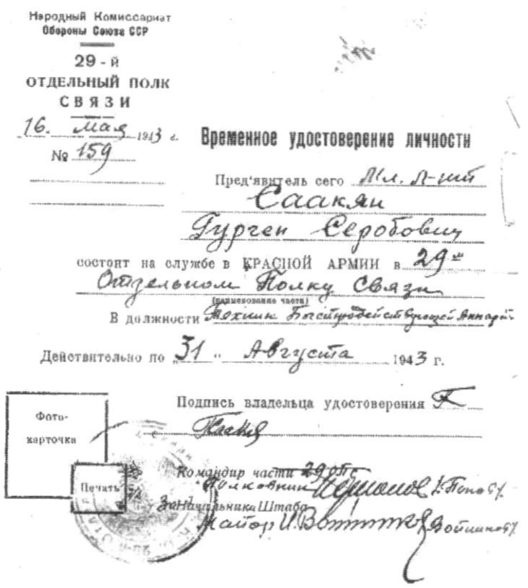

Народный Комиссариат
Обороны Союза ССР

29-й
ОТДЕЛЬНЫЙ ПОЛК
СВЯЗИ

16. мая 1943 г.
№ 159

Временное удостоверение личности

Пред'явитель сего Мл. л-нт
Саакян
Гурген Серобович
состоит на службе в КРАСНОЙ АРМИИ в 29-м
Отдельном Полку Связи
(наименование части)
В должности [illegible]

Действительно по 31 Августа 1943 г.

Фото-карточки

Подпись владельца удостоверения [illegible]

Печать

Командир части [illegible]

Начальник Штаба
Майор [illegible]

Figure 5. The entry on this document witnesses that sub-lieutenant Sahakyan G.S. is a service technician of high-performance equipment. The document was issued on May 16, 1943.

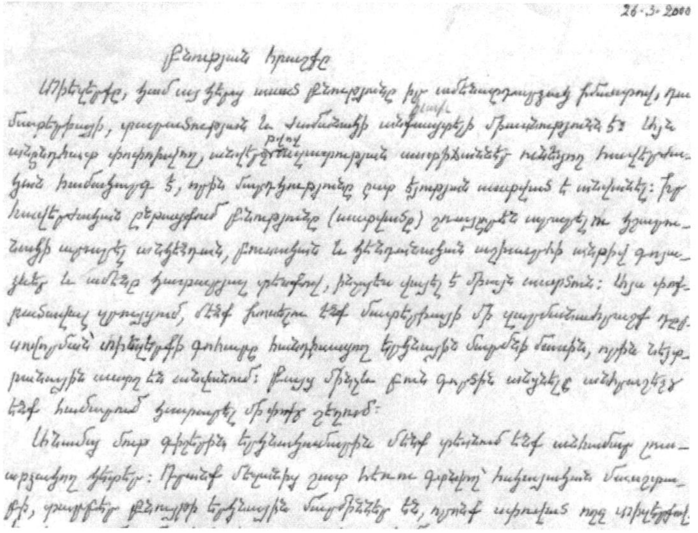

Figure 6.
A fragment of incomplete synopsis of G.S. dated from March 26, 2000. On that day he died.

The unit commander once could not contain himself any longer and told him with irritation:- "Deuce take it! We are here to fight. And every day some of us die, but you are unable to tear yourself away from your physics. Can't you die without this physics?" Such was G.S. at the battle-front and generally everywhere and always.

Gurgen Sahakyan was an unusual post-graduate student. He was enlisted to the post-graduate student of the Physical Institute after P.N. Lebedev (FIAS USSR) in Moscow. The co-supervisors of his post-graduate studies have been the famous physicists D.I. Blokhintsev (Correspondent Member of AS USSR) and I.E. Tamm (Academician of AS USSR announced later on a Nobel Prize Winner).

G.S. was an exceptional man. In figure 6 a fragment of his records on the theory of radio-frequency emission of pulsars is shown. It is unusual by two reasons. First, on that day the Death unexpectedly intervened in his affairs (please, take notice of the date of entries – March 26, 2000) and interrupted the statement of a research. And, second, on those days G.S. was neither more, nor less than 87 years old. We see that G.S. *was faithful to science literally up to the end of his life.*

3. An extraordinary scientific leader

G.S. founded the Chair of Theoretical Physics of YSU in 1951 and has headed that during 34 years. In 1985 he resigned the guidance of Chair to his follower Prof. E.V. Chubaryan that is now an Academician of NAS RA. After that G.S. has been the status leader and the principal scientific associate of CTP for long 15 years (up to the end of his life). In 1972 G.S. was 59 years of age, when he issued his basic monograph [6]. It made him world-famous among the scientists involved in research activities in the field of neutron star physics and it was translated into English and was issued abroad in 2 years period, that was very short in those days. In this monograph a systematic statement of the theory of superdense celestial bodies is given for the first time. It is still widely used by physicists engaged in the research of neutron stars and white dwarfs even at present – after 40 years since its publication.

G.S. achieved *wondrous collaboration* with a distinguished astrophysicist of the XX-th century,- Academician of the Academy of Sciences of USSR, Professor V.A. Ambartsumian.

Photo 7.
Prof. V.A. Ambartsumian, Academician of AS USSR, The founder and Director of the Byurakan Astrophysical Observatory (1946-1988), the President of AS Arm. SSR (1947-1993), a distinguished astrophysicist of the XX-th century.

The collaboration continued for a quarter of century (1959-1985) it being known that V.A and G.S. have jointly published just a few research works (e.g., [2-5]). Though after 1980 there were no joint publications, *the collaboration stayed active and fruitful.* This collaboration made for the publication of the mentioned monograph [6] by G.S. that was known to world-wide scientific community, firstly, and, secondly, for the organization and conduction in Armenia of two Soviet Conferences on GR and

Journal of Physics: Conference Series **496** (2014) 011003 doi:10.1088/1742-6596/496/1/011003

relativistic gravity in 1972 and 1988, and of All-Union symposium in 1985. G.S. supervised the works of both the conferences and symposium. Owing to the collaboration between V.A. and G.S. the CTP headed by G.S. became the leading institute in the USSR on the theory of superdense celestial bodies and gravitation.

G.S. was an extraordinary leader. He supported and encouraged the gifted (hard-working and giving promise) scientific associates. His opinion was always attentively considered, given credence and trusted. But how he achieved that? Maybe because he was extremely obsessed with the science and endlessly devoted to that? Maybe because he worked almost 15 hours a day? Probably because he was always ready to share his ideas and apprehend the ideas of his colleagues. G.S. was capable of initiating an enthusiasm in associates and that brought its fruits.

4. The followers of G.S.

G.S. was the research supervisor or the adviser of approximately 15 Candidate and Doctor Theses of scientists from Armenia, Ukraine, Czeskoslovakia and other countries. In future they proved to be known scientists in the science community of Armenia. The first followers of G.S. were his associates in the Chair:

1. Prof. Edward V. Chubaryan – Honorable Head of the Academician Gurgen Sahakyan's Chair of Theoretical Physics of YSU, Academician of NAS RA
2. Prof. David M. Sedrakian - Head of the Chair of YSU: Academician Victor Ambartsumian's Chair of General Physics and Astrophysics, Academician of NAS RA, Vice-President of the Academy of Sciences of Arm. SSR (1990-1994), Editor of the journal "Astrofizika"
3. Prof. Yuri L. Vardanyan - Dean of the Faculty of Radiophysics of YSU, Corresponding Member of NAS RA
4. Prof. Roland M. Avagyan - Dean of the Faculty of Physics of YSU, Corresponding Member of NAS RA
5. Prof. Vladimir Vl. Papoyan, YSU (17.10.1937-06.05.2004)
6. Prof. Gohar G. Arutyunyan - acting Head of the Academician Gurgen Sahakyan's Chair of Theoretical Physics of YSU
7. Associate Prof. Ashot V. Hovsepyan – YSU.

The last follower of G.S. proved to be

8. Prof. Aram A. Saharian –YSU, Regular Associate, Abdus Salam International Centre for Theoretical Physics, Trieste, Italy.

5. The students of G.S.

G.S. has guided CTP during 34 years (from 1951 to 1985) and was a Dean of the Physics Faculty of YSU in 1967-1972. In that period approximately 3000 students graduated from the Faculty of Physics of YSU, including several hundred theoretical physicists. For the Republic of Armenia these figures were spectacular. Moreover, in 1969-1976 some students from the German Democratic Republic have been taught at the Chair headed by G.S. There were three turn-outs of these students (see photo 8).

These graduates from the Faculty of Physics made weighty contribution to the development of physics in general and especially of theoretical physics in Armenia. Nowadays they successfully work not only in Armenia, but also abroad: Russia, England, Germany, USA, Israel, Italy, Greece, Guinea and other countries.

6. The authority of G.S.

G.S. founded the Chair of Theoretical Physics of YSU in 1951 and has headed that during 34 years. He was a Dean of the Physics Faculty of YSU. He organized and conducted two Soviet Conferences and one All-Union symposium. The CTP headed by G.S. became the leading institute in the USSR on the theory of superdense celestial bodies and gravitation. He was the author of 2 text-books [7,8], 4 monographs and collected articles [1,5,9,10] and of nearly 150 science publications. The activities of

G.S. in different fields: physics of elementary particles, physics of superdense plasma, physics of neutron stars, relativistic gravity, have been invariably fruitful.

Photo 8. The first turn-out of students from Armenia and DDR that were jointly taught at CTP, together with their tutors (1974). **From left to right: (the upper row)** assistant Samvel Shaginayn, Lutz Schimansky-Geier, Stefan Gottloeber, Karen Aramyan, Roland Kirschner, Ian-Peter Muecket, Roland Schepke, Peter Medler, Tenner Bertram, Araik Sedrakyan, **(middle row)** Prof. David M. Sedrakyan, Prof. Gurgen S. Sahakyan, Svetlana Harutyunyan, Prof. Edward V. Chubaryan, assistant Karen Shahabasyan, assistant Araik Avetisyan, **(lower row)** Hrachya Zazyan, Haik Minasyan, Garnik Sargsyan, Gagik Melikyan, Levon Grigoryan, Georgi Savvidy.

G.S. was the only researcher from the Chair of Theoretical Physics that (a) being a post-graduate student he lived in one room with A.D. Sakharov (the world-famous scientist and human rights activist), (b) the research supervisors of which were world-known scientists D.I. Blokhintsev and I.E. Tamm, and that (c) fought on the front during 7 years. G.S. was older than any of his associates in CTP minimum by 20 years.

In photo 9 G.S. is seen in his study room. The free table in the room and a chair beside it were for the associates. During almost 40 years this seat was considered to be the most popular and important place at CTP for all the post-graduate students, all future candidates and even Doctors of Science that many years have been numerous times at this desk for hours of joint research efforts. But that is not the half of it. However early was an associate at this place, G.S. was already working at his desk waiting for him. However late did the associate leave the room, G.S would (most likely to) stay and continue to work.

There was some inscrutability natural to G.S. and he always commanded an esteem to his personality. It is maybe due to these features that his authority was indisputable for all the CTP associates, but maybe because G.S. treated his colleagues, or to be more exact his young colleagues, as his relatives. He usually addressed the colleagues (the students, post-graduates, young scientists and sometimes not only the young ones) with famous "My deary".

The students and colleagues of G.S. follow his example.

Photo 9. All associates of CTP remember G.S. Sahakyan just as it is seen in this photograph.

References

[1] Sahakyan G S 1958 *Energy spectra and nuclear interactions of cosmic ray particles* (YSU, Yerevan) (in Russian)

[2] Ambartsumian V A and Sahakyan G S 1960 *Astron. J, AS USSR* **37** 193 (in Russian)

[3] Ambartsumian V A and Sahakyan G S 1961 *Astron. J. AS USSR* **38** 785 (in Russian)

[4] Ambartsumian V A and Sahakyan G S 1961 *Astron. J. AS USSR* **38** 1016 (in Russian)

[5] Ambartsoumian V A, Mirzoyan L V, Saakyan G S, Vsekhsvyatskij S K and Kazyutinskij V V 1969 *Modern Problems of Ccosmogony* (Nauka, Moscow) (in Russian); 1971 (Mir, Moscou) (in French); 1972 Second revised and enlarged edition (Nauka, Moscow) (in Russian); 1976 (Birkhäuser Verlag, Basel und Stuttgart) (in German); 1980, Second edition (Akademie-Verlag, Berlin) (in German)

[6] Sahakian G S 1972 *Equilibrium Configurations of Degenerate Gaseous Masses* (Nauka, Moscow) (in Russian); 1974 (Wiley, New York) (in English)

[7] Sahakyan G S and Chubaryan E V 1972 *Quantum Mechanics* (Luys, Yerevan) (in Armenian); 1982 Second revised and enlarged edition (in Ruissian); 2009 Revised edition (YSU, Yerevan) (in Armenian)

[8] Sahakyan G S 1985 *Space-Time and Gravitation* (YSU, Yerevan) (in Russian)

[9] Sahakian G S 1995 *Physics of Neutron Stars* (Joint Institute for Nuclear Research, Dubna) (in Russian)

[10] Sahakian G S 1998 *Physics of Neutron Stars* - Second enlarged and revised edition (YSU, Yerevan) (in Russian)

Journal of Physics: Conference Series **496** (2014) 011004 doi:10.1088/1742-6596/496/1/011004

Peer review statement

All papers published in this volume of *Journal of Physics: Conference Series* have been peer reviewed through processes administered by the proceedings Editors. Reviews were conducted by expert referees to the professional and scientific standards expected of a proceedings journal published by IOP Publishing.

Journal of Physics: Conference Series **496** (2014) 012001 doi:10.1088/1742-6596/496/1/012001

Influence of phase-transition scenarios on the abrupt changes in the characteristics of compact stars

G B Alaverdyan

Yerevan State University, Alex Manoogyan str. 1, Yerevan, Armenia

E-mail: galaverdyan@ysu.am

Abstract. We study the abrupt changes in the characteristics of compact stars due to the quark deconfinement phase transition. The hadronic phase is described within the relativistic mean-field theory, including a scalar-isovector δ-meson effective field. To describe the quark phase, we use the MIT bag model, in which the interactions between u, d and s quarks inside the bag are taken into account in the one-gluon exchange approximation. We analyze catastrophic changes of the parameters of the near-critical configuration of compact star and compute the amount of the energy released by a corequake for the two extreme cases of the deconfinement phase transition scenarios. The first one corresponds to the ordinary first-order phase transition (Maxwell scenario) and the second one corresponds to the phase transition calculated using the bulk Gibbs equilibrium conditions and global charge neutrality (Glendenning scenario).

1. Introduction

Clarification of the structure and internal composition of compact stars is one of the main problems in modern physics. The circumstance that the central density of neutron stars is several times larger than the saturation density of nuclear matter makes neutron stars a specific natural laboratory for studying the characteristics of exotic states of matter at extremely high densities. Because of the very large densities achieved in the centers of compact stars various exotic particle species and phases of matter such as, for example, hyperons, deconfined quark plasma of u, d, s quarks, π and K meson condensates can arise. Over the past few decades many researchers have intensively studied various aspects related to the formation of exotic degrees of freedom in neutron stars and proposed observational tests that can confirm the existence in the interiors of compact stars such constituents (for review see, e.g., Refs. [1, 2] and references therein). The authors of several papers attempted to draw a conclusion about the existence of quark matter in compact stars through investigations of the mass-radius relation for such stars relying on the fact that compact stars made of quark matter have smaller radii than their hadronic counterparts of the same mass [3]. The studies of the thermal evolution the quark stars can also provide information about the existence of a quark component in the star [4]. Furthermore, the studies of gravitational radiation from compact stars can provide additional channel of information on existence of the quark matter inside compact stars [5].

Phase transitions accompanied by discontinuities of the thermodynamic potentials are of special interest, because they lead to a dynamical rearrangement of neutron stars. Depending on the value of surface tension σ_s, the phase transition from nuclear matter to quark matter can occur within two scenarios [6, 7]: one corresponds to the ordinary first order phase transition at constant pressure with a density jump (Maxwell construction), the other to formation of mixed hadron-quark matter with a continuous variation of pressure and density. The problem of the formation of a mixed phase taking into account the finite dimensions of the quark structures inside nuclear matter, the Coulomb interaction, as well as the surface energy, has been examined in Refs. [7, 8, 9]. It was shown that the mixed phase is energetically favorable for small values of the surface tension between the quark matter and the nuclear matter. The uncertainty in the value of the surface tension makes it impossible to determine the actual phase transition scenario that will take place. The quark-hadron phase transition can be triggered, for example, by the dynamical process of accretion of matter onto the surface of a hadronic neutron stars. Such accretion can lead to an increase of the central density of the stars and to formation of a new phase containing deconfined quarks in the center of the star. The process of catastrophic rearrangement with the formation of a quark core of finite radius at the star's center will be accompanied by a release of a colossal amount of energy, which is comparable to the energy release during a supernova explosion. Note that a similar process of both restructuring and energy release takes place also in the case of pion condensation in the cores of neutron stars [10, 11, 12, 13, 14].

The recent series of our articles [15, 16, 17] were devoted to a detailed investigation of quark deconfinement phase transition in neutron star matter. The nuclear matter was described within the relativistic mean-field (RMF) theory with the scalar-isovector δ-meson effective field. The bulk calculation results of the mixed phase structure (Glendenning construction) [18] were compared with the results of a usual first-order phase transition (Maxwell construction). Here we investigate the energy release and the change in the integral parameters of compact stars due to a phase transition from hadronic to quark matter in these two extreme scenarios and identify how these depend on the chosen scenario of the phase transition.

2. Equation of state of neutron star matter

2.1. Equation of state of hadronic phase

We use the equation of state (EoS) of Baym, Bethe, and Pethick (BBP) [19] for the description of hadronic phase in the lower density region corresponding to the outer and inner crust of the star. In nuclear and supranuclear density region ($n \geq 0.1$ fm^{-3}) we use the relativistic Lagrangian density of many-particle system consisting of nucleons, p, n, electrons and isoscalar-scalar (σ), isoscalar-vector (ω), isovector-scalar (δ), and isovector-vector (ρ) - exchanged mesons. The Lagrangian of the theory is given by

$$\begin{aligned}
\mathcal{L} = \overline{\psi}_N[\gamma^\mu(i\partial_\mu - g_\omega\omega_\mu - \frac{1}{2}g_\rho\vec{\tau}_N\vec{\rho}_\mu) - (m_N - g_\sigma\sigma - g_\delta\vec{\tau}_N\vec{\delta})]\psi_N \\
+\frac{1}{2}(\partial_\mu\sigma\partial^\mu\sigma - m_\sigma\sigma^2) - \frac{b}{3}\,m_N(g_\sigma\sigma)^3 - \frac{c}{4}\,(g_\sigma\sigma)^4 \\
+\frac{1}{2}m_\omega^2\omega^\mu\omega_\mu - \frac{1}{4}\Omega_{\mu\nu}\Omega^{\mu\nu} + \frac{1}{2}m_\rho^2\vec{\rho}^\mu\vec{\rho}_\mu - \frac{1}{4}\Re_{\mu\nu}\Re^{\mu\nu} \\
+\frac{1}{2}(\partial_\mu\vec{\delta}\partial^\mu\vec{\delta} - m_\delta^2\vec{\delta}^2) + \overline{\psi}_e(i\gamma^\mu\partial_\mu - m_e)\,\psi_e, \qquad (1)
\end{aligned}$$

where σ, ω_μ, $\vec{\delta}$, and $\vec{\rho}^\mu$ are the fields of the σ, ω, δ, and ρ exchange mesons, respectively, m_N, m_e, m_σ, m_ω, m_δ, m_ρ are the masses of the free particles, $\psi_N = \begin{pmatrix} \psi_p \\ \psi_n \end{pmatrix}$ is the isospin doublet for nucleonic bispinors, and $\vec{\tau}$ are the isospin 2×2 Pauli matrices. Antisymmetric tensors of the vector fields ω_μ and $\vec{\rho}_\mu$ are given by

$$\Omega_{\mu\nu} = \partial_\mu\omega_\nu - \partial_\nu\omega_\mu, \qquad \Re_{\mu\nu} = \partial_\mu\vec{\rho}_\nu - \partial_\nu\vec{\rho}_\mu. \qquad (2)$$

Journal of Physics: Conference Series **496** (2014) 012001 doi:10.1088/1742-6596/496/1/012001

In our calculations we take $a_\delta = (g_\delta/m_\delta)^2 = 2.5$ fm^2 for the δ coupling constant, as in Ref. [20]. Also we use $m_N = 938.93$ MeV for the bare nucleon mass, $m_N^* = 0.78\ m_N$ for the nucleon effective mass, $n_0 = 0.153$ fm^{-3} for the baryon number density at saturation, $f_0 = -16.3$ MeV for the binding energy per baryon, $K = 300$ MeV for the incompressibility modulus, and $E_{sym}^{(0)} = 32.5$ MeV for the asymmetry energy. Then the five other constants, $a_i = (g_i/m_i)^2$ ($i = \sigma,\ \omega,\ \rho$), b and c, can be determined numerically: $a_\sigma = (g_\sigma/m_\sigma)^2 = 9.154$ fm^2, $a_\omega = (g_\omega/m_\omega)^2 = 4.828$ fm^2, $a_\rho = (g_\rho/m_\rho)^2 = 13.621$ fm^2, $b = 1.654 \cdot 10^{-2}$ fm^{-1}, $c = 1.319 \cdot 10^{-2}$. The knowledge of the model parameters allows us to solve the set of four equations in a self-consistent way and to determine the (renamed) mean-fields, $\sigma \equiv g_\sigma\bar{\sigma}$, $\omega \equiv g_\omega\bar{\omega}_0$, $\delta \equiv g_\delta\bar{\delta}^{(3)}$, and $\rho \equiv g_\rho\bar{\rho}_0{}^{(3)}$, which depend on baryon number density n and asymmetry parameter $\alpha = (n_n - n_p)/n$. The standard quantum-hadro-dynamics (QHD) procedure allows us to obtain the expressions for energy density $\varepsilon(n,\alpha)$ and pressure $P(n,\alpha)$ (for details see Ref. [15]).

2.2. Equation of state of strange quark phase
To describe the quark phase an improved version of the MIT bag model was used, in which the interactions between u, d, s quarks inside the bag are taken in a one-gluon exchange approximation [21]. We choose $m_u = 5$ MeV, $m_d = 7$ MeV and $m_s = 150$ MeV for quark masses, $B = 60$ MeV/fm^3 for bag parameter and $\alpha_s = 0.5$ for the strong interaction constant.

2.3. Maxwell and Glendenning constructions
Using these EoS for the nucleonic and the quark phases we first calculate the physical parameters of the phase transition in the case of the Glendenning construction (where these phases satisfy the Gibbs condition and are separately electrically charged, but the global electrical neutrality of the system is maintained). We then examine the second case where both phases are separately neutral and the transition is the usual first-order phase transition corresponding to the well-known Maxwell construction. Model EoSs of neutron star matter for Glendenning and Maxwell construction cases are presented in Fig. 1. In case of the Maxwell construction the phase transition occurs at constant pressure $P_0 = 2.11$ MeV/fm^3 and nucleonic matter with the energy density $\varepsilon_N = 114.5$ MeV/fm^3 coexists with the quark matter whose energy density is $\varepsilon_Q = 271.4$ MeV/fm^3. In the case of Glendenning construction the deconfinement phase transition proceed through formation of a mixed hadron-quark phase. The boundaries of the mixed phase are $\varepsilon_N = 72.79$ MeV/fm^3, $P_N = 0.43$ MeV/fm^3 and $\varepsilon_Q = 1280.88$ MeV/fm^3, $P_Q = 327.75$ MeV/fm^3.

It was shown in Ref. [22] that for an ordinary first order phase transition, the density discontinuity parameter $\lambda = \varepsilon_S/(\varepsilon_N + P_0)$ plays a decisive role in the stability of neutron stars with arbitrarily small cores made of the denser phase of matter. Here P_0 is coexistence pressure of the two phases, and ε_N, ε_S are the energy densities of the normal and superdense phases, respectively. Paraphrasing the conclusions of Ref. [22], we can state that the stability criterions for the first-order hadron-quark phase transition: if $\lambda < 3/2$, then a neutron star with an arbitrarily small core of strange quark matter is stable, whereas if $\lambda > 3/2$, neutron stars with small quark cores are unstable. In the latter case, for a stable star there is a nonzero minimum value for the radius of the quark core. In case of the Maxwell construction of deconfinement phase transition discussed in this article, the value of jump parameter is $\lambda = 2.327$.

3. Changes in the Stellar Parameters triggered by the formation of quark phase
Using the neutron star matter EoSs obtained in previous section, we have integrated the Tolman-Oppenheimer-Volkoff (TOV) equations [24, 25] and obtained the gravitational mass M, radius R, baryonic mass $M_0 = m_N N_B$ (m_N is the nucleon mass and N_B the total number of baryons) and moment of inertia I of compact stars for the different values of central pressure P_c.

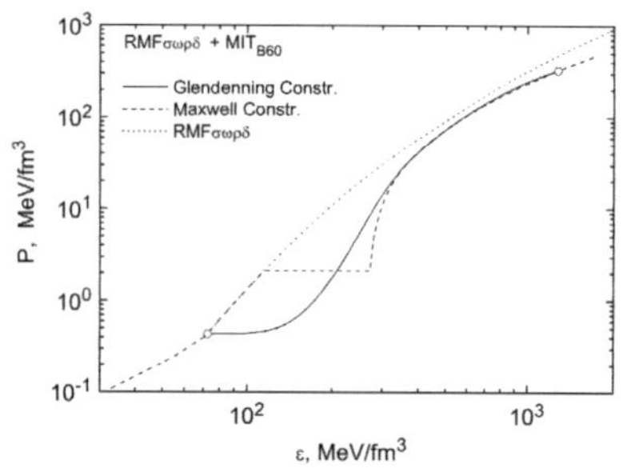

Figure 1. EoS of neutron star matter for the two different hadron-quark phase transition constructions. Solid and dashed lines correspond to the Glendenning and Maxwell constructions, respectively, whereas the dotted line to the pure *npe* matter. Open circles represent the mixed phase boundaries.

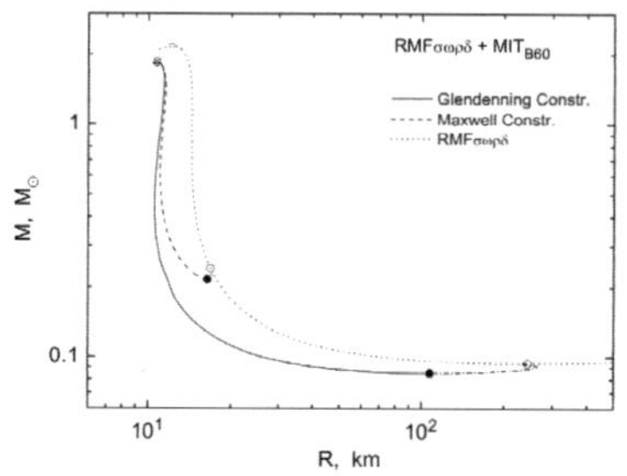

Figure 2. Mass-radius relations for model EoSs presented in Fig. 1. The labeling of the curves is the same as in Fig. 1. Open circles mark the critical configurations, solid circles mark the stable hybrid stars with minimal mass. The dash-dotted line between circles corresponds to the branch of unstable stars.

In Fig. 2 we show the $M(R)$ dependence of compact stars for different quark deconfinement phase transition scenarios. Open circles denote the critical configurations while solid circles denote the stable stars with minimal mass in the center of which there are deconfined quarks. Moreover, in the case of Maxwell construction, this core is composed of a quark-electron plasma, while in the Glendenning construction case it is composed of a mixed quark-hadron matter. We can see that for both phase transition scenarios considered here there are unstable star branches between critical and minimum-mass configurations (dot-dashed line segments). The fact that in the case of Maxwell construction, the $M - R$ relation exhibits such a behavior, is not surprising, since according to the Seidov criterium [22, 23], the infinitesimal core of denser matter in ordinary first order phase transition is unstable when $\lambda > 3/2$. The corresponding EoS of neutron star matter considered here satisfies this condition. The appearance of the unstable branch of compact stars with infinitesimal core consisting of a mixed hadron-quark matter in the case of Glendenning construction is not standard. In fact, in this scenario the energy density is a continuous function of pressure and in most cases leads to a monotonic increase of stars' mass in the domain which corresponds to the lower boundary of the mixed phase. This implies that the configurations with an infinitesimally small core containing the mixed phase are in many cases stable. In the case of Glendenning construction considered here, the branch of unstable compact stars appears near the lower threshold of the mixed phase. Consider the situation where matter is accreted onto the surface of an ordinary neutron star located below the critical configuration. The baryonic mass of the star will increase, the star will reached the mass of the critical configuration at which instance a transition to the configuration with deconfined quark phase will take place.

It is worthwhile to note that the maximum masses of the stars containing deconfined quarks are $M_{max} = 1.853 M_{\odot}$ and $M_{max} = 1.828 M_{\odot}$ for the Glendenning and the Maxwell constructions, respectively. Since the transition of ordinary neutron star to a star containing quark matter occurs at a constant baryon number, it is convenient to consider the star characteristics as functions of baryonic mass M_0. The binding energies $E_{bind} = (M - M_0)c^2$ of compact stars as a function of baryonic mass M_0 for the Maxwell and the Glendenning hadron-quark phase

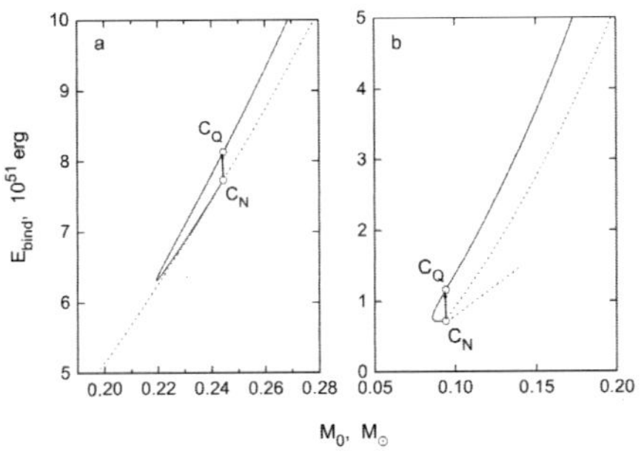

Figure 3. Binding energy of compact stars as a function of baryonic mass M_0 for Maxwell (left panel) and Glendenning (right panel) constructions. Dotted line corresponds to the pure nuclear matter. C_N marks the critical configuration of hadronic star, and C_Q marks the neutron star with same baryonic mass containing quark phase.

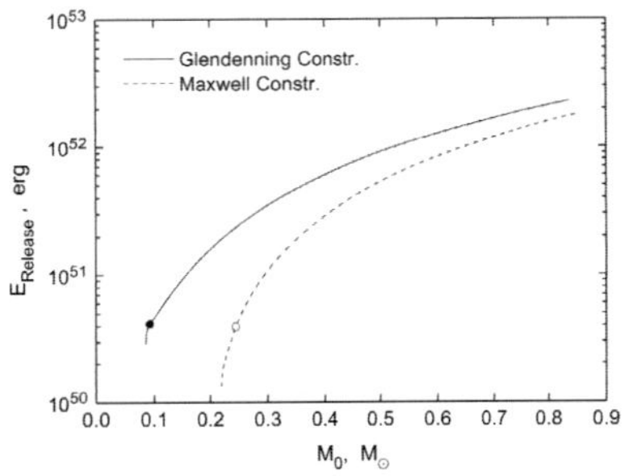

Figure 4. The total energy release due to the hadron-quark phase transition as a function of baryonic mass of neutron star M_0 for the Maxwell and the Glendenning constructions. Circles correspond to the $C_N \to C_Q$ transition shown in Fig. 3.

transition scenarios are shown in Fig. 3. Accretion of matter onto the surface of the critical configuration C_N will lead to a jump-like transition to the configuration C_Q, which will have a finite-size core containing deconfined quark matter. This transition will be accompanied by an enormous release of energy determined by the difference in binding energies of these configurations:

$$E_{release} = E_{bind}(C_Q) - E_{bind}(C_N) = (M(C_N) - M(C_Q))c^2. \tag{3}$$

In Fig. 4 we plot the total energy release as a function of baryonic mass of star. One can see that in both scenarios of the phase transition he released energy increases with increasing baryonic mass M_0. For fixed values of M_0 the energy that is converted in the phase transition is larger in the case of Glendenning construction than in the case of Maxwell construction. In addition, the minimum required baryonic mass for the catastrophic rearrangement of the neutron star, accompanied by the formation of a quark core in the center of the star, is greater in the Maxwell scenario. In the case considered here, the quark deconfinement phase transition in the neutron star interior leads to the energy release of order $10^{50} \div 10^{52}$ erg. Fig. 5 shows the changes in stellar radius $\Delta R = R_Q - R_N$ due to the quark deconfinement phase transition as a function of the baryonic mass M_0 in both the Glendenning and the Maxwell scenarios. It is seen that in both cases the compact stars radii decrease. Fig. 6 shows the fractional changes in the moment of inertia of compact stars, $(I_Q - I_N)/I_N$, for the two types of phase transition. It is seen that the dependence of the quantities ΔR and $\Delta I/I_N$ on the baryonic mass of the stars are clearly distinct in the two alternative scenarios of phase-transition.

4. Conclusion

Starting from relativistic mean-field description of hadronic phase and improved MIT model description of the quark phase we obtained the EoS of compact stars with a quark-deconfinement phase transitions, which was treated using either the Maxwell or the Glendenning construction. We found the dependence of conversion energy on the baryonic mass of neutron stars and analyzed the changes in stellar radii and moments of inertia due to the deconfinement phase

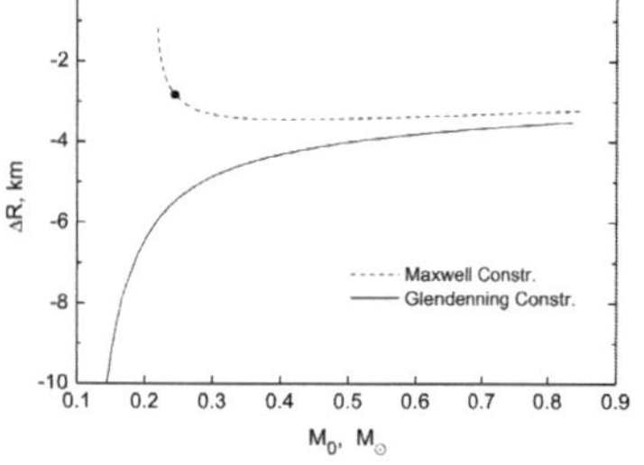

Figure 5. Changes in the stellar radius due to the hadron-quark phase transition as a function of baryonic mass M_0 of a neutron star for both Maxwell and Glendenning scenarios.

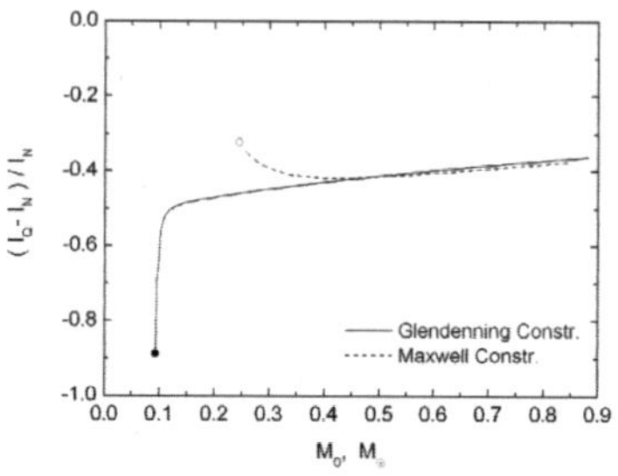

Figure 6. Fractional changes in the stellar moment of inertia $\Delta I/I_N$ as a function of baryonic mass M_0 of a neutron star for both Maxwell and Glendenning scenarios.

transitions. We demonstrated that for a fixed value of the baryonic mass of the star, the energy conversion in the case of the Glendenning construction is greater than in the case of the Maxwell construction. The minimum required baryonic mass for the catastrophic rearrangement of the neutron star and the formation of a quark core in the center of the star is greater in the case of the Maxwell construction as compared to the Glendenning construction. In the cases considered here, the quark deconfinement phase transition in the neutron star interior leads to the energy release of order $10^{50} \div 10^{52}$ erg.

References

[1] Glendenning N K 2000 *Compact Stars. Nuclear Physics, Particle Physics, and General Relativity* (Springer, Berlin)
[2] Haensel P, Potekhin A Y and Yakovlev D G 2007 *Neutron Stars 1. Equation of State and Structure* (Springer, New-York)
[3] Drago A, Lavagno A and Pagliara G 2004 *Phys. Rev.* D **69** 057505
[4] Grigorian H, Blaschke D and Voskresensky D 2005 *Phys.Rev.* C **71** 045801
[5] Andersson N, Jones D I and Kokkotas K D 2002 *Mon. Not. R. Astron. Soc.* **337**, 1224
[6] Heiselberg H Pethick C J and Staubo E F 1993 *Phys. Rev. Lett.*, **70**, 1355
[7] Heiselberg H and Hjorth-Jensen M 2000 *Phys. Rep.* **328** 237
[8] Voskresensky D N, Yasuhira M and Tatsumi T 2003 *Nucl. Phys.* A **723** 291
[9] Maruyama T, Chiba S, Schulze H-J and Tatsumi T 2007 *Preprint* 0708.3277 v1 [nucl.-th]
[10] Migdal A B, Chernoutsan A I and Mishustin I N 1979 *Phys. Lett.* B **83** 158
[11] Haensel P and Proszynski M 1982 *Astrophys. J.* **258** 306
[12] Zdunik J L, Haensel P and Schaeffer R 1987 *Astron.& Astrophys.* **172** 95
[13] Alaverdyan G B and Vartanyan Yu L 1989 *Rep. Acad. of Sci. of Armenia, Physics* **88** 18
[14] Berezhiani Z, Bombaci I, Drago A, Frontera F and Lavagno A 2003 *Astrophys. J.* **568** 1250
[15] Alaverdyan G B 2009 *Astrophysics* **52** 132
[16] Alaverdyan G B 2010 *Research in Astron. Astrophys.* **10** 1255
[17] Alaverdyan A G, Alaverdyan G B, Chiladze A O 2010 *Int. Journal of Modern Phys. D* **19** 1557
[18] Glendenning N K 1992 *Phys. Rev.* D **46** 1274
[19] Baym G, Bethe H and Pethick Ch 1971 *Nucl.Phys.* A **175** 255
[20] Liu B, Greco V, Baran V, Colonna M and Di Toro M 2002 *Phys. Rev.* C **65** 045201
[21] Farhi E and Jaffe R L 1984 *Phys. Rev.* D **30** 2379
[22] Seidov H 1971 *Astron. Zh.* **48** 443
[23] Schaeffer R, Zdunik L and Haensel P 1983 *Astron. Astrophys.* **126** 121
[24] Tolman R 1939 *Phys. Rev.* **55** 364
[25] Oppenheimer J and Volkoff G 1939 *Phys. Rev.* **55** 374

Mass and radius constraints for compact stars and the QCD phase diagram

David B. Blaschke
Institute for Theoretical Physics, University of Wroclaw, 50-204 Wroclaw, Poland
Bogoliubov Laboratory for Theoretical Physics, JINR Dubna 141980 Dubna, Russia
E-mail: blaschke@ift.uni.wroc.pl

Hovik A. Grigorian
Laboratory of Information Technologies, JINR Dubna 141980 Dubna, Russia
Department of Theoretical Physics, Yerevan State University, 0025 Yerevan, Armenia
E-mail: hovik.grigorian@gmail.com

David E. Alvarez-Castillo
Bogoliubov Laboratory for Theoretical Physics, JINR Dubna 141980 Dubna, Russia
Instituto de Física, Universidad Autónoma de San Luis Potosí, 78290 San Luis Potosí, SLP, México
E-mail: alvarez@theor.jinr.ru

Alexander S. Ayriyan
Laboratory of Information Technologies, JINR Dubna 141980 Dubna, Russia
E-mail: ayriyan@jinr.ru

Abstract. We suggest a new Bayesian analysis using disjunct M-R constraints for extracting probability measures for cold, dense matter equations of state. One of the key issues of such an analysis is the question of a deconfinement transition in compact stars and whether it proceeds as a crossover or rather as a first order transition. The latter question is relevant for the possible existence of a critical endpoint in the QCD phase diagram under scrutiny in present and upcoming heavy-ion collision experiments.

1. Introduction

One of the most challenging problems of modern physics concerns the structure of the phase diagram of quantum chromodynamics (QCD). Experimental programs with heavy-ion collisions (HIC) at ultrarelativistic energies and large-scale simulations of lattice QCD at finite temperature are performed to identify the position and the character of the suspected transition from a gas of hadronic resonances to a quark-gluon plasma in these systems characterized by almost perfect symmetry between particles and antiparticles, i.e. vanishing baryon density. It is nowadays established that there is a crossover transition with a pseudocritical temperature of

$T_c = (154 \pm 9)$ MeV [1] from lattice QCD and a chemical freeze-out temperature $T_{\rm fo} = (156 \pm 5)$ MeV [2] for all hadron species from a statistical model analysis of hadron production data at $\sqrt{s} = 2.7$ TeV from the ALICE experiment at CERN LHC.

Both, HIC experiments and ab-initio lattice QCD cannot address, however, the QCD phase diagram at low and vanishing temperatures where the QCD phase transition is eventually of first order so that a critical end point (CEP) of first order transitions would result. The position of such a CEP in the phase diagram would be a landmark for the studies of strongly interacting matter under extreme conditions as it could help to identify the universality class of QCD. However, the beam energy scan programs at RHIC (STAR experiment) or at CERN SPS (NA61 experiment) have not yet brought any evidence for the existence of the CEP. Moreover, some theoretical studies argue that it could be absent at all because of the persistence of repulsive vector meson mean fields in dense matter [3], see also the discussion in [4].

In this situation, where effective model approaches give contradicting results, and new experimental facilities like NICA at JINR Dubna and the CBM experiment at FAIR Darmstadt are not yet operative, a guidance for progress in the field could come from astrophysics of compact objects, namely from the precise mass and radius measurement of pulsars [5]. The main question to be answered is [6, 7]: Can there be a trace of a (strong) first order phase transition in cold nuclear (neutron star) matter in mass-radius data from compact star observations?

In the present contribution, we outline a model study for this case which is based on Bayesian analysis (BA) methods and how they could guide future pulsar observational campaigns.

2. EoS & stars with a QCD phase transition

For this study we follow the scheme suggested by Alford, Han and Prakash [8] for the hybrid EoS with a first order phase transition,

$$p(\epsilon) = p^I(\epsilon)\ \Theta(\epsilon_c - \epsilon) + p^{II}(\epsilon)\ \Theta(\epsilon - \epsilon_c - \Delta\epsilon), \tag{1}$$

where $p^I(\epsilon)$ is given by a pure hadronic EoS and $p^{II}(\epsilon)$ represents the high density nuclear matter introduced here as quark matter given in the bag-like form

$$p^{II}(\epsilon) = c^2_{\rm QM}(\epsilon - \epsilon_0) = c^2_{\rm QM}\epsilon - B, \tag{2}$$

with $c^2_{\rm QM}$ as the squared speed of sound in quark matter, and the bag constant B, or the energy density offset ϵ_0 being synonymous for parametrizing the latent heat $\Delta\epsilon$ of the phase transition, occuring at the critical pressure $p_c = p(\epsilon_c) = p^I(\epsilon_c) = p^{II}(\epsilon_c + \Delta\epsilon)$. It has been shown by Haensel et al. [9] that Eq. (2) describes pretty well the superconducting NJL model derived in [10] and applied for hybrid stars with an extension by a repulsive vector meanfield first in [11], recently revisited and systematically sampled in [12].

For the hadronic EoS we take the well known model of APR [13] that is in agreement with experimental data at densities about nuclear saturation. For this hadronic branch (I) all the relevant thermodynamical variables, energy density ϵ, pressure p, baryon density n and chemical potential μ are well defined and taken as input for determination of the hybrid (hadronic + quark matter) EoS.

The free parametes of the model are the transition density ϵ_c, the energy density jump $\Delta\epsilon \equiv \gamma\epsilon_c$ and $c^2_{\rm QM}$. For the present study we will use all hybrid EoS of the given type which are obtained when varying these three parameters within the limits: $400 < \epsilon_c[\mathrm{MeV/fm^3}] < 1000$, $0 < \gamma < 1.0$ and $0.3 < c^2_{\rm QM} < 1.0$. The resulting EoS are shown in the upper left panel of Fig. 1 and demonstrate which part of the pressure versus energy density plane is covered by our three-dimensional parameter sampling. For the present study we use a set of 1000 EoS corresponding to a coverage of the parameter space by 10 values in each dimension.

Using the above set of hybrid EoS one calculates the corresponding set of neutron star sequences by solving the Tolman-Oppenheimer-Volkoff (TOV) equations [14]. Of particular interest for the comparison with observational data are the gravitational mass vs. radius ($M-R$) and gravitational mass vs. baryon mass ($M-M_B$) diagrams, shown in Fig. 1 in the middle and lower left panels, respectively. One can read off the following correlations with the hybrid EoS parameters: (i) the higher the critical energy density (ϵ_c) the higher the onset mass for hybrid star configurations, (ii) the larger the jump in energy density at the transition (γ) the stronger the effect of compactification of the hybrid star configuration, which is reflected also in a larger gravitational binding energy (mass defect) for the rightmost curves in the $M-M_B$ diagram (light color). Increasing γ eventually leads to an instability, indicated by an increasing mass with increasing radius, and (iii) the increase of the maximum mass for a star sequence which results from increasing the stiffness of quark matter, i.e. the speed of sound $c^2_{\rm QM}$.

In the next step, we would like to apply observational constraints to the obtained 1000 compact star sequences and filter the most likely parameter sets by applying Bayesian methods.

3. Observational constraints and Bayesian analysis

We want to compare the theoretical results with suitable constraints from neutron star observations in order to conclude for the likeliness of the underlying EoS model. A pioneering study of this kind has been performed recently in [15]. However, as there are suspicions that the analysis of the burst sources used in [15] may have a systematic bias towards smaller radii (cf. [16, 17]), we use here instead three statistically independent constraints, none of them related to bursting sources:

- a maximum mass constraint from PSR J0348+0432 [18],
- a radius constraint from the nearest millisecond pulsar PSR J0437-4715 [19], which both are shown in the middle panel of Fig. 1, and
- a constraint on the gravitational binding energy from the neutron star B in the binary system J0737-3039 (B) [20], see the bottom panel of Fig. 1. The M_B from this full hydro simulation is 1% smaller than the one by [21] where mass loss was neglected.

The above constraints are shown with their respective $1\sigma-$, $2\sigma-$, and $3\sigma-$ confidence regions in the lower panels of Fig. 1. For the BA, we have to sample the above defined parameter space and to that end we introduce a vector of the parameter values

$$\pi_i = \vec{\pi}\left(\epsilon_c(k), \gamma(l), c^2_{\rm QM}(m)\right), \tag{3}$$

where $i = 0 \ldots N-1$ (here $N = N_1 \times N_2 \times N_3$) as $i = N_1 \times N_2 \times k + N_2 \times l + m$ and $k = 0 \ldots N_1 - 1$, $l = 0 \ldots N_2 - 1$, $m = 0 \ldots N_3 - 1$. Here N_1, N_2 and N_3 denote the number of parameters ϵ_c, γ and $c^2_{\rm QM}$, respectively. The goal is to find the set π_i corresponding to an EoS and thus a sequence of configurations which contains the most probable one based on the given constraints using BA. For initializing the BA we propose that *a priori* each vector of parameters π_i has a probability equal to unity: $P(\pi_i) = 1$ for $\forall i$. Then one proceeds as follows.

3.1. Mass constraint

We propose that the error of mass measurement is normal distributed $\mathcal{N}(\mu_A, \sigma_A^2)$, where $\mu_A = 2.01$ M$_\odot$ and $\sigma_A = 0.04$ M$_\odot$, according to the mass measurements for the massive pulsar PSR J0348+0432 [18]. Using this assumption we can calculate the conditional probability of the event E_A that the mass of a neutron star corresponds to this measurement

$$P(E_A | \pi_i) = \Phi(M_i, \mu_A, \sigma_A), \tag{4}$$

where M_i - maximal mass constructed by π_i and $\Phi(x, \mu, \sigma)$ is the cumulative distribution function for the normal distribution.

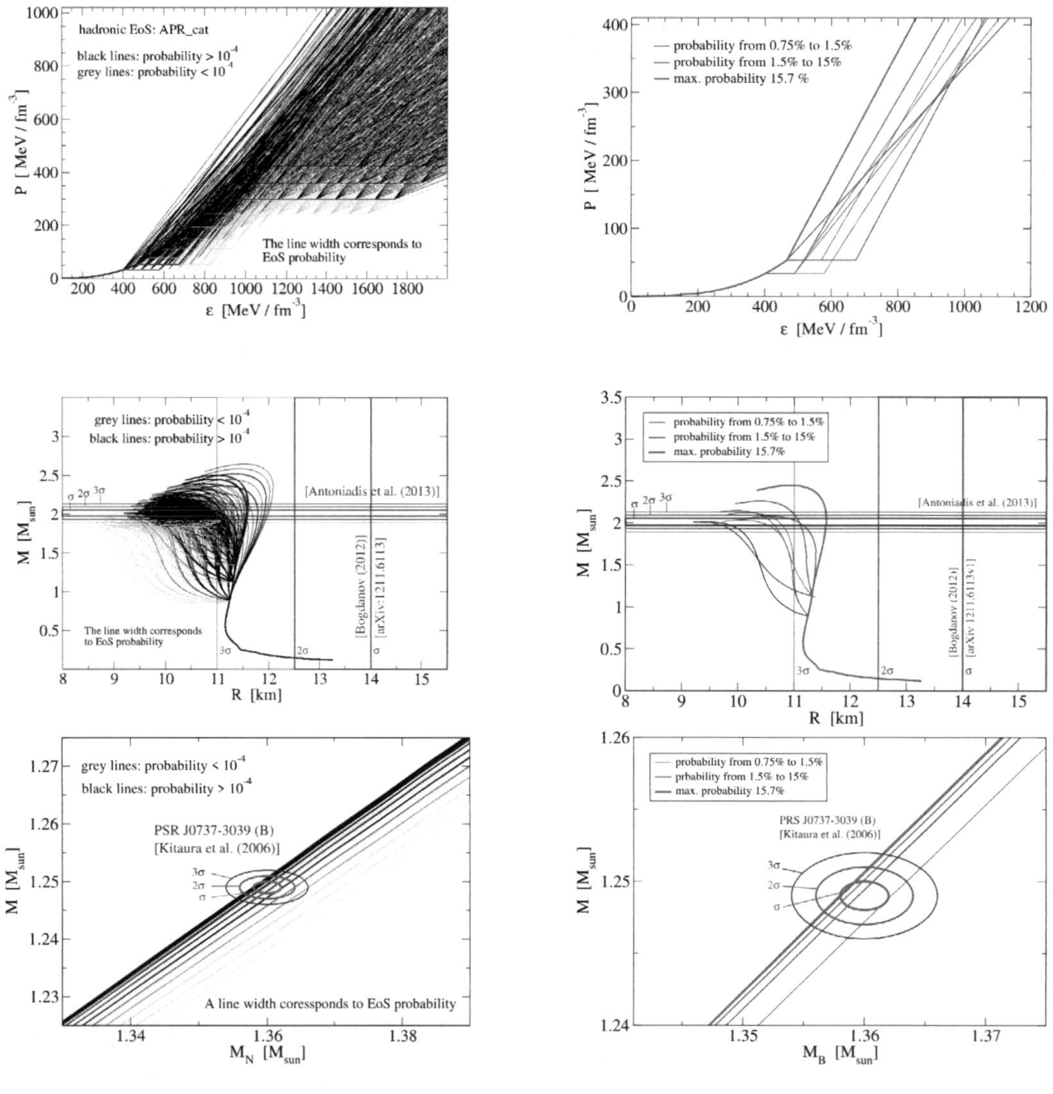

Figure 1. Hybrid EoS scheme for sets of three parameters $(\epsilon, \gamma, c_s^2)$ before (left panels) and after (right panels) the Bayesian analysis.

3.2. Radius constraint

From an analysis of the timing of the nearest millisecond pulsar PSR J0437-4715 Bogdanov [19] extracts a radius of $\mu_B = 15.5$ km at a mass of 1.7 $M_\odot$ with a variance of $\sigma_B = 1.5$ km. We will consider this value mass independent, neglecting the mild variation given in Ref. [19] since it is inessential for the present study. Now it is possible to calculate conditional probability of the event E_B that the radius of a neutron star corresponds to the given measurement

$$P\left(E_B \left|\pi_i\right.\right) = \Phi(R_i, \mu_B, \sigma_B) \; . \tag{5}$$

3.3. Gravitational binding ($M - M_B$) constraint

This constraint gives a region in the $M - M_B$ plane. For our analysis we use the mean values $\mu = 1.249\ M_\odot$, $\mu_B = 1.360\ M_\odot$ and the standard deviations $\sigma_M = 0.001\ M_\odot$ and $\sigma_{M_B} = 0.002\ M_\odot$ which are given in [20].

We need to estimate the probability for the closeness of a theoretical point $M_i = (M_i, M_{Bi})$ to the observed point $\mu = (\mu, \mu_B)$. The required probability can be calculated using the following formula

$$P\left(E_K\left|\pi_i\right.\right) = \left[\Phi\left(\xi\right) - \Phi\left(-\xi\right)\right] \cdot \left[\Phi\left(\xi_B\right) - \Phi\left(-\xi_B\right)\right], \tag{6}$$

where $\Phi\left(x\right) = \Phi\left(x, 0, 1\right)$, $\xi = \sigma_M / d_M$ and $\xi_B = \sigma_{M_B} / d_{M_B}$, with d_M and d_{M_B} being the absolute values of components of the vector $\mathbf{d}_{\mathrm{i}} = \mu - \mathbf{M}_i$, where $\mu = (\mu, \mu_B)^T$ is given in [20] and $\mathbf{M}_i = (M_i, M_{Bi})^T$ is the solution of the TOV equations using the ith vector of EoS parameters π_i. Note that formula (6) does not correspond to the multivariate normal distribution.

3.4. Calculation of a posteriori probabilities

Note, that these measurements are independent of each other. This means that we can calculate the complete conditional probability of an event described by one of the objects in the star sequence addressed by π_i corresponds to a product of the probabilities for all three constraining measurements

$$P\left(E\left|\pi_i\right.\right) = P\left(E_A\left|\pi_i\right.\right) \times P\left(E_B\left|\pi_i\right.\right) \times P\left(E_K\left|\pi_i\right.\right). \tag{7}$$

4. Results and conclusions

As a result of the BA for the three-parameter EoS (1) each of the $10 \times 10 \times 10 = 1000$ points in the three-dimensional parameter space has been assigned a probability value according to (7). We have selected the six most likely parameter sets and show the corresponding equations of state in the upper right panel of Fig. 1. The corresponding $M - R$ and $M - M_B$ plots are given in the lower right panels of Fig. 1, together with the corresponding observational constraints.

Alternativey, one can present the probabilities as histograms in the parameter space. Here, we are restricted to a two-dimensional subset (LEGO plots). Therefore we choose the three classes of speed of sound values for which we found the maximum probabilities and depict for each of them the LEGO plot of probabilities in the subspace of critical energy density (ϵ_c) and strength of the first-order transiition (γ) in Fig. 2. It is interesting to note that for the stiffest

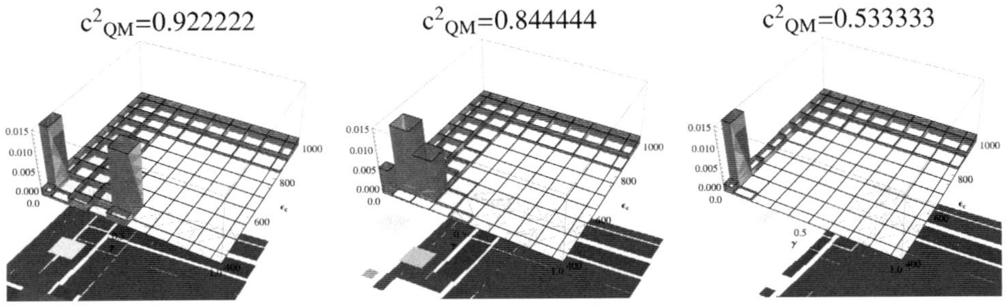

Figure 2. LEGO plots for Bayesian analysis in the plane of two parameters (ϵ_c, γ) for three fixed values of the third parameter c_s^2.

high-density EoS ($c^2_{\mathrm{QM}} = 0.922$), we obtain a bimodality of the probability distribution. A crossover behaviour ($\gamma = 0$) has the same probability as a strong first order transition ($\gamma = 0.5$).

We conclude that the Bayesian Analysis we have presented here presents an alternative to the previously developed one in Ref. [15]. It is based on three statistically independent constraints which are extremely selective and result in maximum probabilities of about 1% for the most likely EoS. Thus this method will have a sufficient selective power when applied to a broader class of EoS. This concerns in particular EoS being stiffer (with larger radii) on the hadronic side and with microscopically founded EoS on the high density (quark matter) side. To test the interesting possibility that a strong first order phase transition might occur in massive neutron stars of 2 $M_\odot$ [18, 22] and lead to the appearance of a "third family" (mass twins) of hybrid stars at this high mass, one should perform radius measurements for these massive neutron stars. If it might turn out that the corresponding radii might be significantly different, e.g., by $1 - 2$ km with a standard deviation of about 500 m, then one would be able to disselect EoS without a phase transition, given the very narrow range of uncertainty in the mass of these objects. This possibility offers bright prospects for future observational campaigns and bears the chance to "prove" the existence of a critical point [6, 7] in the QCD phase diagram from astrophysical observations!

Acknowledgements

D.B., H.G. and D.E.A.-C. acknowledge the stimulating and inspiring atmosphere at the workshop in Yerevan, as well as support by the Volkswagen Foundation. The authors are grateful for clarifying discussions with many colleagues about this topic, in particular with T. Fischer, T. Klähn, R. Lastowiecki, J. Lattimer, M.C. Miller, J. Trümper, F. Weber. The work was supported by the NCN "Maestro" programme under contract number UMO-2011/02/A/ST2/00306. A.S.A., D.E.A-C. and H.G. are grateful for support by the Bogoliubov-Infeld programme. A.A. thanks to JINR grant No. 14-602-01 and H.G. acknowledges support by the Volkswagen Foundation under grant No. 85 182. D.E.A-C. is grateful for support from funds of the collaboration between South Africa and JINR Dubna for his participation at the School on Bayesian Analysis in Physics and Astronomy at the University of Stellenbosch and the SKA Conference at the STIαS.

References

[1] A. Bazavov, *et al.*, Phys. Rev. D **85**, 054503 (2012).
[2] F. Becattini, *et al.*, Phys. Rev. Lett. **111**, 082302 (2013).
[3] N. M. Bratovic, T. Hatsuda and W. Weise, Phys. Lett. B **719**, 131 (2013).
[4] G. A. Contrera, A. G. Grunfeld and D. B. Blaschke, arXiv:1207.4890 [hep-ph].
[5] For a recent review, see M. C. Miller, arXiv:1312.0029 [astro-ph.HE].
[6] D. Blaschke, D. E. Alvarez-Castillo and S. Benic, PoS CPOD **2013**, 063 (2013); [arXiv:1310.3803 [nucl-th]].
[7] D. E. Alvarez-Castillo and D. Blaschke, arXiv:1304.7758 [astro-ph.HE].
[8] M. G. Alford, S. Han and M. Prakash, Phys. Rev. D **88**, 083013 (2013).
[9] J. L. Zdunik and P. Haensel, Astron. & Astrophys., **551**, A61 (2013); [arXiv:1211.1231 [astro-ph.SR]].
[10] D. Blaschke, *et al.*, Phys. Rev. D **72**, 065020 (2005).
[11] T. Klähn, *et al.*, Phys. Lett. B **654**, 170 (2007).
[12] T. Klähn, D. B. Blaschke and R. Łastowiecki, Phys. Rev. D **88**, 085001 (2013).
[13] A. Akmal, V. R. Pandharipande, D. G. Ravenhall, Phys. Rev. C **58**, 1804 (1998).
[14] N. K. Glendenning, *Compact stars*, Springer, New York (1997).
[15] A. W. Steiner, J. M. Lattimer and E. F. Brown, Astrophys. J. **722**, 33 (2010).
[16] S. Bhattacharyya, M. C. Miller, D. K. Galloway, Mon. Not. R. Astron. Soc. **401**, 2 (2010).
[17] J. E. Trümper, Prog. Part. Nucl. Phys. **66**, 674 (2011).
[18] J. Antoniadis, *et al.*, Science **340**, 6131 (2013).
[19] S. Bogdanov, Astrophys. J. **762**, 96 (2013).
[20] F. S. Kitaura, H.-Th. Janka, W. Hillebrandt, Astron. & Astrophys. **450**, 345 (2006).
[21] P. Podsiadlowski, *et al.*, Mon. Not. Roy. Astron. Soc. **361**, 1243 (2005).
[22] P. Demorest, T. Pennucci, S. Ransom, M. Roberts and J. Hessels, Nature **467**, 1081 (2010).

Equation of state of hypernuclear matter: tuning hyperon–scalar-meson couplings

Giuseppe Colucci and Armen Sedrakian

Institute for Theoretical Physics, J. W. Goethe-University, D-60438 Frankfurt-Main, Germany

E-mail: colucci@th.physik.uni-frankfurt.de, sedrakiani@th.physik.uni-frankfurt.de

Abstract. We discuss to which extent the modifications of the hyperon–scalar-meson coupling constants affect the equation of state (EoS) hypernuclear matter. The study is carried out within a relativistic density functional theory. The nucleonic matter is described in terms of a density-dependent parametrization of nucleon-meson couplings, whereas the hyperon–meson couplings are deduced from the octet model. We identify the parameter space of hyperon-meson couplings for which massive stellar configurations with $M \leq 2.25 M_\odot$ exist. We also discuss the EoS at finite temperatures with and without of a trapped neutrino component and show that neutrinos stiffen the EoS and change qualitatively the composition of stellar matter.

1. Introduction

The recent observations of two-solar-mass pulsars in binary orbits with white dwarfs [1, 2] spurred an intensive discussion of the phase structure of dense matter, which is consistent with the implied observational lower bound on the maximum mass of any sequence of compact stars based on the unique equation of state (hereafter EoS) of dense matter. In this article we review and summarize the key result of our study of hypernuclear matter in the context of these observations of massive compact stars [3].

Large central densities achieved in massive compact stars may require substantial population of heavy baryons (hyperons), because these become energetically favorable once the Fermi energy of neutrons becomes of the order of their rest mass. The onset of hyperons (and more generally any new constituent) reduces the degeneracy pressure of a cold thermodynamic ensemble. Therefore the EoS becomes softer than in the absence of the hyperons (or any other constituent). This decreases the maximum mass of a compact stars to values which contradict the observation of massive compact stars in nature. The controversy between the theory and observations is the "hyperonization puzzle" in compact stars.

Hyperons in dense nuclear matter have been studied using a number of methods, including Lagrangian based relativistic density functional methods [4, 5, 6, 7, 8, 9, 10, 11, 12] with coupling parameters fixed by the nuclear phenomenology, as well as models based on hyperon-nucleon potentials (for recent work see [13, 14]). While the potential models fail to produce heavy enough stars and are most likely accurate at densities not much larger than the nuclear saturation density, the relativistic Lagrangian based models are suitable candidates for extrapolation to high density regime.

The hypernuclear EoS was investigated [3] by us recently in the framework of the density-dependent relativistic density functional method. In particular, we focus on the sensitivity of

Journal of Physics: Conference Series **496** (2014) 012003 doi:10.1088/1742-6596/496/1/012003

the EoS of hypernuclear matter to the unknown hyperon–scalar-meson couplings. These are constrained only by imposing SU(6) symmetry breaking and the nonet mixing. Within this framework, we argue that the parameters can be tuned such that two-solar massive hyperonic compact stars can exist. The EoS and composition of matter were also studied at finite temperatures relevant for the hot proto-neutron star stage of evolution and it was shown that the neutrino component stiffens the EoS of hypernuclear matter and substantially changes the composition of matter [3].

2. Theoretical model and choice of couplings

The relativistic Lagrangian density of our model reads

$$\begin{aligned}\mathcal{L} &= \sum_B \bar{\psi}_B\left[\gamma^\mu\left(i\partial_\mu - g_{\omega B}\omega_\mu - \frac{1}{2}g_{\rho B}\boldsymbol{\tau}\cdot\boldsymbol{\rho}_\mu\right) - (m_B - g_{\sigma B}\sigma)\right]\psi_B \\ &+ \frac{1}{2}\partial^\mu\sigma\partial_\mu\sigma - \frac{1}{2}m_\sigma^2\sigma^2 - \frac{1}{4}\omega^{\mu\nu}\omega_{\mu\nu} + \frac{1}{2}m_\omega^2\omega^\mu\omega_\mu - \frac{1}{4}\boldsymbol{\rho}^{\mu\nu}\boldsymbol{\rho}_{\mu\nu} + \frac{1}{2}m_\rho^2\boldsymbol{\rho}^\mu\cdot\boldsymbol{\rho}_\mu \\ &+ \sum_\lambda \bar{\psi}_\lambda(i\gamma^\mu\partial_\mu - m_\lambda)\psi_\lambda - \frac{1}{4}F^{\mu\nu}F_{\mu\nu},\end{aligned} \tag{1}$$

where the B-sum is over the $J^P = \frac{1}{2}^+$ baryon octet, ψ_B are the baryonic Dirac fields with masses m_B. The meson fields σ, ω_μ and $\boldsymbol{\rho}_\mu$ mediate the interaction among baryon fields, $\omega_{\mu\nu}$ and $\boldsymbol{\rho}_{\mu\nu}$ represent the field strength tensors of vector mesons and m_σ, m_ω, and m_ρ are their masses. The baryon-meson coupling constants are denoted by g_{mB}. The last line of Eq. (1) stands for the contribution of the free leptons, where the λ-sum runs over the leptons e^-, μ^-, ν_e and ν_μ with masses m_λ. The last term is the electromagnetic energy density. The contribution of neutrinos is included at temperatures above those at which the neutrinos decouple from matter, which is of order of several MeV.

The nucleon–meson coupling constants have been taken according to the DD-ME2 density-dependent parameterization [15]. The density dependence of the couplings implicitly takes into account many-body correlations among nucleons which are beyond the mean-field approximation. The nucleon-meson coupling constants are parametrized as $g_{iN}(\rho_B) = g_{iN}(\rho_0)h_i(x)$, for $i = \sigma, \omega$, and $g_{\rho N}(\rho_B) = g_{\rho N}(\rho_0)\exp[-a_\rho(x-1)]$ for the $\boldsymbol{\rho}_\mu$-meson, where ρ_B is the baryon density, ρ_0 is the saturation density, $x = \rho_B/\rho_0$ and the explicit form of the functions $h_i(x)$ and the values of couplings can be found elsewhere [3, 15]. The pressure and energy density of the model is further supplemented from the contribution coming from the so-called rearrangement self-energy [16], which guarantees the thermodynamical consistency.

In order to fix the hyperon–meson couplings we consider the SU(3)-flavor symmetric octet model. Due to the universal coupling of the $\boldsymbol{\rho}_\mu$ meson to the isospin current [17] and the ideal mixing between the ω and ϕ mesons[18], the couplings between hyperons and vector mesons are as follows

$$\begin{aligned} g_{\Xi\rho} &= g_{N\rho}, \qquad g_{\Sigma\rho} = 2g_{N\rho}, \qquad g_{\Lambda\rho} = 0, \\ g_{\Xi\omega} &= \frac{1}{3}g_{N\omega}, \quad g_{\Sigma\omega} = g_{\Lambda\omega} = \frac{2}{3}g_{N\omega}. \end{aligned} \tag{2}$$

Within the octet model the baryon-scalar mesons couplings of the scalar octet can be expressed in terms of only two parameters, the nucleon–a_0 meson coupling constant g_S and the $F/(F+D)$ ratio of the scalar octet [19]. By considering the mixing with the scalar singlet state, one is then left with the following relation between the coupling of the baryons with the σ-meson [3]:

$$2(g_{N\sigma} + g_{\Xi\sigma}) = 3g_{\Lambda\sigma} + g_{\Sigma\sigma}, \tag{3}$$

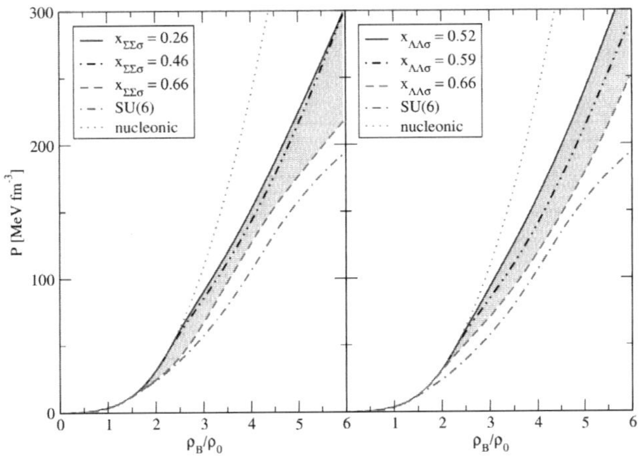

Figure 1. Equations of state of hypernuclear matter for a range of values of hyperon-σ meson couplings defined in terms of $x_{H\sigma} = g_{H\sigma}/g_{N\sigma}$, $H \in \Lambda, \Sigma$. The nucleonic EoS (dotted line, magenta online) and hyperonic EoS with SU(6) quark model couplings (dot-dashed line, green online) are shown as a reference. The nucleonic coupling constants correspond to the DD-ME2 parametrization [15]; the hyperon-vector meson couplings are fixed as explained in the text. Left panel: we assume $x_{\Lambda\sigma} = 0.58$, as in the NSC potential model, and a range $0.26 \leq x_{\Sigma\sigma} \leq 0.66$ which generates the shaded area. Right panel: we assume $x_{\Sigma\sigma} = 0.448$ and a range $0.26 \leq x_{\Lambda\sigma} \leq 0.66$. The cases $x_{\Sigma\sigma} = 0.46$ (left panel) and $x_{\Lambda\sigma} = 0.59$ (right panel), shown by dash-double-dotted (black online) lines, fit the depth of the potentials of the Σ^- and Λ hyperons in nuclear matter at saturation.

We assume that the hyperon coupling constants must be positive and less than the nucleon coupling constant. Then, by solving Eq. (3) for one of the dependent hyperon-σ meson coupling constant, say $g_{\Xi\sigma}$, we obtain

$$g_{N\sigma} \leq \frac{1}{2}(3g_{\Lambda\sigma} + g_{\Sigma\sigma}) \leq 2g_{N\sigma}. \tag{4}$$

We further proceed by first fixing the value of $g_{\Lambda\sigma}$ coupling constat at the value provided by the Nijmegen Soft Core (NSC) hypernuclear potential [20] and varying the range of couplings $g_{\Sigma\sigma}$ within the limits provided by Eq. (4) and then we repeat the calculations by interchanging the role of Σ and Λ hyperons. We also studied the case where one of couplings is fixed to the depth of the potential of the Σ^- and Λ hyperons in nuclear matter at the saturation density.

3. Results

The dependence of the EoS on the variation of the hyperon–scalar meson coupling at $T = 0$ is shown in Fig. 1 for a range of parameter space (for details see the figure caption). It is clearly seen that the hyperonization of matter softens the nucleonic EoS. The softening is smallest for the lowest possible values of the couplings of the hyperons to the scalar mesons.

Fig. 2 shows the particle fractions of fermions at zero temperature, for the limiting cases of the SU(6) quark model couplings (left panel) and the stiffest hypernuclear EoS (right panel).

Journal of Physics: Conference Series **496** (2014) 012003 doi:10.1088/1742-6596/496/1/012003

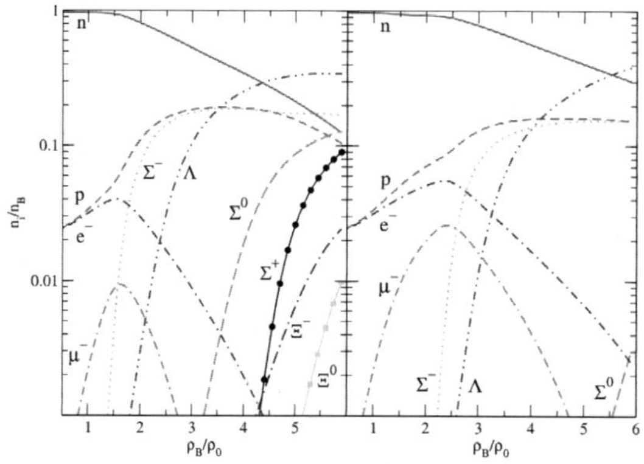

Figure 2. Particle fractions in hypernuclear matter at $T = 0$. Left panel: hyperon–scalar meson couplings are fixed as in the SU(6) symmetric quark model. Right panel: a stiff hypernuclear EoS from our parameter space with $x_{\Sigma\sigma} = 0.448$ and $x_{\Lambda\sigma} = 0.52$.

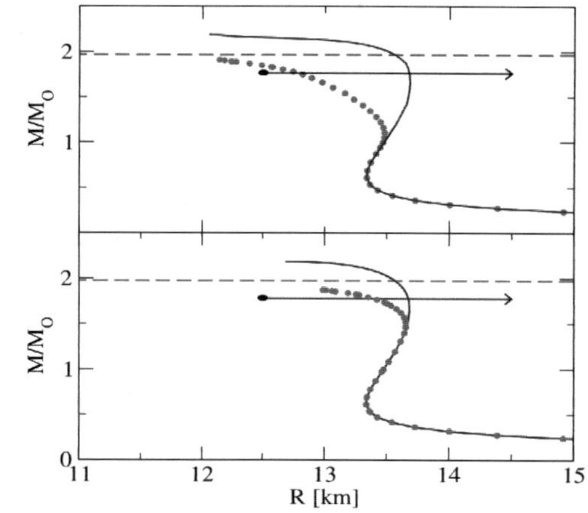

Figure 3. The mass–radius relations for compact hypernuclear stars at zero temperature. The solid (blue) lines show the cases $x_{\Sigma\sigma} = 0.26$ and $x_{\Lambda\sigma} = 0.58$ (upper panel) and $x_{\Sigma\sigma} = 0.52$ and $x_{\Lambda\sigma} = 0.448$ (lower panel). The (red) dots show the cases $x_{\Sigma\sigma} = 0.66$ and $x_{\Lambda\sigma} = 0.58$ (upper panel) and $x_{\Sigma\sigma} = 0.66$ and $x_{\Lambda\sigma} = 0.448$ (lower panel). The dash-dotted (green) line shows the observational lower limit on the maximum mass $1.97M_{\odot}$. The arrow shows the mass-radius constraint of Ref. [21] at 2σ level, which is $M = 1.76M_{\odot}$ and $R \geq 12.5$ km.

The first EoS is characterized by large hyperon–scalar-meson couplings, whereas the second by small ones. Thus, the larger are the hyperon–scalar-meson couplings the more favorable is the formation of hyperons and the softer is the EoS. The mass radius relation for these EoS is shown in Fig. 3, which demonstrates the large enough masses can be achieved within the

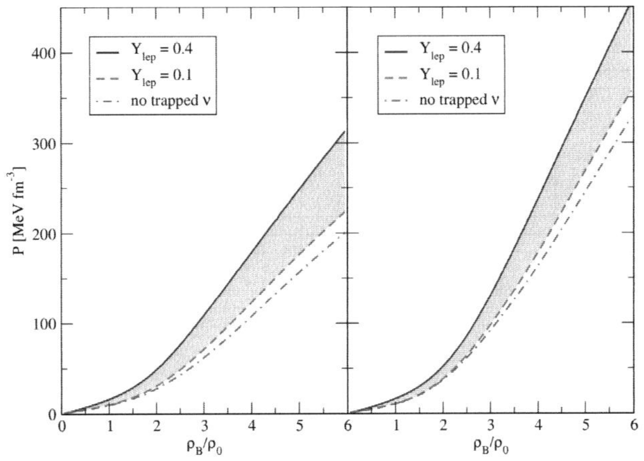

Figure 4. Equation of state in the DD-ME2 parametrization [15] at finite temperature, $T = 50$ MeV. Left panel: the hyperon–scalar meson coupling constants are fixed by the quark model. In this figure the dependence on the presence of trapped neutrinos is shown. The dot-dashed line (green online) corresponds to the case without neutrinos. The presence of neutrinos make the EoS stiffer. The dashed region represents the variation of the EoS with the lepton fraction. The dashed line (red online) corresponds to a lepton fraction $Y_L = 0.1$ and the full line (blue online) to $Y_L = 0.4$. Right panel: same as left, in which the coupling constants are the one of the stiffest case considered, corresponding to $g_{\Sigma\sigma} = 0.448 g_{N\sigma}$, fixed by the NSC potential and $g_{\Lambda\sigma} = 0.52 g_{N\sigma}$.

present model that are consistent with the observational lower bound on the maximum mass of a compact star. Fig. 4 shows the EoS of finite temperature hypernuclear matter with trapped neutrinos; the lepton fraction is varied in the range $0.1 \leq Y_L \leq 0.4$ (shaded area in Fig. 4). This corresponds to the shaded area. The left and right panels correspond to the softest and the stiffest hypernuclear EoS described above. This compared to the neutrino-less case ($Y_\nu \neq 0$). Clearly neutrinos stiffen the EoS matter - the larger is the fraction of neutrinos, i.e. Y_L, the stiffer is the EoS. The stiffening of the EoS can be attributed to the fact that the thermal population of neutrinos adds its contribution to the pressure of matter.

The obtained EoS can be conveniently represented by piecewise polytropic EoS of the form [3] [1]

$$P = \sum_{i=1}^{4} K_i (\rho/\rho_0)^{\Gamma_i} \theta(\rho - a_i\rho_0)\theta(b_i\rho_0 - \rho), \tag{5}$$

where Γ_i is the polytropic index, K_i is a dimensionful constant, $[K_i] = \text{MeV fm}^{-3}$, and ρ_0 is the saturation density; the values of the fit parameters can be found in [3].

[1] We use the occasion to correct a misprint in the original formula (Eq. (32) of [3]), where the density normalization by ρ_0 is missing.

4. Conclusions

Because the information on the properties of hypernuclear matter is far less extensive than for nucleons it is currently impossible to exclude hyperons as constituents of densest regions of compact stars. Our study [3] confirms this within a specific relativistic density functional approach to hypernuclear matter with tuned hyperon–scalar-meson couplings. We find that hyperonization in massive stars is favored for small ratios of the hypernuclear-to-nuclear couplings; in particular, hyperons need to be coupled to scalar mesons weaker than predicted by the SU(6) quark model. For certain values of the hyperon–scalar meson couplings hypernuclear EoS can still produce stellar equilibrium configurations of compact stars compatible with the two-solar-mass pulsar observations.

Neutrino trapping leads to strong modification in the population of hyperons and to a shift in the threshold density at which they first appear. As a consequence, a stiffening of the EoS in the early stage of the neutron star formation is observed. Instead of deleptonization with increasing density, seen in neutrino-less matter, the abundances of charged leptons remain constant, which among other things leads to inversion of the density thresholds for appearance of charged Σ's.

Acknowledgments

The work of GC was supported by the HGS-HIRe graduate program at Frankfurt University.

References

[1] Demorest P B, Pennucci T, Ransom S M, Roberts M S E and Hessels J W T 2010 *Nature* **467** 1081–1083 (*Preprint* 1010.5788)
[2] Antoniadis J, Freire P C C, Wex N, Tauris T M, Lynch R S, van Kerkwijk M H, Kramer M, Bassa C, Dhillon V S, Driebe T, Hessels J W T, Kaspi V M, Kondratiev V I, Langer N, Marsh T R, McLaughlin M A, Pennucci T T, Ransom S M, Stairs I H, van Leeuwen J, Verbiest J P W and Whelan D G 2013 *Science* **340** 448 (*Preprint* 1304.6875)
[3] Colucci G and Sedrakian A 2013 *Phys. Rev. C* **87** 055806 (*Preprint* 1302.6925)
[4] Glendenning N K 1982 *Physics Letters B* **114** 392–396; Glendenning N K 1985 *Astrophys. J.* **293** 470–493; Glendenning N K 1987 *Zeitschrift fur Physik A Hadrons and Nuclei* **326** 57–64; Glendenning N K 1987 *Zeitschrift fur Physik A Hadrons and Nuclei* **f 327** 295–300
[5] Weber F and Weigel M K 1989 *Nuclear Physics A* **493** 549–582
[6] Kapusta J I and Olive K A 1990 *Physical Review Letters* **64** 13–15
[7] Ellis J, Kapusta J I and Olive K A 1991 *Nuclear Physics B* **348** 345–372
[8] Glendenning N K and Moszkowski S A 1991 *Physical Review Letters* **67** 2414–1417
[9] Ellis P J, Knorren R and Prakash M 1995 *Physics Letters B* **349** 11–17 (*Preprint* arXiv:nucl-th/9502033)
[10] Schaffner J and Mishustin I N 1996 *Phys. Rev. C* **53** 1416–1429 (*Preprint* arXiv:nucl-th/9506011)
[11] Weber F (ed) 1999 *Pulsars as astrophysical laboratories for nuclear and particle physics*
[12] Sedrakian A 2007 *Progress in Particle and Nuclear Physics* **58** 168–246 (*Preprint* arXiv:nucl-th/0601086)
[13] Schulze H J and Rijken T 2011 *Phys. Rev. C* **84** 035801
[14] Logoteta D, Vidaña I, Providência C, Polls A and Bombaci I 2012 *Journal of Physics Conference Series* **342** 012006
[15] Lalazissis G A, Nikšić T, Vretenar D and Ring P 2005 *Phys. Rev. C* **71** 024312
[16] Fuchs C, Lenske H and Wolter H H 1995 *Phys. Rev. C* **52** 3043–3060 (*Preprint* arXiv:nucl-th/9507044)
[17] Sakurai J J 1960 *Annals of Physics* **11** 1–48
[18] KLOE Collaboration 2009 *Journal of High Energy Physics* **7** 105 (*Preprint* 0906.3819)
[19] de Swart J J 1963 *Reviews of Modern Physics* **35** 916–939
[20] Maessen P M M, Rijken T A and de Swart J J 1989 *Phys. Rev. C* **40**(5) 2226–2245 URL http://link.aps.org/doi/10.1103/PhysRevC.40.2226
[21] Bogdanov S 2013 *Astrophys. J.* **762** 96 (*Preprint* 1211.6113)

Journal of Physics: Conference Series **496** (2014) 012004 doi:10.1088/1742-6596/496/1/012004

Nuclear constraints on the equation of state and rotating neutron stars

A E L Dieperink[1] **and C Korpa**[2]

[1] Kernfysisch Versneller Instituut, NL-9747AA Groningen, The Netherlands

[2] Department of Theoretical physics, University of Pécs, Ifjúság útja 6, 7624 Pécs, Hungary

E-mail: `dieperink@kvi.nl`

Abstract. In this contribution nuclear constraints on the equation of state for a neutron star are discussed. A combined fit to nuclear masses and charge radii leads to improved values for the symmetry energy and its derivative at nuclear saturation density, $S_{\rm v} = 31$ MeV and $L = 68 \pm 8$ MeV. As an application the sensitivity of some properties of rotating supramassive neutron stars on the EoS is discussed.

1. Nuclear Constraints on the Equation of State

Despite numerous efforts to tighten the nuclear constraints on the EoS there remains considerable uncertainty. The spreading in the pressure at nuclear saturation density as summarized by Lattimer [1] a decade ago was roughly a factor six; results of present day mean field calculations vary by about a factor four [2]. Therefore it remains a challenge to try to improve the situation. The pressure as a function of density is given by $P(\rho) = \rho d\epsilon/d\rho - \epsilon(\rho)$.
In neighborhood of saturation density ρ_s, with $u = \rho/\rho_s$ and x proton fraction one has

$$\epsilon(u,x) = B + K/18(u-1)^2 + S_A(u)(1-2x)^2 + .., \tag{1}$$

where K the compressibility and S_A the symmetry energy (SE). Hence the pressure near ρ_s is

$$P(\rho \sim \rho_{\rm s}) \sim u^2\rho_{\rm s}[\frac{K}{9}(u-1) + \frac{dS_A}{du}(1-2x)^2 + ..]. \tag{2}$$

In practice the leading contribution comes from the last term, the derivative of the SE; the latter is usually parameterized in the liquid drop model (LDM) as

$$S_A = \frac{(N-Z)^2}{A}\frac{S_{\rm v}}{1+yA^{-1/3}}, \tag{3}$$

where $S_{\rm v}$, $S_{\rm s}$ denote the volume and surface SE, and $y = S_{\rm v}/S_{\rm s}$.
The quantity of interest for the EoS, the derivative $L = 3\rho_s dS/d\rho|_s$, can in good approximation [1] be related to $S_{\rm v}$ and y: $y \sim 0.646 + S_{\rm v}/98\text{MeV}+0.436L/S_{\rm v} + 0.087(L/S_{\rm v})^2$.
In practice the values of $S_{\rm v}, y$ when fitted to masses using the LDM appear to be strongly correlated [3], and the same is true for L, $S_{\rm v}$ (see the 1-σ confidence ellips in fig. 1), and a similar correlation is found in microscopic (mean field) models.

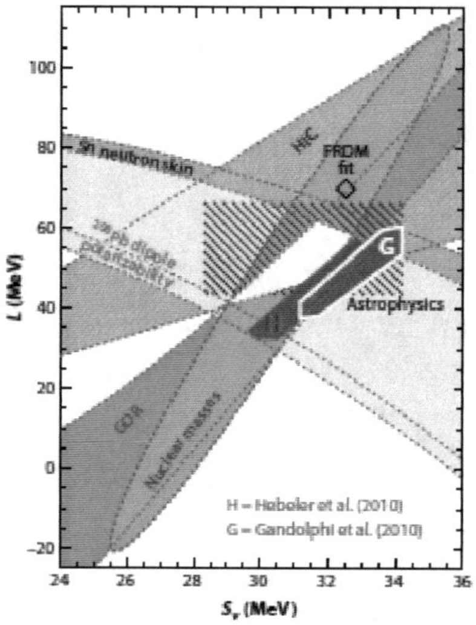

Figure 1. Constraints on L *vs* S_{v}, the filled ellipsoid corresponds to the constraint from fitting nuclear masses, from [1].

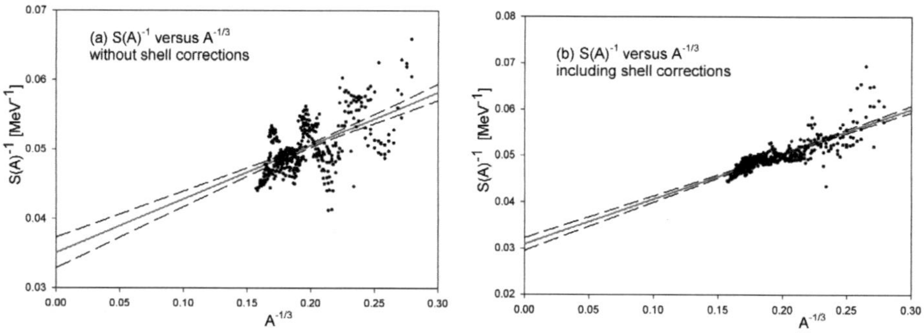

Figure 2. SE vs $A^{-1/3}$ without (left) and with (right) shell corrections, from [4].

However, one can improve the situation sketched above in several ways. As a first step one may consider differentials of masses with respect to $N - Z$ (rather than a global fit), which allows one to fit the parameters in $\frac{S_{\mathrm{v}}}{1+yA^{-1/3}}$ in isolation of other terms [4, 5].

By plotting $1/S$ vs $A^{-1/3}$ (see fig. 2) one obtains the value of $1/S_{\mathrm{v}}$ from the crossing of the fit line with y-axis ($A = \infty$) and the slope $y = 2.6 \pm 0.8$. From the figure the correlation between slope and S_{v} is evident.

As a second step an appreciable increase in the accuracy can be achieved by including shell corrections [4] shown in the right part of the figure. In passing we note that in ref. [6] a quite accurate result for S_{v}, S_{s} is reported by using double differences of masses. However, their results are obtained for a parametrization of the SE different from eq. (3), and moreover depend on the choice of the Wigner energy.

As a final step one can improve the situation further by using information from charge radii, which mainly depends only on the ratio y. (In fig. 1 this is indicated by the band labeled "skins

of Sn", but we consider the result of this particular analysis rather model dependent).
In the spirit of the LDM and distinguishing proton and neutron radii we decompose [4]

$$R_{\rm p}(N,Z) = R_0(N,Z) + \frac{N}{A}R_{\rm np}(N,Z) + \delta R_{\rm C}(N,Z), \tag{4}$$

where the isoscalar term $R_0 = (NR_{\rm n} + ZR_{\rm p})/A \sim r_0 A^{1/3}$, and the isovector term (essentially the neutron skin)

$$R_{\rm np} = R_{\rm n} - R_{\rm p} = \frac{2r_0}{3}\frac{N-Z}{A}\frac{1}{1+A^{1/3}/y}, \tag{5}$$

and R_C the Coulomb contribution [3]. The point is that $R_{\rm np}$ depends only on $y = S_{\rm v}/S_{\rm s}$ (apart from the Coulomb contribution).
To determine y from data one can envision the following options
(i) measure the neutron skin using parity violating electron scattering (PREX). However, the first experiment [7] on ^{208}Pb yielded a rather large error $R_{\rm np} \sim 0.33 \pm 0.17$ fm.
(as a side remark: atomic parity violation, in progress, appears a promising alternative tool, with a possible precision of about 1% in the skin in Ra isotopes),
(ii) fit to observed charge radii using the expression (4). (As an alternative one may consider fitting differences like isobar shifts, $R_{\rm p}(N,Z) - R_{\rm p}(N-i, Z+i)$; the latter are independent of R_0, but in general have larger experimental uncertainties.)
The values for $S_{\rm v}$ and y from a combined fit of masses and radii are given in table 1, which are compared to some other results from fits and microscopic approaches. Note that phenomenology

Table 1. Results for the parameters $S_{\rm v}$, y and L obtained from fits and microscopic approaches

$S_{\rm v}$(MeV)	y	L(MeV)	$R_{\rm np}(^{208}{\rm Pb})$ (fm)	ref	model
		fit to masses			
32.5 ± 0.5	1.98	70 ± 15		[8]	FRLDM
32.1 ± 0.3	1.9			[6]	double diff
31.1	2.31 ± 0.38	66 ± 13		[10]	LDM
32	3.0	94		[5]	analysis IAS
31	2.5 ± 0.4	80 ± 15		[4]	LDM+shell corr
30.5	2.35 ± 0.20	68 ± 8	0.185 ± 0.015	present	masses+ charge radii
		Microscopic	approaches		
30		58 ± 15		[9]	Skyrme+skin Sn isotopes
32 ± 1.5		45 ± 15		[11]	QMC
31 ± 1	1.85 ± 0.25	46 ± 10		[13]	EFT
31 ± 1		64 ± 8	0.195 ± 0.02	[12]	EDF

seems to favor larger L values than most microscopic models; this trend is not understood yet.

2. Rotating supramassive neutron stars and the EoS

Several energetic observations can be associated with formation of neutron stars (NS) or black holes (BH), supernovae, gamma ray bursts (GRB). Some short GRB's (~1s) have been attributed to NS mergers.
Recently the observation [14] of bright radio pulses was reported, with radio flux ~ Jy at GHz frequencies and $\Delta T \leq 1$ms, which do not repeat, while no γ- or x-rays were observed.

Journal of Physics: Conference Series **496** (2014) 012004 doi:10.1088/1742-6596/496/1/012004

Falcke and Rezzolla [15] proposed the following interpretation: a supramassive rotating NS (i.e., a NS with a mass larger than the maximum mass of a static NS, *e.g.* created by accretion in binary system) slows down due to magnetic braking, and at critical point collapses into Kerr BH (see fig. 2). The created event horizon will hide star's surface, hence only emission from the detached magnetosphere can be observed; the estimated timescale (freefall) $\tau \sim 0.04 R_{10}^{3/2} M_2^{-1/2}$ ms appears to be consistent with the observation.
For a non-rotating star the mass vs radius relation (given the EoS) is obtained by solving the TOV equation; a rotating star requires a more general approach to general relativity. The scenario proposed in ref. [15] seems not unrealistic but some questions remain. For example the authors took a very simple representation for the EoS: a single polytrope $P = K\rho^\gamma$ with $\gamma = 2$ and K adjusted such that $M(0)_{\max} = 2.1 M_\odot$.
Naturally one may ask how large is the sensitivity to EoS? To investigate this we took a 3-polytrope EoS , $P(\epsilon) = K_i \epsilon^{\Gamma_i}$. It fits $P(\rho_s)$ and has the proper low-density behavior; specifically, $\Gamma_1 = 1.5$ for $\epsilon < 67$ MeV/fm^3, $\Gamma_2 = 2.68$, for $67 < \epsilon < 650$, and $\Gamma_3 = 1.41$ for $\epsilon > 650$. The K_i values are fixed by continuity of the pressure P and the normalization $P(\epsilon = 650) = 180$ MeV/fm^3. It yields $M_{\max} = 2.2 M_\odot$.
Using the rns code [17] the mass as a function of the equatorial radius or the central density, and the critical frequency (the Keppler or mass shedding limit) have been computed, see fig. 4. Qualitatively the main features of fig. 2 are confirmed, i.e., $M(\Omega_K)$ increases by 20%, $R(\Omega_K)$ increases by 50%. However, Ω_K itself turns out to be more sensitive to the EoS [18, 16].
Finally we point out that an observation of a high rotation frequency of a pulsar can lead to a constraint on the mass-radius diagram. Namely for a Newtonian uniformly rotating rigid star with mass M and radius R one has

$$\tau_{\min} = 2\pi\sqrt{\frac{R^3}{GM}} = 0.545 \left(\frac{M_\odot}{M}\right)^{1/2} \left(\frac{R}{10\text{km}}\right)^{3/2} \text{ms}. \tag{6}$$

Using general relativity a similar empirical relation (valid for $M < 0.9 M_{\max}(0)$ with a weak dependence on the EoS) has been derived [18] (M, R refer to the static star)

$$\tau_{\min} = C\ (M_\odot/M)^{1/2} (R/10\text{km})^{3/2} \text{ms}, \tag{7}$$

where $C \sim 0.92 \pm 0.04$; with our EoS we find $C = 0.90$. At present, the fastest rotating pulsar has $\nu_{obs} = 716$ Hz. Obviously, the Keplerian frequency of any neutron star must satisfy $\nu_K \geq \nu_{\text{obs}}$. This inequality constraints a region on the M *vs* R diagram as shown in the left panel of fig. 4.

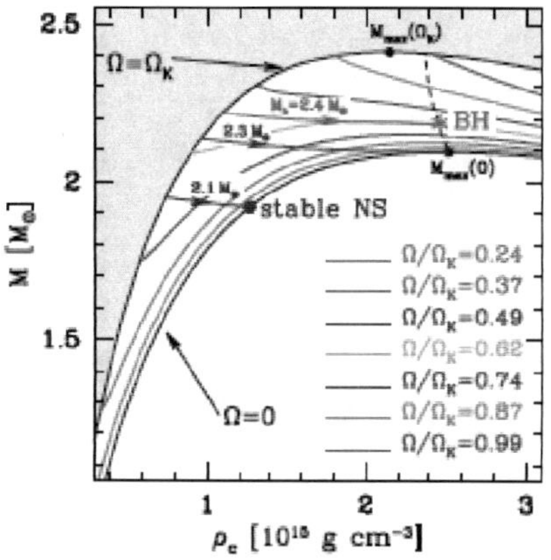

Figure 3. Gravitational mass *vs* central density for various ratios $f = \Omega/\Omega_K$, where Ω_K is taken as the critical Ω belonging to star with maximum mass; arrows represent tracks of NS's slowing down due to magnetic braking, from [15].

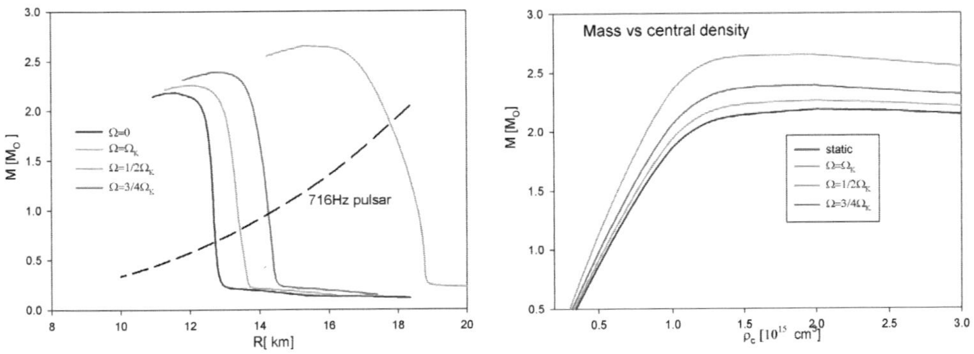

Figure 4. Gravitional mass *vs* radius (left) and *vs* central density (right) for various Ω. In the left panel the broken line denotes the lower limit on the mass of a static star from eq. (7).

References

[1] Lattimer J M (2012) *Ann. Rev. Nucl. Part. Sci.* **62** 485
[2] Steiner A and Gandolfi S (2012) *Phys. Rev. Lett.* **108** 081102
[3] Danielewicz P (2003) *Nucl. Phys.* **A727** 203
[4] Dieperink A and van Isacker P (2009) *Eur. J. Phys.* **A32** 11
[5] Danielewicz P and Lee J (2009) *Int. J. Mod. Phys.* **E18** 892
[6] Jiang H *et al.* (2012) *Phys. Rev.* **C85** 024301
[7] Abrahamyan S *et al.* (2012) *Phys. Rev. Lett.* **108** 112502
[8] Möller P *et al.* (2012) *Phys. Rev. Lett.* **108** 052501
[9] Chen L W *et al.* (2010) *Phys. Rev.* **C82** 024321
[10] Liu M *et al.* (2010) *Phys. Rev.* **82** 064306
[11] Gandolfi S *et al.* (2012) *Phys. Rev.* **C85** 032801(R)
[12] Agrawal B K *et al.* (2013) *arXiv:* nucl-th/1305.5336
[13] Hebeler K *et al.* (2010) *Phys. Rev. Lett.* **105** 160102
[14] Thornton D *et al.* (2013) *Science* **341** 53
[15] Falcke H and Rezzolla L (2013) *arXiv:* astro-ph/1307.1409
[16] Lo K-W and Lin L-M (2011) *Astrophys. J.* **728** 12
[17] Stergioulas N and Friedman J L (1995) *Astrophys. J.* **444** 306
[18] Haensel P *et al.* (2009) *A.&A.* **502** 605

2nd International Symposium on the Modern Physics of Compact Stars and Relativistic Gravity IOP Publishing
Journal of Physics: Conference Series **496** (2014) 012005 doi:10.1088/1742-6596/496/1/012005

On the radii of hot strange stars

A G Alaverdyan, G S Hajyan

Yerevan State University, Alex Manoogian str. 1, Yerevan, 0025, Armenia

E-mail: gevorg.hajyan@gmail.com

Abstract. The equilibrium states of hot strange stars are studied. The equation of state of hot strange quark matter is determined on the basis of the MIT bag model. It is shown that for hot strange quark stars the mass-central density and mass-radius relations do not depend on the central temperature if quarks are ultrarelativistic. For these relations the weight of the thermal energy is of fundamental significance. That explains the nature of change in the radius of a strange star when it cools down.

1. Introduction

The dependence of the radii of neutron stars on their temperature is generally small and is frequently neglected. The studies of the evolution of the hot white dwarfs (WD) and neutron stars (NS) assume that their radii are constant in time [1, 2, 3, 4, 5]. For central temperatures $T_c < 10^{10}$K the constant radius approximation is valid. The cooling time of a NS from an initial central temperature $T_c = 10^{10}$K to a value $T_c = 10^9$K is about a year [2, 3, 5]. The cooling time from $T_c = 10^{11}$K to $T_c = 10^{10}$K is even less – of the order of from several minutes to several hours. Initial temperatures of neutron stars and hot strange stars (hereafter HSS) can be of the order of $T_c \sim 10^{12}$K if the major part of the gravitational energy of the progenitor star is converted into heat.

In Ref. [6] sequences of isothermal strange stars (SS) were computed. The equation of state (EoS) of hot strange quark matter (HSQM) was taken to be the one of the MIT bag model [7]. For simplicity, the presence of electrons in Ref. [6] was ignored, and the quark-gluon interaction constant was assumed to vanish. The calculations were performed for the cases where the bag parameter B depends on the quark density and where it does not depend on this density. From these calculations it follows that the higher the temperature the smaller is the maximum mass for any given isothermal sequence (see Figs. 7 and 8 in Ref. [6]). However, because the pressure of matter at constant density of particles is always increasing with the increase of temperature, the temperature dependence of the mass maximum found in Ref. [6] appears to be strange. Similar strange behavior is observed for the radii of some configuration as a function of the temperature.

In Ref. [6] the EoS of HSQM is determined in a complicated way. We believe that for this equation of state it was sufficient to use the known results for the thermal corrections to the energy and pressure of a degenerate Fermi gas [8]. The absence in Ref. [6] of the explanation of the strange temperature dependences of the maximum mass and radius of HSS, as well as calculations for stars with a fixed baryonic number, motivated us to reconsider the problem of determining of the integral parameters of strange quark star at non-zero temperature.

2. Problem statement, the equation of state

The purpose of this study is to determine

(i) the equilibrium configurations of hot strange stars,
(ii) the dependence of the radius and mass of an SS, with fixed baryonic number on temperature
(iii) the amount of energy lost during the cooling of HSS.

To solve these problem we use the MIT bag model of quark matter [7] in its simplest form: quarks are ultrarelativistic, the constant of the quark-gluon interaction $\alpha_c = 0$, and the bag parameter B is constant. The energy density and pressure within this model, including the leading order thermal corrections to the zero-temperature ideal gas result [8] are given by

$$\begin{aligned}
\epsilon &= \sum \epsilon_i(\mu_{0i}, T) + B, && (1)\\
P &= \sum P_i(\mu_{0i}, T) - B, && (2)\\
\epsilon_i &= 3\mu_i^4 \, [\, 1 + (m_i/\mu_{0i})^2 + (2\pi^2/3)\cdot(T/\mu_{0i})^2 \,] \,/\, 4\pi^2, && (3)\\
P_i &= \mu_i^4 \, [\, 1 - (m_i/\mu_{0i})^2 + (2\pi^2/3)\cdot(T/\mu_{0i})^2 \,] \,/\, 4\pi^2, && (4)\\
n_i &= (\mu_{0i}^2 - m_i^2)^{3/2} \,/\, \pi^2. && (5)
\end{aligned}$$

Here m_i is the mass, μ_i is the chemical potential, ϵ_i and P_i are the energy density and pressure and n_i is the density of quarks at $T = 0$. The index i labels the quarks with fixed discrete quantum numbers (i.e. color, flavor). Here and below we work with natural units where $\hbar = k = c = 1$.

Because the quarks are assumed to be ultrarelativistic, we can set in Eqs. (1)-(3) $m_i = 0$. This approximation is well satisfied for the light u and d flavors, but holds only approximately for the s quark ($m_s \approx 150$) MeV. The non-zero mass the s quark leads to a violation of symmetry in the chemical composition. It is entirely due to this that in the local electrical neutrality implies the presence of electrons in SQM. In Ref. [6] the mass of the s-quark is taken into account but at the same time it is assumed that electrons are absent. The electrical neutrality is ensured by the condition

$$n_u = (n_d + n_s)/2. \tag{6}$$

We consider such an approach to be inconsistent and erroneous. In SS the Fermi energy of quarks reaches values of the order of $500 - 600$ MeV, which is 3-4 times larger than m_s. Therefore, in a first approximation m_s may be neglected. In this approximation electrons are absent, so the condition of beta-equilibrium between quarks follows from the equality of their chemical potentials $\mu_i(n_i, T)$:

$$\mu_u(n_u, T) = \mu_d(n_d, T) = \mu_s(n_s, T). \tag{7}$$

It is easy to show that these conditions are equivalent to the conditions:

$$\mu_{0u}(n_u) = \mu_{0d}(n_d) = \mu_{0s}(n_s). \tag{8}$$

In the state of thermodynamic equilibrium, the chemical potentials depend on the temperature and on baryonic number density $n = (n_u + n_d + n_s)/3$. To summarize, our models assumes that $\alpha_c = 0$, $B = 0$, and $m_s = 0$,, therefore the concentrations and the chemical potentials of all the quarks are equal. Introducing $\mu \equiv \mu_{0i}$ we obtain from (1)-(4)

$$P = (\epsilon - 4B), \tag{9}$$

i. e., within our approximations the pressure depends only on the energy density of HSQM.

3. The complete system of equations of the equilibrium state of HSS

The equilibrium structure of HSS is determined by the equations of hydrostatic equilibrium [the Tolman-Oppenheimer-Volkoff (TOV) equations]

$$\frac{dP}{dr} = -\frac{G}{r^2}\frac{(P+\varepsilon)(m+4\pi r^3 P)}{1-2G\,m/r}, \tag{10}$$

$$\frac{dm}{dr} = 4\pi r^2 \varepsilon, \tag{11}$$

where G is the gravitational constant, r is the radial coordinate, m is the mass within the sphere of the radius r. These equations should be supplemented by the equation for the baryonic charge $N(r)$ in the sphere of the radius r

$$\frac{dN}{dr} = 4\pi r^2 n \cdot (1-2Gm/r)^{-1/2}, \tag{12}$$

as well as by the condition of isothermality

$$T \cdot \exp(\nu/2) = const. \tag{13}$$

where g_{00} is the temporal metric coefficient in the Schwarzschild metric. In Ref. [6] it was assumed that T =const. in the entire star. However, in a strong gravitational field the quantity that remains constant within the star (either NS or SS) is $T\sqrt{g_{00}}$. Equations (12)-(13) should be supplemented by the equation for the field [9]

$$\frac{d\nu}{dP} = -\ 2/(\epsilon + P). \tag{14}$$

In our case from (9), (13) and (14) we obtain for $T(r)$

$$T(r) = T_c \left[\frac{P(r)+B}{P_c+B}\right]^{1/4}, \tag{15}$$

where T_c is the central temperature.

4. Numerical results and discussion

For given T_c and P_c we solved Eqs. (10)-(12) to obtain sequences of isothermal HSS. In our numerical calculations we assume the value of the bag constant $B = 80$ MeV. Since within the approximations outlined above the HSQM pressure depends only on the energy density the dependence of the mass of stellar configurations $M(\rho_c)$ on the central density $\rho_c = \epsilon_c/c^2$ of matter is independent of the central temperature of the star. This statement applies also to the radius-mass relationship $R(M)$. In contrast, the full baryonic number N_0 depends on the central energy density as well as the central temperature T_c. Figure 1 shows the dependence of the mass on the central density ρ_c in g/cm^3 and Fig. 2 shows the dependence of the radius R on the mass. According to the static stability criterion, the configurations on iso-entropic sequences are stable up to the first maximum. At the maximum of these curves the stability is lost [10, 11]. Since all the HSS including the iso-entropic ones are on the same $M(\rho_c)$ curve (Fig.1), all the configurations up to the maximum are stable. For the isothermal series from Ref. [6] this stability criterion is not applicable. To determine the point of loss of stability on these curves more research is needed. Below, only stable HSS will be considered. Let us find out how the radius of HSS changes when it cools. The dependence of the radius of SS and NS on their mass are very different. Self-confining SQM at zero pressure has a density $\rho_0 = 4B/c^2 = 5.7 \cdot 10^{14} g/cm^3$ (see

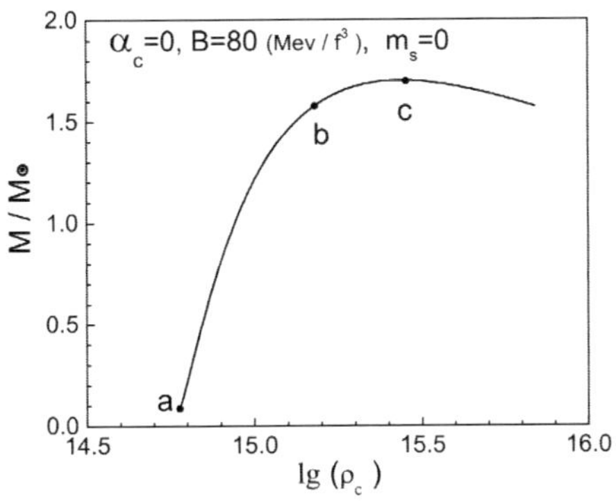

Figure 1. The masses M of hot strange stars as a function of the central density ρ_c in g/cm^3.

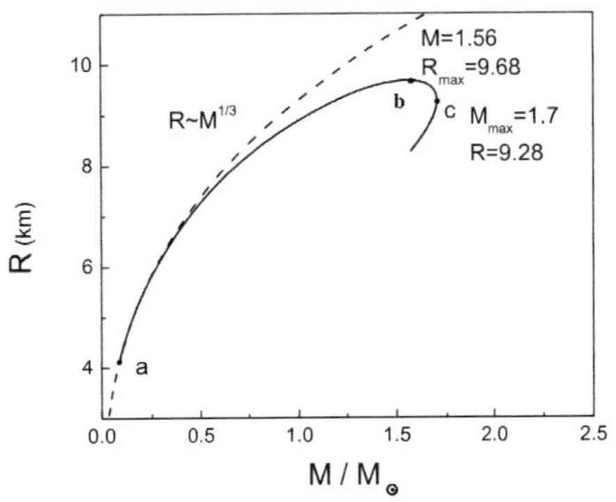

Figure 2. The radii R of hot strange stars as a function of the mass (see the explanations in text).

[1, 2, 3, 4]). For small masses $R \sim M^{1/3}$, as the gravitational forces are weak for a significant compression of SQM. In Fig. 2 this dependence is shown by the dashed curve. With the increase of mass, gravitational forces are growing and the radius SS grows slower than $R \sim M^{1/3}$. For the masses $M > 1.56M_\odot$ gravity is already so strong that the familiar from neutron stars form of the $R(M)$ curve is restored. In Fig. 2 the point b ($M = 1.56M_\odot$) separates these two areas. For different models of SQM and variants of the EoS the $R(M)$ dependence for cold SS was obtained in many studies, in particular in Refs. [12, 13, 14, 15]. As it has already been stated for the hot and cold SS $R(M)$ curves coincide. When a HSS cools its total baryonic number N_0 remains constant, and the mass decreases as the star loses energy. If at the initial moment HSS is in the region ab of the $R(M)$ curve, the cooling star will move down along this curve towards the point a. The final position of the SS on the $R(M)$ curve is determined only by the numerical value N_0. For small masses (the area close to the point a in Fig. 2) $R \sim (N_0/n_0)^{1/3}$, where n_0 is the baryonic number of SQM at $P = 0$. From (5) and (11) it follows that as the temperature decreases n_0 increases. Consequently, the radius of the HSS decreases. If the HSS is located close to the point b in Fig. 2 and its mass $M < 1.56M_\odot$ it mass will decrease as it cools down.

Although for these masses the gravitational forces are decreasing, the decrease of pressure and the increase of n_0 are dominant which again leads to compression of the star. Thus, the HSS located in the region ab in Fig. 2 will get compressed as they cool.

Consider a HSS which is born in a supernova explosion and has a high temperature. If the HSS at the formation moment is in the region bc of the $R(M)$ curve (see Fig. 2) it will expand as it cools, although its temperature and pressure will fall. At first glance, it is strange that when the pressure is reduced the star expands. In such cases, as a rule, stars compress, as in the case for example for hot white dwarfs. However, in the case of HSS the role of mass reduction due to the loss of energy during the cooling is much more important than for ordinary stars. The weakening of gravity is so significant that even with a decrease of pressure the star expands. If these HSS evolve along the $R(M)$ curve up to the point b, then they only expand. Otherwise, after passing the point b, the expansion will turn into compression.

Figures 3, 4 and 5 we show the dependence of the change of the radius on the central temperature T_c for HSS with baryonic numbers $N_0 = (0.65;\ 2;\ 2.4) \cdot 10^{57}$. In these figures we also show the initial and final values of the masses (in units of the solar mass) and the radii (in kilometers). For example a HSS with baryonic number $N_0 = 6.5 \cdot 10^{56}$ and initial parameters

$M = 0.77M_\odot$, $T_c = 80$ MeV $\simeq 9.28 \cdot 10^{11}$ K and $R = 8.27$ km loses 38 percent of its mass and monotonically compresses to $R_0 = 7.23$ km in the course of cooling down to $T_c = 0$ (see Fig. 3). The two other configurations ($N_0 = 2 \cdot 10^{57}$ and $N_0 = 2.4 \cdot 10^{57}$) with initial temperatures 47.5

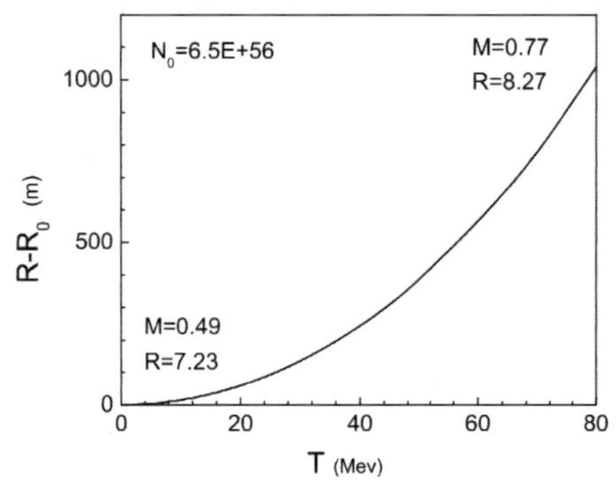

Figure 3. The change of the radius $R - R_0$ of the hot strange star with the barionic number $N_0 = 6.5 \cdot 10^{56}$ at cooling (explanations in text).

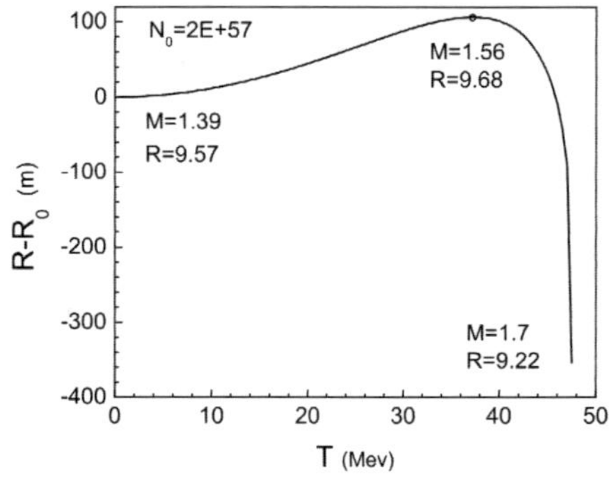

Figure 4. The change of the radius $R - R_0$ of a hot strange star with the baryonic number $N_0 = 2 \cdot 10^{57}$ during the cooling (see for explanation explanations in text).

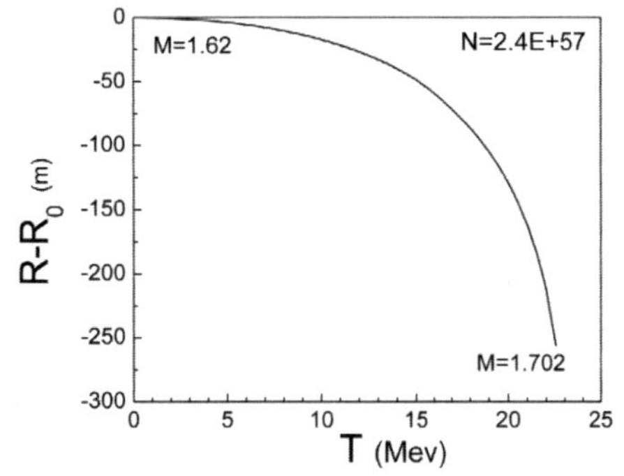

Figure 5. The change of the radius $R - R_0$ of a hot strange star with the baryonic number $N_0 = 2.4 \cdot 10^{57}$ during cooling.

MeV and 22.5 MeV respectively have the maximum masses $M = 1.7M_\odot$. The configuration with $N_0 = 2 \cdot 10^{57}$ during the cooling at first expands by ~ 100 m, and then compresses by ~ 450m. The third configuration only expands. We conclude that the strange behavior of the radius during the cooling of the HSS with $M > 1.56M_\odot$ is due to the weakening of the gravitational field caused by the reduction of the mass of the star.

5. Comparison with the results of Ref. [6]

There is a fundamental difference between our results and those of Ref. [6]. While in this study all isothermal sequences have the same $M = M(\rho_c)$, in Ref. [6] not only they don't coincide but also have significant differences in the numerical values (Figs. 7-12 and Table 1 and 2 of Ref. [6]). The reason for these differences may be only the non-zero value of the mass of the strange quark m_s. In Ref. [6] the value of m_s used in numerical calculations is not given. However, when $m_s = 0$ the conditions of chemical equilibrium (8) and electro-neutrality (6) [Eqs. (8), (9), (12) in Ref. [6]] are incompatible. Indeed, according to (5) and (8) $n_u = n_d$, because $m_u = m_d = 0$, and from (5) and (6) $n_u = n_s$, which at the same chemical potentials and $m_s \neq 0$ is impossible.

The obtained results qualitatively remain valid also in the case when the bag constant B depends on the concentration of barionic number. Surely the duration of the high temperature stage of HSS is very short. Neutrino energy losses, bremsstrahlung of electrons on the surface of the SS, electron-positron pair emission, etc. quickly cool the star [16]. However, we should keep in mind that if all the cold SS were formed by the above mentioned scenario, the maximum mass of the existing cold SS will be below the maximum value $M(\rho_c)$ .The HSS with a high temperature and maximum mass loses mass during cooling. From Fig. 4 and 5 it follows that the difference between the highest possible value of cold SS and maximum$M(\rho_c)$ can be significant. We plan to study this problem in a separate study.

Acknowledgments
The authors would like to thank Professors Yu. Vartanyan, G. Alaverdyan and the participants of the seminars of the Chair of Theory of Wave Processes of the Department of Radiophysics and of the Chair of Theoretical Physics of the Department of Physics at Yerevan State University for discussions.

References

[1] Schwarzschild H 1958 *Structure and Evolution of the Stars* **Prinseton University Press**
[2] Tsuruta S and Cameron A G W 1966 *Canad.J.Phys.* **44** 1863
[3] Hajyan G S and Vartanyan Yu L 1971 *AStrophysics* **7** 237
[4] Vartanyan Yu L, Hajyan G S and Harutyunyan A S 1973 *Soviet Astronomy* **50** 305
[5] Yakovlev D G, Gnedin O Y, Kaminker A D and Potekhin A Y 2007 *arXiv:0710.2047v2*
[6] Bordbar G H, Poostforush A and Zamani A 2011 *Astrophysics* **54** 277, arXiv:1103.2436v1
[7] Chodos A, Jaffe R, Johnson K, Thorne C B and Weisskopf V F 1974 *Phys. Rev.* D **9** 3471
[8] Landau L D and Lifshitz E M 1980 *Statistical Physics* **Butterworth-Heineman**
[9] Landau L D and Lifshitz E M 1975 *The Classical Theory of Fields* **Butterworth-Heineman**
[10] Zel'dovich Ya B 1963 *Voprosy kosmologii Izd. AN SSSR* **9** 36
[11] Bisnovatiy-Kogan G S 1974 *Astron. Astrophys.* **31** 3910
[12] Alcock C, Farhi E and Olinto A 1986 *Astrophys. J.* **310** 261
[13] Hansel P, Zdunik J L and Shaefer R 1986 *Astron. Astrophys* **160** 121
[14] Weber F, Glendenning N K and Weigel 1991 *Astrophys. J.* **373** 579
[15] Vartanyan Yu L, Harutyunyan A R and Grigoryan A K 1994 *Astrophysics* **37** 499
[16] Aksenov A G, Milgrom M and Usov V V 2005 *arXiv:astro-ph/0506647 v1*

Journal of Physics: Conference Series **496** (2014) 012006 doi:10.1088/1742-6596/496/1/012006

Delayed Pion Spectroscopy of Hypernuclei

Amur Margaryan[1,9], Patrick Achenbach[2], Robert Ajvazyan[1], John Annand[3], Franco Garibaldi[4], Nersik Grigoryan[1], Yuu Fujii[5], Osamu Hashimoto[5,10], Lubomir Majling[6], Satoshi N Nakamura[5], Josef Pochodzalla[2], Joerg Reinhold[7], Liguang Tang[8], Kyo Tsukuda[5], Henrik Vardanyan[1], Simon Zhamkochyan[1]

[1]Yerevan Physics Institute, 0036 Yerevan, Armenia
[2]Institut fur Kernphysik, Johannes Gutenberg-Universität, 55099 Mainz, Germany
[3]Department of Physics and Astronomy, University of Glasgow, Scotland, UK
[4]Istituto Superiore di Sanita Dipartmento TESA and INFN-Rome 1, Italy
[5]Department of Physics, Tohoku University, 980-8578 Sendai, Japan
[6]Nuclear Physics Institute, Academy of Sciences, Rez, Czech Republic
[7]Department of Physics, Florida International University, Miami, FL 33174, USA
[8]Department of Physics, Hampton University, Hampton, VA 23668, USA

E-mail: mat@mail.yerphi.am

Abstract. New possibilities of hypernuclear studies at modern electron accelerators based on recently developed radio frequency photomultiplier tubes are discussed.

1. Introduction

The studies of hypernuclear systems are of crucial importance for our understanding of the physics of neutron stars. The very early discussion by Ambartsumyan and Sahakyan [1] based on the model of free (non-interacting) hadron and lepton gases made a very plausible case for the existence of a non-vanishing hyperonic charge in neutron stars. Dense matter in β-equilibrium in the absence of strong interactions starts to populate Σ-hyperons already at $4n_0$, where n_0 is the normal nuclear matter density, whereas the lighter Λ-hyperons appear at $8n_0$. The inclusion of nuclear forces generically reduces these critical densities by substantial amount. Theoretically predicted thresholds for appearance of various hyperons are around $2n_0$ [2]. Because of the importance of interactions, high quality hypernuclear data are essential for the understanding not only the nuclear physics but also the physics of the compact stars (see [3] and references therein).

The binding energies of light hypernuclei provide the most valuable experimental information that constrains various models of YN interaction. Table 1, taken from Ref. [4], lists the results for the Λ separation energies obtained from theoretical ab initio calculations using different YN interactions, along with the existing experimental results [5]. It is seen that the precise experimental measurements of binding energies of light hypernuclei can discriminate various models of YN interactions.

The binding energies of light hypernuclei have been measured in emulsion exclusively from π^--decays [5]. In addition to the quoted statistical errors, they have also systematic errors of about 0.04

[9] Corresponding author
[10] Deceased

MeV. Monochromatic π^- mesons from weak decays of hypernuclei have been studied by magnetic spectrometers in fragmentation reactions with K^- beams on targets in the range from ^{7}Li to ^{16}O [6]. This group was the first to detect the decayed discrete π^-mesons from $^4_\Lambda H$ hyperfragments as a "*delayed-particle*" in stopped K^- reactions by magnetic spectrometer (see Ref. [7] for details). This result is a clear demonstration for a new "*delayed-* π^-" spectroscopy of nuclear matter.

Table 1. Λ separation energies, B_Λ given in units of MeV, of A = 3-5 Λ hypernuclei for different YN interactions.

YN	$B_\Lambda(^3_\Lambda H)$	$B_\Lambda(^4_\Lambda H)$	$B_\Lambda(^4_\Lambda H^*)$	$B_\Lambda(^4_\Lambda He)$	$B_\Lambda(^4_\Lambda He^*)$	$B_\Lambda(^5_\Lambda He)$
SC97d(S)	0.01	1.67	1.20	1.62	1.17	3.17
SC97e(S)	0.10	2.06	0.92	2.02	0.90	2.75
SC97f(S)	0.18	2.16	0.63	2.11	0.62	2.10
SC89(S)	0.37	2.55	Unbound	2.47	Unbound	0.35
Experiment	0.13±0.05	2.04±0.04	1.00±0.04	2.39±0.03	1.24±0.04	3.12±0.02

In 2007 the usage of magnetic spectrometers to measure the momenta of pions from weak two-body decays of electroproduced hyperfragments was proposed for Jefferson Lab [8]. The method is aimed at the determination of ground state masses of light Λ-hypernuclei off the stability line with high precision. Since the hypernuclei are produced from fragments, their nucleon and atomic numbers could be widely distributed, and hyperisotopes, e.g. A = 3-5 Λ hypernuclei, can be studied which are hardly accessible in missing-mass experiments. The proposed investigations are based on the unique time structure of the radio frequency (RF) driven electron accelerators, which provide a few ps electron bunches in every few ns, continuously, and are enabled by (1) the use of kaon spectrometer (HKS), (2) the precise magnetic spectrometer for hypernuclear decayed pions (HπS), and (3) the development of an ultra high resolution timing technique, based on recently developed novel photon detector RF Photo-Multiplier Tube (RF PMT) [9, 10].

The project was defined with two types of experiments as follows:

- Experimental studies based on simultaneous detection of kaons and decay pions by using HKS and HπS. In this case produced kaons are detected in coincidence with decayed π^- and used as a positive tag for strangeness production [11]. Similar experimental program was started at MAMI, Mainz [12, 13].
- Experimental studies which use HπS and ultra high resolution timing technique.

In this paper we are discussing perspectives of the decayed pion spectroscopy at RF driven accelerators by using HπS and ultra high resolution timing technique based on RF PMTs.

2. Delayed π^- spectroscopy

In the first type of experiments the produced kaons and decay pions were detected simultaneously by using two magnetic spectrometers, HKS and HπS (Fig. 1). The detection probability $R_{K^+}(HKS)$ of produced K$^+$ in the momentum acceptance $P_0 \pm \Delta P$ of the HKS magnetic spectrometer is: $R_{K^+}(HKS) = [\Delta\Omega/(4\pi)] \times \varepsilon_s^K \times \varepsilon_{ef} \approx 10^{-3}$, were $\Delta\Omega = 20-30$ msr is the acceptance of HKS, $\varepsilon_s^K \cong 0.5$ is the survival probability of K$^+$, $\varepsilon_{ef} \cong 0.5$ is the detection efficiency of the detector package, $\Delta P \cong 100$ MeV. Similar figure we have for the detection probability of decayed pions:

$R_{\pi^-}(H\pi S) \approx 10^{-3}$. These two events are correlated in time but not in space, consequently their simultaneous detection probability is: $R_{K^+\&\pi^-} = R_{K^+}(HKS) \times R(\pi^-) \times R_{\pi^-}(H\pi S)$, where $R(\pi^-) \leq 10^{-2}$ is the probability of decay pion formation for each produced K^+ [8]. Therefore, for the first type of experiment the useful event probability for each produced K^+ is $R_{K^+\&\pi^-} \approx 10^{-8}/K^+$.

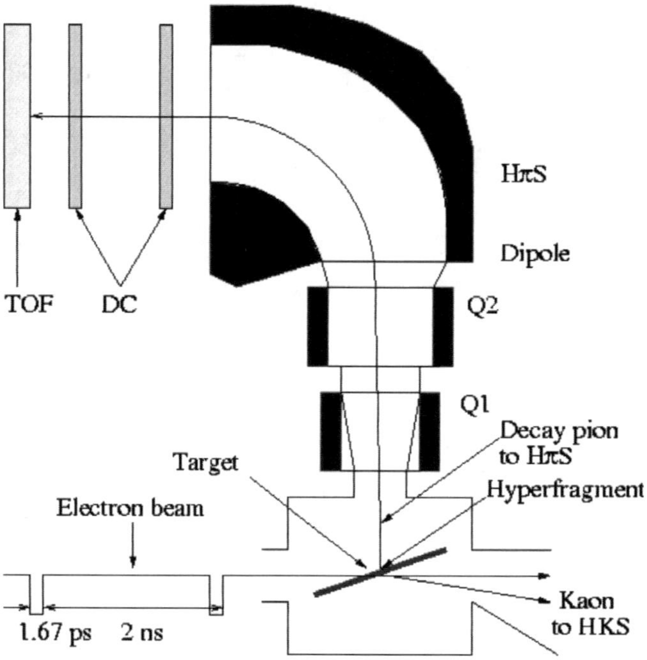

Figure 1. The scheme of the decayed pion experiment.

The useful event rate can be increased by several orders of magnitude by using the oncept of "*delayed-π^-*" spectroscopy [14]. The experimental setup in this case consists only of the decay pion spectrometer, HπS (see Fig. 1). The tracking detector package of HπS is the same as before, but as a timing technique, we propose to use RF PMT based Cherenkov detectors [15]. The expected time resolution of such detectors is better than 10 ps, FWHM. The expected reconstructed transit-time spread of pions in HπS is about or less than 20 ps, FWHM, and consequently the expected total time resolution is about or less than 30 ps, FWHM. With CEBAF at Jefferson Lab and MAMI at Mainz RF driven electron beams, the production time is about a few ps in every few ns. By using such a high resolution timing technique the decay pions can be separated from the huge amount of promptly produced background by using the time information only, i.e. without detecting K^+ mesons as a positive tag of strangeness production. Indeed, the probability to find promptly produced pions in the region with times larger than 100 ps is less than 10^{-5}, and ~70% of decay pions from Λ or hyperfragments (lifetime 260 ps) are delayed more than 100 ps. Therefore, the HπS with the RF PMT based Cherenkov detectors at RF driven electron beams, allows to carry out "delayed π^- spectroscopy", similar to "delayed γ-ray spectroscopy". The useful decay pion event probability in the "delayed π^-spectroscopy" experiment is: $R_{\pi^-} = 0.7 \times R_S \times R(\pi^-) \times R_{\pi^-}(H\pi S)$, where R_S is the probability of strangeness production. In this case, all two-body reactions, in which photons produce strangeness contributed. These reactions are: $1{:}\,\gamma + p \rightarrow \Lambda + K^+$; $2{:}\,\gamma + p \rightarrow \Sigma^0 + K^+$; 3: $\gamma + p \rightarrow \Sigma^+ + K^0$; $4{:}\,\gamma + n \rightarrow \Lambda + K^0$; $5{:}\,\gamma + n \rightarrow \Sigma^- + K^+$; 6: $\gamma + n \rightarrow \Sigma^0 + K^0$. In addition all useful virtual photon spectra take part, unlike the first type of experiment, were only K^+ production reactions, and a part of virtual photons, that produce K^+ mesons with momentum laying in HKS momentum acceptance ($\Delta P \cong 100$ MeV), contributed. Due to this $R_{\pi^-} \geq 10^3 \times R_{K^+\&\pi^-}$, i.e. the expected rates of useful events in case of "delayed π^-spectroscopy" is 10^3 times higher than rates expected in the first type of experiments.

3. Absolute calibration of the HπS

The key point of the proposed experimental program is determination of the binding energies of light hypernuclei by decay π^- momentum measurement. These moments are in the range around 100 MeV/c and could be measured precisely by using high resolution magnetic spectrometer $H\pi S$. However, method of absolute calibration of the magnetic spectrometer within 10^{-4} precision remains still open. Absolute calibration of the HπS can be realized by time-of-flight (TOF) measurement of promptly produced pions and electrons. We propose to use the high resolution and highly stable TOF system with RF PMT based Cherenkov detectors located on the focal plane of $H\pi S$ [16]. Indeed, the TOF of particle with mass m and momentum p and a flight path L is:

$$t_\pi = \frac{L}{\beta_\pi c} = \frac{L}{c}\left(1 + \frac{m_\pi^2 c^2}{p^2}\right)^{1/2} \quad (1)$$

$$t_e = \frac{L}{\beta_e c} = \frac{L}{c}\left(1 + \frac{m_e^2 c^2}{p^2}\right)^{1/2} \quad (2)$$

where c is the speed of light, $\beta = v/c$. From these equations we have:

$$\frac{L}{c} = \left(\frac{t_e m_\pi^2 - t_\pi m_e^2}{m_\pi^2 - m_e^2}\right)^{1/2} \quad (3)$$

$$p_\pi = \frac{L}{c}\frac{m_\pi c}{[t_\pi^2 - (L/c)^2]^{1/2}} \quad (4)$$

Therefore absolute calibration of the HπS can be performed by measuring flight times of pions, t_π and electrons, t_e, and determining L/c and p_π by using equations (3) and (4). The Monte Carlo simulations demonstrated that precision better than 10 keV can be achieved. This method of absolute calibration can be used continuously during experiment. It is assumed that for a fixed configuration of the system, the flight path length L and the magnetic rigidity of the spectrometer stay stable with in relative precision better than 10^{-4}. Similar technique is established for precise measurements of masses of exotic nuclei (see e.g. [17]).

Tagged-Weak π-Method

Hypernuclear investigations provide unique information on many-body hadronic systems by utilizing new "strangeness" degrees of freedom. A hyperon added to a nuclear system does not feel the Pauli exclusion. It also experiences nuclear forces that are different from those among nucleons in a nucleus. Therefore, a hyperon introduced in a nucleus may give rise to various changes of the nuclear structure. It may affect the size and the shape of a nucleus, change its cluster structure, lead to emergence of new symmetries, change the collective motions, etc. An example of how one can modify a nucleus by adding to it a distinguishable baryon is given by the experiment on γ-spectroscopy of $^{7}_{\Lambda}$Li [18]. When a Λ in 1s orbit is added to a loosely bound nucleus such as ^{6}Li, the nucleus is expected to shrink into a more compact system due to the attractive force between Λ and nucleons ("glue-like" role of the Λ). This effect can be verified from the E2 transition probability B(E2), which contains information of the nuclear size. Experimentally B(E2) is derived from the lifetime of the $5/2^+$ state. The expected lifetime ($\sim 10^{-11}$ sec) is of the same order as the stopping time of the recoil $^{7}_{\Lambda}$Li in lithium in the case of the (π^+, K^+) reaction at 1.05 GeV/*c*. It was determined to be $5.8^{+0.9}_{-0.7} \pm 0.7$ ps with the Doppler Shift Attenuation Method (DSAM). B(E2) was then derived to be 3.6 ± 0.5 e^2fm^4. This result, compared with the B(E2) = 10.9 ± 0.9 e^2fm^4 of the core nucleus ^{6}Li ($3^+ \rightarrow 1^+$), indicates about 20% shrinkage of the $^{7}_{\Lambda}$Li size vs. ^{6}Li. The DSAM can be applied to measure lifetimes in the range of $10^{-12} - 10^{-11}$ s. For the E2 transition with energy around ~1 MeV the expected lifetimes will be too long ($\sim 10^{-10}$ s) for DSAM and the γ transition competes with weak decay. For this case a new "γ-weak coincidence method" has been proposed [19] for the future J-PARC experiment.

Recently a new "tagged-weak π-method" has been proposed for measurements of electromagnetic rates of hypernuclear states with lifetimes of $\sim 10^{-10}$ s, tagged by weak decayed pions [20]. By using "tagged-weak π-method" and RF PMT based Cherenkov detectors, the electromagnetic rates of hypernuclear states with lifetimes down to 10^{-11} s could be measured. It is worthy to mention that unlike DSAM, "tagged-weak π-method" can be used for hyperfragments as well.

4. Conclusions

The recently developed high resolution RF PMTs opens new opportunities for hypernuclear studies. The RF PMT based Cherenkov detectors allow the realization of a new type nuclear spectroscopy, the "*delayed-*π^-" spectroscopy at modern electron accelerators. By using "*delayed-*π^-" spectroscopy, binding energies of hypernuclear ground and low lying states can be measured with precision of about 10 keV. In addition by measuring lifetimes of the low lying states and by applying "tagged-weak π-method", electromagnetic rates of hypernuclear states with lifetimes down to 10^{-11}s could be investigated. The experimental information on these characteristics of the ΛN interaction is essential for the discrimination and improvement of baryon-baryon interaction models, which are needed for a unified, comprehensive understanding of the strong force that determines the properties of the high-density nuclear matter with hyperons.

Acknowledgments
This work is supported in part by the State Committee of Science of Republic of Armenia (Grant Number 9-8/AB).

References

[1] Ambartsumyan V A and Saakyan G S 1960 The degenerate superdense gas of elementary particles *Sov. Astron.* **4** p 187
[2] Sahakian G S and Vartanian Y L 1963 On the possible phase states of matter at extremely high densities *Nuovo Cimento* **30** p 82
[3] Schaffner-Bielich J. 2008 Hypernuclear physics for neutron stars *Nucl. Phys.* A **804** Issues 1–4 pp 309–321 (*Preprint* astro-ph/0801.3791v1)

[4] Nemura H, Akaishi Y and Suzuki Y 2002 Ab initio approach to s-shell hypernuclei ${}^{3}_{\Lambda}H, {}^{4}_{\Lambda}H, {}^{4}_{\Lambda}He, and\ {}^{5}_{\Lambda}He$ with a $\Lambda N - \Sigma N$ interaction *Phys. Rev. Lett.* **89** 142504-1

[5] Davis D H, Pniewski J 1965 Hypernuclei *Contemp. Phys.* **27** p 91

[6] Tamura H et al. 1989 Formation of ${}^{4}_{\Lambda}H$ hypernuclei from K^{-}absorption at rest on light nuclei *Phys. Rev.* C **40** p R479

[7] Yamazaki T et al. 1988 New aspects and new tools in hypernuclear studies: experiments with a superconducting toroidal spectrometer *Preprint* INS-Rep.-728

[8] Margaryan A, Hashimoto O, Majewski S, Tang L 2007 Study of hypernuclei by pionic decay at JLab *Leter of intent to Jlab PAC31* LOI-07-001 Jefferson Lab

[9] Margaryan A et al 2006 Radio frequency picosecond phototube *Nucl. Instr. and Meth.* A **566** p 321

[10] Margaryan A, Ajvazyan R, Annand J, K. Gynashyan, I. Cox, S. Zhamkochyan 2012 Single photon THz timer with radio frequency photomultiplier tube *International Workshop on New Photon-detectors, LAL, Orsay, France June 13-15 (Proceeding of Science,* http://pos.sissa.it)

[11] Tang L, Margaryan A, Nakamura S N, Reinhold J, Yuan L (cospokespersons) 2007 Study of light hypernuclei by pionic decay at JLab *Exp. Proposal* E-08-012 Jefferson Lab

[12] Tsukada K et al. 2013 Decay pion spectroscopy of electro-produced hypernuclei *Few-Body Syst.* **54** p 375

[13] Esser A et al. 2013 Prospects for hypernuclear physics at Mainz: from KAOS@MAMI to PANDA@FAIR *Nucl. Phys.* A **914** pp 519-529

[14] Margaryan A 2007 RF picosecond timing technique and new possibilities for hypernuclear studies *The European Physical Journal A –Hadrons and Nuclei* **33** n3 p 47

[15] Margaryan A, Hashimoto O, Majewski S, Tang L 2008 RF Cherenkov timing technique for high energy physics applications *Nucl. Instr. and Meth.* A **595** p 274

[16] Margaryan A 2009 Radio frequency phototube, optical clock and precise measurements in nuclear physics *Preprint* arXiv: 0910.3011

[17] Estrade Alfredo et al 2008 Time-of-flight mass measurements of neutron rich nuclides, *10th Symposium on Nuclei in the Cosmos, July 27 - August 1 2008, Mackinac Island, Michigan, USA, Proceeding of Science,* PoS (NIC) p 184

[18] Tanida K et al 2001 Measurement of the B(E2) of ${}^{7}_{\Lambda}Li$ and shrinkage of the hypernuclear size *Phys. Rev. Lett.* **86** p 1982

[19] Tamura H 2001 High resolution spectroscopy of Λ hypernuclei: present status and perspectives *Nucl. Phys.* A **691** p 86c

[20] Margaryan A, Hashimoto O, Kakoyan V, Knyazyan S, Tang L 2011 Tagged-weak π-method *Physics of Atomic Nuclei* **74** Issue 2 p 216

Protons in High Density Neutron Matter

Misak M. Sargsian

Department of Physics, Florida International University, Miami, FL 33199, USA

E-mail: sargsian@fiu.edu

Abstract. We discuss the possible implication of the recent predictions of *two new properties* of high momentum distribution of nucleons in asymmetric nuclei for neutron star dynamics. The *first* property is about the approximate scaling relation between proton and neutron high momentum distributions weighted by their relative fractions (x_p and x_n) in the nucleus. The *second* is the existence of inverse proportionality of the high momentum distribution strength of protons and neutrons to $x_{p/n}$. Based on these predictions we model the high momentum distribution functions for asymmetric nuclei and demonstrate that it describes reasonably well the high momentum characteristics of light nuclei. We also extrapolate our results to heavy nuclei as well as infinite nuclear matter and calculate the relative fractions of protons and neutrons with momenta above k_F. Our results indicate that for neutron stars starting at *three* nuclear saturation densities the protons with $x_p = \frac{1}{9}$ will populate mostly the high momentum tail of the momentum distribution while only 2% of the neutrons will do so. Such a situation may have many implications for different observations of neutron stars which we discuss.

1. Introduction

One of the exciting recent results in the studies of short-range correlations (SRCs) in nuclei is the observation of the strong (by factor of 20) dominance of the pn SRCs, relative to the pp and nn correlations, for nuclear internal momenta of $\sim 300 - 600$ MeV/c[1, 2].

This observation is understood[1, 3, 4] based on the dominance of the tensor forces in NN interaction at the above mentioned momenta corresponding to average nucleon separations of $\sim$ 1 Fm. At these distances the dominating NN central potential crosses the zero due to transition from attractive to repulsive interaction allowing tensor forces to dominate in this transition range. The tensor interaction projects the NN SRC part of the wave function into the isosinglet - relative angular momentum, $L = 2$ state, almost identical to the D-wave component of the deuteron wave function. As a results pp and nn components of the NN SRC will be strongly suppressed since they are dominated by the central NN potential with relative $L = 0$. The *resulting picture* for the nuclear matter consisting of protons and neutrons at densities in which inter-nucleon distances are $\sim$ 1 Fm is rather unique: it represents a system with suppressed pp and nn but enhanced pn interactions.

The goal of our study is to understand the implication of the above described conditions on the the momentum distribution of protons and neutrons in high density nuclei matter.

2. New Relation between High Momentum p- and n-distributions in Nuclei:

Due to short range nature of NN interaction the nuclear momentum distribution, $n^A(p)$, for momenta, p, exceeding the characteristic nuclear Fermi momentum k_F is predominantly defined by the momentum distribution in the SRCs. There is a rather large experimental body of information indicating that for the range of $k_F < p \leq 600$ MeV/c the SRCs are dominated by 2N correlations, which consist of mainly the pn pairs (for recent reviews see [5, 6]).

In recent work[7] based on the dominance of the pn SRCs we predicted two new properties for the nuclear momentum distributions at $\sim k_F < p < 600$:
(i) There is an approximate equality of p- and n- momentum distributions weighted by their relative fractions in the nucleus $x_p = \frac{Z}{A}$ and $x_n = \frac{A-Z}{A}$:

$$x_p n_p^A(p) \approx x_n n_n^A(p), \tag{1}$$

with $\int n_{p/n}^A(p) d^3p = 1$.
(ii) The probability of proton or neutron being in high momentum NN SRC is inverse proportional to their relative fractions and can be related to the momentum distribution in the deuteron $n_d(p)$ as:

$$n_{p/n}^A(p) \approx \frac{1}{2x_{p/n}} a_2(A, y) \cdot n_d(p), \tag{2}$$

where $a_2(A, y)$ is interpreted as a per nucleon probability of finding 2N SRC in the given A nucleus[8, 9, 10] and the nuclear asymmetry parameter is defined as $y = |1 - 2x_p| = |x_n - x_p|$.

The above two properties are obtained assuming no contributions from pp, nn as well as higher order SRCs. They follow from the assumption that the whole strength of nuclear high momentum distribution as well as per nucleon probability of proton and neutron to be in the SRC is defined by the *same pn* correlation.

These properties can be checked for light nuclei by direct calculations of momentum distribution of nucleons in asymmetric nuclei (see below).

Predictions can be made also for heavy nuclei and infinite nuclear matter if one estimates the a_2 parameters entering in Eq.(2) for large A and extrapolate for infinite nuclear matter. Since SRCs are defined by local properties of nuclei, one expects that the A dependence of a_2 is related to the nuclear density, i.e. $a_2(A, y) = a_2(\rho, y)$. This could allow us to evaluate the high momentum part of the nucleon momentum distribution not only for finite[7] but also for infinite nuclear matter.

3. High Momentum Features of Light Nuclei

One can check the validity of the above two (Eqs.(1) and (2)) observations for light nuclei for which it is possible to perform realistic calculations based on the Faddeev equations for A=3 systems[11], Correlated Gaussian Basis(CGB) approach[12] as well as Variational Monte Carlo method(VMC) for light nuclei A (recently being available for up to $A = 12$ [13]).

The validity of Eq.(1) is checked in Fig.1 for ^3He nucleus, based on the solution of Faddeev equation[11], and for ^{10}Be based on VMC calculations[13]. The solid lines with and without squares in Fig.1(a) represent neutron and proton momentum distributions for both nuclei weighted by their respective x_n and x_p factors.

As one can see for ^3He the proton momentum distribution dominates the neutron momentum distribution at small momenta reflecting the fact that in the mean field the probability of finding proton is larger than neutron just because there are twice as much protons in ^3He than neutrons. The same is true for ^{10}Be for which now the neutron momentum distribution dominates at small momenta. However at ≥ 300 MeV/c for both nuclei, the proton and neutron momentum distributions become close to each other up to the internal momenta of 600 MeV/c. This is the region dominated by tensor interaction. As Fig.1(a) shows the high momentum distribution modeled according to Eq.(2) (dotted lines) agrees reasonably well with the realistic calculations.

Journal of Physics: Conference Series **496** (2014) 012007 doi:10.1088/1742-6596/496/1/012007

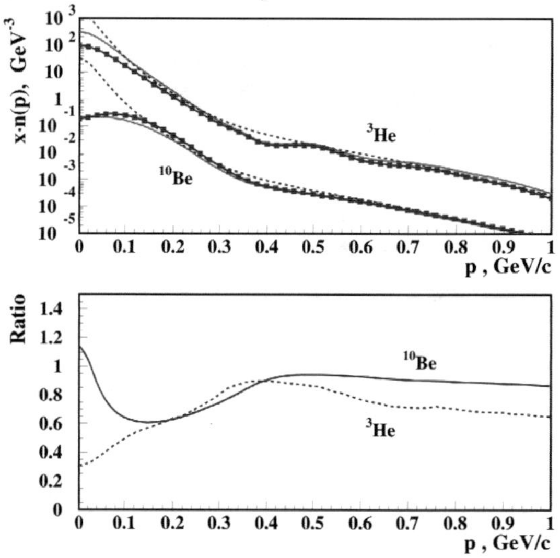

Figure 1. (a) The momentum distributions of proton and neutron weighted by x_p and x_n respectively. The doted lines represent the prediction for the momentum distribution according to Eq.(2). (b) The $x_{p/n}$ weighted ratio of neutron to proton momentum distributions. See the text for details.

Table 1. Kinetic energies (in MeV) of proton and neutron

A	y	E^p_{kin}	E^n_{kin}	$E^p_{kin} - E^n_{kin}$
^{8}He	0.50	30.13	18.60	11.53
^{6}He	0.33	27.66	19.06	8.60
^{9}Li	0.33	31.39	24.91	6.48
^{3}He	0.33	14.71	19.35	-4.64
^{3}He[11]	0.33	13.70	18.40	-4.7
^{3}He[12]	0.33	13.97	18.74	-4.8
^{3}H	0.33	19.61	14.96	4.65
^{8}Li	0.25	28.95	23.98	4.97
^{10}Be	0.2	30.20	25.95	4.25
^{7}Li	0.14	26.88	24.54	2.34
^{9}Be	0.11	29.82	27.09	2.73
^{11}B	0.09	33.40	31.75	1.65

This effect is more visible for the ratios of weighted n- to p- momentum distributions in Fig.1(b), demonstrating that the approximation of Eq.(1) in the range of 300 – 600 MeV/c is good on the level of 15%. Note that the similar features present for all other asymmetric nuclei calculated within the VMC method in Ref.[13].

The prediction of Eq.(2) can be checked by comparing the kinetic energies of proton and neutron, in which case we expect that per nucleon kinetic energy of the lesser component to be larger. As it can be seen from Table 1 this prediction is confirmed too by realistic calculations.

4. High Momentum Properties of Heavy Nuclei

Presently, no realistic calculations exist for asymmetric heavy nuclei for the predictions of Eqs.(1) and (2) to be checked. The main predictions of Eq.(2) is that high momentum protons and neutrons became increasingly *unbalanced* with an increase of the nuclear asymmetry, y. To quantify this, using Eq.(2) one can calculated the fraction of the nucleons having momenta $\geq k_F$ as:

$$P_{p/n}(A,y) \approx \frac{1}{2x_{p/n}} a_2(A,y) \int\limits_{k_F}^{\infty} n_d(p) d^3p, \tag{3}$$

where the parameter $a_2(A,y)$ can be taken from the experimental analysis of inclusive $A(e,e')$ data at kinematics dominated by SRC[10, 14, 15, 16]. The results of the estimates of $P_{p/n}(A,y)$ for medium to heavy nuclei are presented in Table 2.

Table 2. Fractions of high momentum ($\geq k_F$) protons and neutrons in nuclei A

A	$P_p(\%)$	$P_n(\%)$	A	$P_p(\%)$	$P_n(\%)$
12	20	20	56	27	23
27	24	22	197	31	20

As it follows from the table, as the asymmetry increases the imbalance between the high momentum fractions of proton and neutron grows. For example, in the Gold, the relative fraction of high momentum ($\geq k_F$) protons is 50% more than that of the neutrons.

Such a dominance of the proton high momentum component in heavy neutron rich nuclei can be checked in high momentum transfer single proton knock-out processes in which long-range nuclear effects are well controlled[17, 18]. The first such experimental verification for heavy nuclei is currently underway in quasi elastic $A(e,e,p)X$ measurement at Jefferson Lab, where the ratio of high momentum fractions of nucleons in ^{56}Fe and ^{208}Pb to that of ^{12}C is extracted. The results[19] are in reasonably good agreement with the prediction of Eq.(3) (Table 2) and they are being prepared for publication.

5. High Momentum Properties of Asymmetric Nuclear Matter

To estimate the fractions of energetic protons and neutrons in asymmetric nuclear matter using Eq.(3) one needs to extrapolate $a_2(A,y)$ for infinite nuclear matter. This was achieved in Ref.[20] where based on the local property of SRCs the $a_2(A,y)$ parameters were represented as a function of local nuclear density ρ and asymmetry y: $a_2(\rho,y)$. Then all the available data on inclusive $A(e,e')X$ scattering[21] at SRC kinematics were used to fit and extrapolate a_2 magnitudes for saturation and above saturation densities of nuclear matter at given asymmetry y. The accuracy of the fit was checked for symmetric nuclear matter at the saturation density, for which the obtained value of $a_2(\rho_0,0) \approx 7.03 \pm 0.41$ is in reasonable agreement with other estimates of a_2 for symmetric nuclear matter[22].

For the case of asymmetric nuclear matter we considered a neutron star matter, in which for asymmetry parameter y, we used the threshold value of $x_p = \frac{1}{9}$ ($y = \frac{7}{9}$) below of which the direct URCA processes:

$$n \to p + e^- + \bar{\nu}_e, \quad p + e^- \to n + \nu_e \tag{4}$$

will stop in the standard model of superdense nuclear matter consisting of degenerate protons and neutrons[23]. Estimating the Fermi momenta of protons and neutrons in Eq.(3) with $k_{F,N} = (3\pi^2 x_N \rho)^{\frac{1}{3}}$, in Fig.2 we present the off-Fermi-shell fractions of protons and neutrons as a function of nuclear density. The most interesting result of these estimates is that in the equilibrium, pn SRCs move the large fraction of protons above the Fermi-shell: at $3\rho_0$ densities half of the protons will be off-Fermi-shell while at $\rho \geq 4.5\rho_0$ all the protons will populate the high momentum tail of the momentum distribution. The situation however is not as dramatic for neutrons, with only about 2 % of neutrons populating the high momentum part of the momentum distribution.

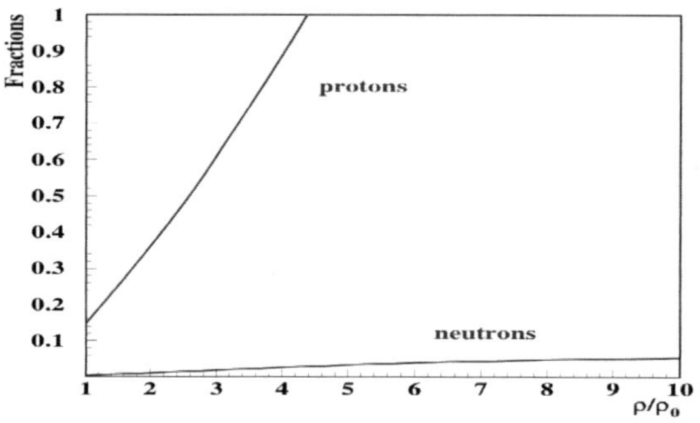

Figure 2. Density dependence of the fraction of off-Fermi-shell nucleons in $x_p = \frac{1}{9}$ matter.

It is worth mentioning that the present result of strong modification of proton momentum distribution in high density asymmetric matter is in qualitative agreement with the nuclear matter calculation based on Green function method[25].

6. Possible Implications for Nuclei and Neutron Stars:

Our main observation is that with an increase of nuclear asymmetry the lesser component become more energetic. This is confirmed[7] for light nuclei by direct estimates of the average kinetic energies of proton and neutron using realistic wave function calculations (Table 1). In the case of neutron rich heavy nuclei ($A \geq 40$), a larger fraction of protons will occupy the high momentum tail of the momentum distributions (Table 2). This may have several verifiable implications for large A nuclear phenomena[7]. Particularly our estimates show that it may explain[26] the large A part of the recently observed correlation between the strengths of the EMC (nuclear partonic distribution modification) and SRC effects[27].

The energetic protons in neutron rich nuclei will result also to the stronger nuclear modification of u-quarks as compared to d-quarks and the effect will grow with A. This provides an explanation[28] of the NuTeV anomaly[29]. The predicted effect also can be checked in parity violating deep inelastic scattering off heavy nuclei.

However, our observation may have more dramatic implications for the dynamics of neutron stars. Some of them are:

- *Cooling of a Neutron Star:* Large concentration of protons above the Fermi momentum will allow the condition for Direct URCA processes $p_p + p_e > p_n$[23] to be satisfied even if $x_p < \frac{1}{9}$. This will allow a situation in which intensive cooling of the neutron stars continues well beyond the critical point $x_p = \frac{1}{9}$ (see also Ref.[5]).
- *Superfluidity of Protons:* Transition of protons to the high momentum tail will smear out the energy gap which will remove the superfluidity condition for the protons.
- *Protons in the Neutron Star Cores:* The concentration of protons in the high momentum tail will result in proton densities $\rho_p \sim p_p^3 \gg k_{F,p}^3$. This will favor an equilibrium condition with "neutron skin" effect in which large concentration of protons populate the core rather than the crust of the neutron star. This and the proton superfluidity condition violation may provide different dynamical picture for generation of magnetic fields in the stars.

- *Isospin locking and the stiff equation of state of neutron stars:* With an increase in density more and more protons move to the high momentum tail where they are in short range tensor correlations with neutrons. In this case one would expect that high density nuclear matter to be dominated by configurations with quantum numbers of tensor correlations ($S = 1, I = 0$). In such a scenario protons and neutrons at large densities will be locked in the NN iso-singlet state. This will double the threshold of inelastic excitation from $NN \to N\Delta$ to $NN \to \Delta\Delta(NN^*)$ transition thereby stiffening the equation of state which is favored by the recent large neutron star mass observation[24].

7. Possible Universality of the Obtained Results

Our observation is relevant to any asymmetric two-component Fermi system in which the interaction within each component is suppressed while the mutual interaction between two components is enhanced. It is interesting that the similar situation is realized for two-fermi-component ultra-cold atomic systems[30] but with the mutual s-state interaction. One of the most intriguing aspects of such systems is that in the asymmetric limit they exhibit very rich phase structure with indication of the strong modification of the small component of the mixture[31]. In this respect our case is similar to that of ultra-cold atomic systems with the difference that the interaction between components has a tensor nature.

This work is supported by U.S. DOE grant under contract DE-FG02-01ER41172.

References

[1] E. Piasetzky, M. Sargsian, L. Frankfurt, M. Strikman and J. W. Watson, Phys. Rev. Lett. **97**, 162504 (2006).
[2] R. Subedi *et al.*, Science **320**, 1476 (2008).
[3] M. M. Sargsian, T. V. Abrahamyan, M. I. Strikman and L. L. Frankfurt, Phys. Rev. C **71**, 044615 (2005).
[4] R. Schiavilla, R. B. Wiringa, S. C. Pieper and J. Carlson, Phys. Rev. Lett. **98**, 132501 (2007).
[5] L. Frankfurt, M. Sargsian and M. Strikman, Int. J. Mod. Phys. A **23**, 2991 (2008).
[6] J. Arrington, D. W. Higinbotham, G. Rosner, M. Sargsian, Prog. Part. Nucl. Phys. **67**, 898 (2012).
[7] M. M. Sargsian, arXiv:1210.3280 [nucl-th].
[8] L. L. Frankfurt and M. I. Strikman, Phys. Rept. **76**, 215 (1981).
[9] L. L. Frankfurt and M. I. Strikman, Phys. Rept. **160**, 235 (1988).
[10] L. L. Frankfurt, M. I. Strikman, D. B. Day and M. M. Sargsian, Phys. Rev. C **48**, 2451 (1993).
[11] A. Nogga, H. Kamada and W. Glockle, Nucl. Phys. **A689**, 357 (2001).
[12] H. Feldmeier, W. Horiuchi, T. Neff and Y. Suzuki, Phys. Rev. C **84**, 054003 (2011).
[13] R. B. Wiringa, R. Schiavilla, S. C. Pieper and J. Carlson, arXiv:1309.3794 [nucl-th].
[14] K. S. Egiyan *et al.*, Phys. Rev. C **68**, 014313 (2003).
[15] K. S. Egiyan *et al.*, Phys. Rev. Lett. **96**, 082501 (2006).
[16] N. Fomin *et al.*, Phys. Rev. Lett. **108**, 092502 (2012).
[17] L. L. Frankfurt, M. M. Sargsian and M. I. Strikman, Phys. Rev. C **56**, 1124 (1997).
[18] M. M. Sargsian, Int. J. Mod. Phys. E **10**, 405 (2001).
[19] E. Piasetzky, http://www.int.washington.edu/talks /WorkShops/int_13_52W/People/Piasetzky_E/Piasetzky.pdf.
[20] M. McGauley, M. M. Sargsian, arXiv:1102.3973 [nucl-th].
[21] O.Benhar, D.Day and I.Sick, [arXiv:1104.1196 [nucl-ex]].
[22] C.Ciofi degli Atti, E. Pace, G.Salme, Phys. Rev. C **43**, 1155 (1991).
[23] J. M. Lattimer, M. Prakash, C. J. Pethick and P. Haensel, Phys. Rev. Lett. **66**, 2701 (1991).
[24] P. Demorest, T. Pennucci, S. Ransom, M. Roberts and J. Hessels, Nature **467**, 1081 (2010).
[25] A. Rios, A. Polls and W. H. Dickhoff, Phys. Rev. C **79**, 064308 (2009).
[26] M. M. Sargsian, arXiv:1209.2477 [nucl-th].
[27] L. B. Weinstein, *et al.*, Phys. Rev. Lett. **106**, 052301 (2011).
[28] M. Sargsian, "Neutrino interactions in the nuclear environment", APS Meeting, April, 2013; Denver, Co.
[29] G. P. Zeller *et al.* [NuTeV Collaboration], Phys. Rev. Lett. **88**, 091802 (2002).
[30] Y. Shin, M. W. Zwierlein, C. H. Schunck, A. Schirotzek and W. Ketterle, Phys. Rev. Lett. **97**, 030401 (2006).
[31] A. Bulgac and M. M. Forbes, Phys. Rev. A **75**, 031605 (2007).

Journal of Physics: Conference Series **496** (2014) 012008 doi:10.1088/1742-6596/496/1/012008

Inhomogeneous condensates in dilute nuclear matter and BCS-BEC crossovers

Martin Stein, Armen Sedrakian
Institute for Theoretical Physics, J. W. Goethe-University, D-60438 Frankfurt-Main, Germany
E-mail: mstein@th.physik.uni-frankfurt.de, sedrakian@th.physik.uni-frankfurt.de

Xu-Guang Huang
Physics Department & Center for Particle Physics and Field Theory, Fudan University, Shanghai 200433, China
E-mail: huangxuguang@fudan.edu.cn

John W. Clark
Department of Physics & McDonnell Center for the Space Sciences, Washington University, St. Louis, MO 63130, USA, and Center for Mathematical Sciences, University of Madeira, Funchal, Portugal
E-mail: jwc@wuphys.wustl.edu

Gerd Röpke
Institut für Physik, Universität Rostock, Universitätsplatz 3, D-18051 Rostock, Germany
E-mail: gerd.roepke@uni-rostock.de

Abstract. We report on recent progress in understanding pairing phenomena in low-density nuclear matter at small and moderate isospin asymmetry. A rich phase diagram has been found comprising various superfluid phases that include a homogeneous and phase-separated BEC phase of deuterons at low density and a homogeneous BCS phase, an inhomogeneous LOFF phase, and a phase-separated BCS phase at higher densities. The transition from the BEC phases to the BCS phases is characterized in terms of the evolution, from strong to weak coupling, of the condensate wavefunction and the second moment of its density distribution in r-space. We briefly discuss approaches to higher-order clustering in low-density nuclear matter.

1. Introduction

The two-nucleon interactions in nuclear matter at sub-saturation densities (below about half the saturation density) are well constrained by the phase-shift data and analysis of elastic nucleon-nucleon collisions, facilitating a quantitative many-body description of nuclear matter in this regime. However, a significant theoretical challenge is posed by the complexity of behavior arising from the dominant attractive part of the nuclear interaction, notably the formation of nuclear clusters and the emergence of nucleonic pair condensates of the Bardeen-Cooper-Schrieffer (BCS) type. Such a description is of direct relevance for the matter found in supernovae and neutron stars. These two settings differ somewhat in the regions of parameter space (density, temperature, isospin asymmetry) involved. For example, in supernovae the nuclear matter is at finite isospin asymmetry which, however, is small compared to that of cold β-catalyzed neutron star matter. In this contribution we review some recent progress in establishing the nature of pairing in low-density nuclear matter.

2. Pairing

If we restrict ourselves to only two-body correlations, nuclear matter at extremely low density is a mixture of quasi-free nucleons and deuterons, the only effect of the interaction being to renormalize the mass of the constituents. As bosons, the deuterons may also condense and form a Bose-Einstein condensate (BEC) even without interactions, provided, of course, the temperature is sufficiently low [1, 2, 3, 4, 5, 6]. Increasing the density of the system while keeping its temperature constant serves to increase the degeneracy of the system. As the density ρ approaches the saturation density $\rho_0 = 2.8 \times 10^{14}$g cm^{-3} of symmetrical nuclear matter, the abundance of deuterons is reduced by Pauli blocking of the phase space available to nucleons (for a recent discussion see [7]). However, because at high densities the nucleons fill a Fermi sphere and because the interaction between nucleons is attractive, there emerges a BCS-type coherent state – a condensate of paired nucleons.

The dominant pairing interaction comes from the $^3S_1$-$^3D_1$ (SD) partial wave, i.e., the same interaction channel that binds the deuteron. Therefore, we may anticipate that the nuclear matter undergoes a BEC-BCS phase transition as the density increases. This effect has been conjectured to occur in the context of intermediate-energy heavy-ion collisions [1, 2] and more recently in the context of supernovae [8, 9].

The theoretical framework developed by Nozières and Schmitt-Rink [10] for description of the BCS-BEC transition in the condensed-matter context can be transposed to nuclear matter. Importantly, however, isoscalar neutron-proton (np) pairing is disrupted by the isospin asymmetry induced by weak interactions in stellar environments and expected in exotic nuclei, since the mismatch in the Fermi surfaces of protons and neutrons suppresses the pairing correlations [11]. The standard Nozières-Schmitt-Rink theory of the BCS-BEC crossover has been modified to account for this fact [12]. The emergence of two mismatched Fermi surfaces in isospin-asymmetrical nuclear matter introduces a new scale into the problem, namely the shift $\delta\mu = (\mu_n - \mu_p)/2$ in the chemical potentials μ_n and μ_p of neutrons and protons (up or down) from their otherwise common value $\bar{\mu} = (\mu_n + \mu_p)/2$. If sufficiently large, this shift can destroy the condensate.

In the isospin symmetrical case (at $\delta\mu = 0$) the condensate is characterized by a gap Δ_0 in the SD channel of order several MeV. A sequence of unconventional phases has been conjectured to occur with increasing isospin symmetry, i.e., as $\delta\mu$ increases from zero to values of order Δ_0. One of these is a neutron-proton condensate whose Cooper pairs have non-zero center-of-mass (CM) momentum [13, 14, 15]. This phase is the analogue of the Larkin-Ovchinnikov-Fulde-Ferrell (LOFF) phase in electronic superconductors [?, 16]. Another possibility is phase separation into superconducting and normal components, proposed in the context of cold atomic gases [17]. At large isospin asymmetry, where SD pairing is strongly suppressed, a BCS-BEC crossover may also occur in the isotriplet $^1S_0$ pairing channel. Finally, it has been conjectured that spontaneous deformation of Fermi surfaces is an alternative to the LOFF phase and can in fact dominate the latter in nuclear systems [14].

Recently, the concepts of unconventional SD pairing and the BCS-BEC crossover were unified in a model of isospin-asymmetric nuclear matter [9] by including some of the phases mentioned above. A phase diagram for superfluid nuclear matter was constructed over wide ranges of density, temperature, and isospin asymmetry. The coupled equations for the gap and the densities of constituents (neutrons and protons) were solved allowing for the ordinary BCS state,

Journal of Physics: Conference Series **496** (2014) 012008 doi:10.1088/1742-6596/496/1/012008

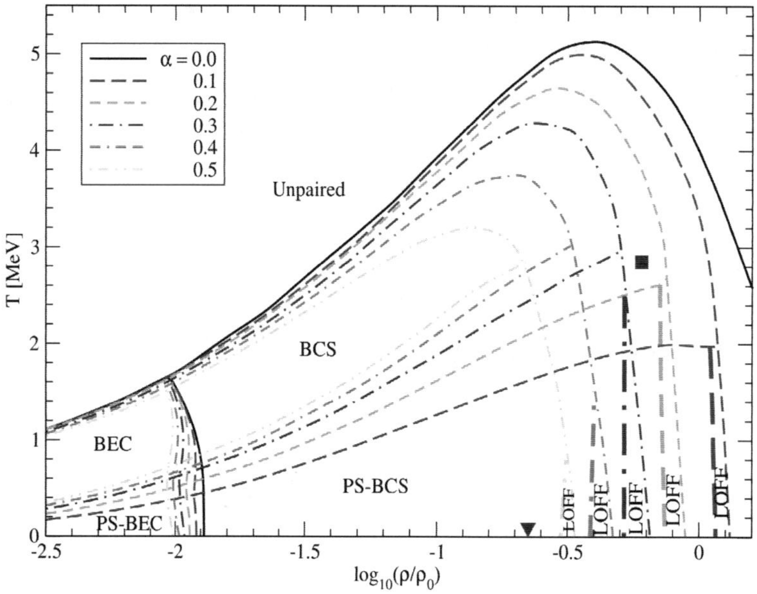

Figure 1. Phase diagram of dilute nuclear matter in the temperature-density plane for several isospin asymmetries α (from [9]). Included are four phases: unpaired phase, BCS (BEC) phase, LOFF phase, and PS-BCS (PS-BEC) phase. For each asymmetry there are two tri-critical points, one of which is always a Lifshitz point [?]. For special values of asymmetry these two points degenerate into single tetra-critical point at $\log(\rho/\rho_0) = -0.22$ and $T = 2.85$ MeV (shown by a square dot) for $\alpha_4 = 0.255$. The LOFF phase disappears at the point $\log(\rho/\rho_0) = -0.65$ and $T = 0$ (shown by a triangle) for $\alpha = 0.62$. The boundaries between BCS and BEC phases are identified by the change of sign of the average chemical potential $\bar{\mu}$.

its low-density asymptotic counterpart BEC state, and two phases that owe their existence to the isospin asymmetry: the phase with a moving condensate (LOFF phase) and the phase in which the normal fluid and superfluid occupy separate spatial domains. The latter phase is referred to as the phase-separated BCS (PS-BCS) phase and, in the strong-coupling regime, the phase-separated BEC (PS-BEC) phase. In this phase the asymmetry is accumulated in the normal domains, whereas the superfluid domain is perfectly isospin symmetric.

The gap equation to be solved has the form

$$\Delta_l(k,Q) \;=\; \frac{1}{2}\sum_{l'}\int\frac{d^3k'}{(2\pi)^3}V_{l,l'}(k,k')\sum_{a,r}\frac{\Delta_{l'}(k',Q)}{2\sqrt{E_S(k')^2+\Delta_{l'}^2(k',Q)}}[1-2f(E_a^r)], \tag{1}$$

where the gap $\Delta(Q)$ in the quasiparticle spectrum is a function of the modulus Q of the total momentum of the pair of particles, the l summation is over the coupled partial waves, $V(\boldsymbol{k},\boldsymbol{k}')$ is the neutron-proton interaction potential, $f(x) = 1/[\exp(x/T)+1]$, $E_r^a = [E_S^2+\Delta^2]^{1/2}+r\delta\mu+aE_A$, and $a, r \in \{+,-\}$, with $E_S = \left(Q^2/4+k^2\right)/2m^* - \bar{\mu}$, $E_A = \boldsymbol{k}\cdot\boldsymbol{Q}/2m^*$, and $\Delta^2 = \sum_l \Delta_l^2$. Note that the gap in the SD state is angle averaged in the denominator of the gap equation; this approximation can be avoided [15, ?]. Here $\bar{\mu}$ is the average of the chemical potentials μ_n and μ_p of neutrons and protons introduced above, and m^* is the effective mass of the nucleons. (Note that we do not distinguish between the effective masses of neutrons and protons.)

The resulting phase diagram of dilute nuclear matter is shown in Fig 1 for several values of isospin asymmetry $\alpha = (n_n - n_p)/(n_n + n_p)$, where n_n and n_p are the number densities of neutrons and protons. Four different phases of matter are present. (a) The unpaired normal phase is always the ground state at sufficiently high temperatures $T > T_{c0}$, where $T_{c0}(\rho)$ is the critical temperature of the normal/superfluid phase transition at $\alpha = 0$. (b) The LOFF phase is the ground state in a narrow temperature-density strip at low temperatures and high densities. (c) The domains of phase separation (PS) appear at low temperatures and low densities, while the isospin-asymmetric BCS phase is the ground state at intermediate temperatures for densities above the crossover to a BEC phase and below the density where the condensate vanishes in the isospin symmetrical limit.

In the extreme low-density and strong-coupling regime, the BCS superfluid phases have two counterparts: the BCS phase evolves into the BEC phase of deuterons, whereas the PS-BCS phase evolves into the PS-BEC phase, in which the superfluid fraction of matter is a BEC of deuterons. The superfluid/unpaired phase transitions and the phase transitions between the superfluid phases are of second order (thin solid lines in Fig. 1), with the exception of the PS-BCS to LOFF transition, which is of first order (thick solid lines in Fig. 1). The BCS-BEC transition and the PS-BCS to PS-BEC transition are smooth crossovers. At non-zero isospin asymmetry the phase diagram features two tri-critical points where the simpler pairwise phase coexistence terminates and three different phases coexist.

Consistent with the earlier studies of the BCS-BEC crossover, one observes in the phase diagram of Fig. 1 a smooth crossover to an asymptotic state corresponding to a mixture of a Bose condensate of deuterons and a gas of excess neutrons. This however occurs at moderate temperatures, where the unconventional phases do not appear. The new ingredient of the nuclear phase diagram is the crossover seen at very low temperatures, where the heterogeneous superfluid phase is replaced by a heterogeneous mixture of a phase containing a deuteron condensate and a phase containing neutron-rich unpaired nuclear matter.

The transition to the BEC regime of strongly-coupled neutron-proton pairs, which are asymptotically identical with deuterons, occurs at low densities. The criterion for the transition from BCS to BEC is that either the average chemical potential $\bar{\mu}$ changes its sign from positive to negative values, or the coherence length ξ of a Cooper pair becomes comparable to the interparticle distance, i.e., ξ becomes of order $d \sim \rho^{-1/3}$ as it ranges from $\xi \gg d$ to $\xi \ll d$. The coherence length is related to the wave function of the condensate $\Psi(\boldsymbol{r})$ by

$$\xi = \sqrt{\langle r^2 \rangle}, \qquad \langle r^2 \rangle = \int d^3r \, r^2 |\Psi(r)|^2, \tag{2}$$

where the wave-function is defined in terms of the kernel of the gap equation according to

$$\Psi(\boldsymbol{r}) = \sqrt{N} \int \frac{d^3p}{(2\pi)^3} [K(\boldsymbol{p}, \Delta) - K(\boldsymbol{p}, 0)] e^{i\boldsymbol{p}\cdot\boldsymbol{r}}, \qquad \int d^3r |\Psi(\boldsymbol{r})|^2 = N^{-1}. \tag{3}$$

Thus the change in the coherence length is related to the change of the condensate wave-function across the BCS-BEC crossover. In the case of neutron-proton pairing, the criteria for the BCS-BEC transition are fulfilled, i.e., $\bar{\mu}$ changes sign and the mean distance between the pairs becomes larger than the coherence length of the superfluid.

In Fig. 2 we display the integrand $r^2|\Psi(r)|^2$ of Eq. (2) as a function of radial distance for BCS and PS states at various densities. In its construction, the PS state corresponds to the symmetrical BCS state insofar as we are concerned with the nature of the Cooper pairs; hence there are no effects from the asymmetry in this case; the asymmetry affects the PS condensate indirectly via the number of particles available for pairing. In the weak-coupling (high-density-limit) case, the integrand function has an oscillatory form which extends over

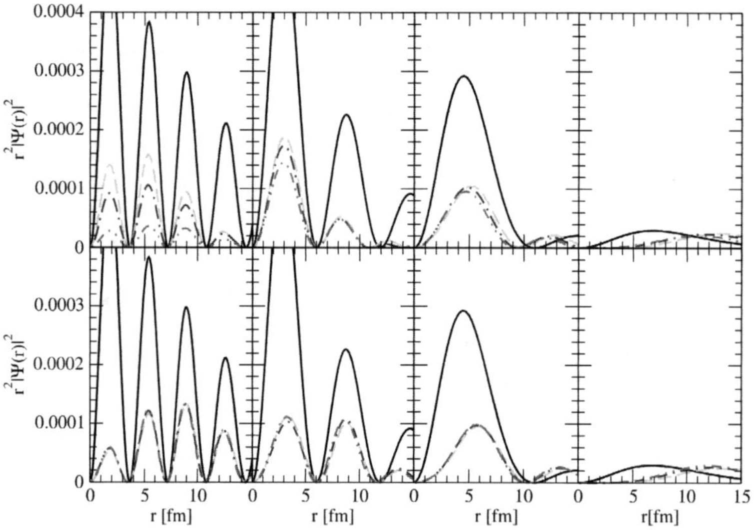

Figure 2. The quantity $r^2|\Psi(r)|^2$ as a function of radial coordinate r for $T = 0.5$ MeV and densities (from left to right) $\log_{10}\rho/\rho_0 = -0.5, -1., -1.5, -2.5$ plotted for the asymmetrical BCS state (upper panel) and the PS phase (lower panel). The values of asymmetry are $\alpha = 0.0$ (solid lines), 0.1 (dashed lines), 0.2 (dash-dotted lines), and 0.3 (dash-double-dotted lines).

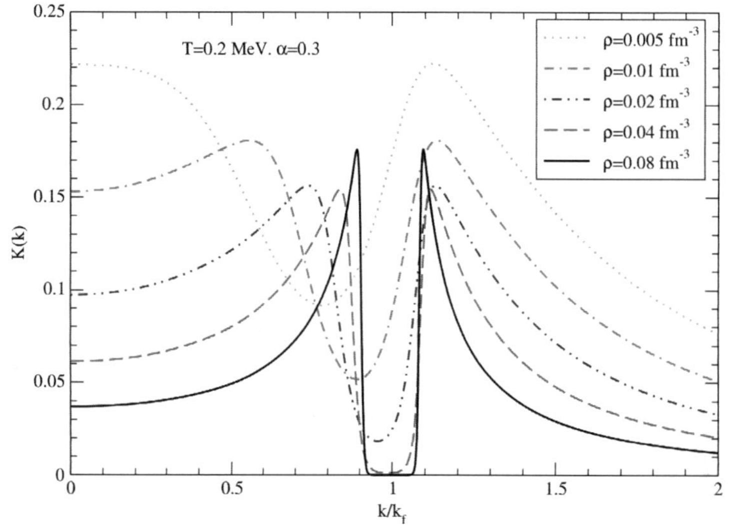

Figure 3. Kernel $K(k)$ of the gap equation (1) as a function of momentum, in units of the relevant Fermi momentum k_F for fixed $\alpha = 0.3$ and $T = 0.2$ MeV at various densities.

many periods of the interparticle distance. This behavior reflects the coherence associated with the condensate, in which the spatial correlations are characterized by scales much larger than the interparticle distance. For smaller densities (larger couplings) the integrand of $\langle r^2 \rangle$ is increasingly concentrated at the origin, with at most a few periods of oscillation. At low density (strong coupling) the pairs are well localized in space within a radius which is small compared to the interparticle distance; there is only one spike in the integrand function. This regime has BEC character, in that the pair correlations just extend over distances comparable to the interparticle distance. In the case of the BCS superfluid it is seen that isospin asymmetry affects only the amplitude of the oscillations, the period being unaffected in the weak-coupling limit and defined by the magnitude of the inverse Fermi momentum.

Further insight into the BCS-BEC crossover can be gained from examining the kernel $K(p)$ of the gap equation (1), which is defined by the sum appearing to the right of the interaction V in this equation. Physically, $K(p)$ can be interpreted as the wave function of the Cooper pairs, since it obeys a Schrödinger-type eigenvalue equation in the limit of extremely strong coupling. The prefactor of the Pauli operator $1-2f$ is a smooth function of momentum with a maximum at the Fermi surface, where E_S vanishes. The momentum ranges whose contributions are important in the gap equation in different regimes of the phase diagram can be identified from Fig. 3. In the BCS regime, $K(p)$ has two sharp maxima which are separated by a depression (of width $\sim \delta\mu$) around the Fermi momentum. Because of strong localization in momentum space, the Cooper pairs have an intrinsic structure which is broad in real space, implying a large coherence length. This is characteristic of the BCS regime. The picture is reversed in the strong-coupling (low-density) limit, where $K(p)$ is a broad function of momentum, corresponding to bound states (deuterons) well-localized in the real space. This is characteristic of the BEC regime. In addition, as the density decreases, the lower peak moves toward $k = 0$, due to the fact that $\bar{\mu}$ changes its sign from positive to negative at the transition from the BCS to the BEC regime. As a consequence, the prefactor of the Pauli operator $1 - 2f$ peaks at $k = 0$ in the BEC regime, rather at the Fermi surface as in the BCS regime.

3. Higher-order clustering

In supernovae the matter is at finite isospin asymmetry (but with small values compared to those of cold neutron stars) and at relatively high temperatures. These conditions could allow for a substantial presence of light clusters, notably deuterons, tritons, ^{3}He nuclei, and alpha particles [8, 18, 19, 20, 21, 22, 23]. The composition of nuclear matter in thermodynamic equilibrium at low density is given by the mass action law, the so-called nuclear statistical equilibrium (NSE). At given temperature T and total baryonic density ρ, the partial densities of the components c (whether nucleons or nuclei) are determined by a phase-space sum (integral) over their Fermi or Bose distribution functions as

$$n_c = g_c \sum_P \left\{ \exp\left[\left(\frac{P^2}{2A_c m} + E_c - Z_c\mu_p - N_c\mu_n \right) \Big/ T \right] \pm 1 \right\}^{-1}, \tag{4}$$

where $\boldsymbol{P}$ is the center-of-mass momentum, Z_c and N_c are the proton and neutron numbers of component c, $A_c = N_c + Z_c$ is its mass number, and g_c is its degeneracy factor (accounting for spin and excitations). A number of studies [8, 18, 19, 20, 21, 22, 23] clearly demonstrate the importance of two- and four-body correlations. The NSE approach is valid at low-densities, where the free-space binding energies of bound states are not affected by the nuclear environment. As the baryon density increases, the energies of bound states are modified due to the strong and electromagnetic interactions. These modifications can be incorporated by solving an "effective"

Schrödinger equations for an A-particle cluster (see [24] and references therein)

$$\left[\sum_{i=1}^{A} E_i^{\rm mf} - E\right]\psi(1\ldots A) + \sum_{i<j}^{A}[1-f(i)-f(j)]\sum_{1'\ldots A'} V_{ij}(1\ldots A,1'\ldots A')\psi(1'\ldots A') = 0\,, \quad (5)$$

where the indices i, j label nucleons, $\psi(1\ldots A)$ is a bound-state wave function, E is its corresponding energy eigenvalue, V_{ij} is the pairwise interaction potential, and $f(i)$ is the distribution function of nucleon i. The mean-field corrections to the quasiparticle energies are included in $E_1^{\rm mf}$ and can be evaluated from the Skyrme or the relativistic mean-field functionals. The modifications of energies of higher-order bound states depend on the center-of-mass momentum $\boldsymbol{P}$ as well as the thermodynamic parameters of the medium. These have been computed in detail recently in Ref. [24], where fitting formulas are also given. The main physical effect of these modifications is to suppress the abundances of clusters upon asymptotic approach to the nuclear saturation density.

To demonstrate the utility of the "effective" Schrödinger equation (5), we consider the example of a four-nucleon (α-like) cluster and assume for simplicity that both T and α are zero. Bound states can appear below the continuum of scattering states, whose edge is given simply by the sum of the energies E_F of two neutrons and two protons at the tops of their respective Fermi seas. In the zero-density limit, α particles are formed with bound-state energy $E_\alpha^0 = -28.3$ MeV. The energy of the center-of-mass motion vanishes at $\boldsymbol{P} = 0$. At finite density of the surrounding nuclear matter, the energy of the four-nucleon bound state is shifted due to Pauli blocking, so that $E_4[T,\rho] = E_\alpha^0 + \Delta E_4^{\rm Pauli}(T,\rho)$. The shift $E_4^{\rm Pauli}$ has been found from Eq. (5) within a variational approach using a Gaussian form for the wave function and a Gaussian separable interaction, adjusted to reproduce the binding energy and the r.m.s. radius of the α particle [25]. The density dependence of the shift (at vanishing T and α) is well approximated by $\Delta E_4^{\rm Pauli}(\rho) = 4515.9\,\rho - 100935\,\rho^2 + 1202538\,\rho^3$, where the energy is given in units of MeV and the density in units of fm^{-3}. Using this formula, it is easy to deduce that the binding energy of an α particle crosses the continuum $4E_F$ at a density $\rho \sim 0.03$ fm^{-3}, above which no α particle bound state can be formed. Analogous fits to in-medium binding energies for other light clusters can be found in [19]. These compare well with exact solutions of the three- and four-body quantum mechanical problems in the medium (see, for example, [3, ?, 26] and references therein). A useful feature of Eq. (4) is that it has the character of a Bose distribution function of a cluster for even A which implies that the physics of BEC of clusters (notably the critical temperature) can be deduced starting from NSE or its modifications that include density- and temperature-dependent binding energies.

4. Outlook

It is apparent from the discussion above that low-density and moderately warm nuclear matter can have a rather complicated phase diagram at finite, but small, isospin asymmetries. It is also apparent that higher-order clusters need to be included in the phase diagram to make the discussion complete. Furthermore, the α particles will form a Bose-Einstein condensate at sufficiently low temperatures (see [27, 28, 29] and references therein). The extent to which these clusters modify the structure of the phase diagram developed here remains to be studied. The diverse microscopic aspects of superfluid, asymmetrical nuclear matter portend diverse ramifications for the astrophysics of supernovae and hot compact stars.

Acknowledgments

The work of MS was supported by the HGS-HIRe graduate program at Frankfurt University. XGH acknowledges support through Fudan University grants EZH1512519 and EZH1512600. JWC acknowledges research support from the McDonnell Center for the Space Sciences and expresses his appreciation for the hospitality of the Center of Mathematical Sciences, University of Madeira.

References

[1] Alm T, Friman B L, Röpke G and Schulz H 1993 *Nuclear Physics A* **551** 45–53
[2] Baldo M, Lombardo U and Schuck P 1995 *Phys. Rev. C* **52** 975–985
[3] Sedrakian A and Clark J W 2006 *Phys. Rev. C* **73** 035803 (*Preprint* nucl-th/0511076)
[4] Mao S, Huang X and Zhuang P 2009 *Phys. Rev. C* **79** 034304 (*Preprint* 0812.0674)
[5] Huang X G 2010 *Phys. Rev. C* **81** 034007 (*Preprint* 1002.0060)
[6] Jin M, Urban M and Schuck P 2010 *Phys. Rev. C* **82** 024911 (*Preprint* 1005.1782)
[7] Röpke G, Bastian N U, Blaschke D, Klähn T, Typel S and Wolter H H 2013 *Nuclear Physics A* **897** 70–92 (*Preprint* 1209.0212)
[8] Heckel S, Schneider P P and Sedrakian A 2009 *Phys. Rev. C* **80** 015805 (*Preprint* 0902.3539)
[9] Stein M, Huang X G, Sedrakian A and Clark J W 2012 *Phys. Rev. C* **86** 062801 (*Preprint* 1208.0123)
[10] Nozières P and Schmitt-Rink S 1985 *Journal of Low Temperature Physics* **59** 195–211
[11] Sedrakian A and Lombardo U 2000 *Physical Review Letters* **84** 602–605 (*Preprint* nucl-th/9907076)
[12] Lombardo U, Nozières P, Schuck P, Schulze H J and Sedrakian A 2001 *Phys. Rev. C* **64** 064314 (*Preprint* nucl-th/0109024)
[13] Sedrakian A 2001 *Phys. Rev. C* **63** 025801 (*Preprint* nucl-th/0008052)
[14] Müther H and Sedrakian A 2003 *Phys. Rev. C* **67** 015802 (*Preprint* nucl-th/0209061)
[15] Shang X and Zuo W 2013 *ArXiv e-prints* (*Preprint* 1308.0364)
[16] Fulde P and Ferrell R A 1964 *Physical Review* **135** 550–563
[17] Bedaque P F, Caldas H and Rupak G 2003 *Physical Review Letters* **91** 247002 (*Preprint* cond-mat/0306694)
[18] Sumiyoshi K and Röpke G 2008 *Phys. Rev. C* **77** 055804 (*Preprint* 0801.0110)
[19] Typel S, Röpke G, Klähn T, Blaschke D and Wolter H H 2010 *Phys. Rev. C* **81** 015803 (*Preprint* 0908.2344)
[20] Hempel M and Schaffner-Bielich J 2010 *Nuclear Physics A* **837** 210–254 (*Preprint* 0911.4073)
[21] Raduta A R and Gulminelli F 2010 *Phys. Rev. C* **82** 065801 (*Preprint* 1009.2226)
[22] Oertel M, Fantina A F and Novak J 2012 *Phys. Rev. C* **85** 055806 (*Preprint* 1202.2679)
[23] Gulminelli F and Raduta A R 2012 *Phys. Rev. C* **85** 025803 (*Preprint* 1110.2034)
[24] Röpke G 2011 *Nuclear Physics A* **867** 66–80 (*Preprint* 1101.4685)
[25] Röpke G 2009 *Phys. Rev. C* **79** 014002 (*Preprint* 0810.4645)
[26] Beyer M, Sofianos S A, Kuhrts C, Röpke G and Schuck P 2000 *Physics Letters B* **488** 247–253 (*Preprint* nucl-th/0003071)
[27] Clark J W and Wang T P 1966 *Annals of Physics* **40** 127–152
[28] Sedrakian A, Müther H and Schuck P 2006 *Nuclear Physics A* **766** 97–106 (*Preprint* nucl-th/0407020)
[29] Yamada T, Funaki Y, Horiuchi H, Röpke G, Schuck P and Tohsaki A 2012 Nuclear Alpha-Particle Condensates *Lecture Notes in Physics, Berlin Springer Verlag* (*Lecture Notes in Physics, Berlin Springer Verlag* vol 848) ed Beck C p 229 (*Preprint* 1103.3940)

Journal of Physics: Conference Series **496** (2014) 012009 doi:10.1088/1742-6596/496/1/012009

Stability of strange dwarfs: a comparison with observations

Yu L Vartanyan, G S Hajyan, A K Grigoryan, T R Sarkisyan

Yerevan State University, Alex Manoogian str. 1, Yerevan, 0025, Armenia

E-mail: yuvartanyan@ysu.am

Abstract. We study the stability of strange dwarfs, superdense stars with a small self-confining core ($M_{core} < 0.02M_{\odot}$) containing strange quark matter and an extended crust consisting of atomic nuclei and degenerate electron gas. The mass and the radius of these stars are of the same order of magnitude as those of ordinary white dwarfs. It is shown that any study of their stability must examine the dependence of the mass on two variables, which can, for convenience, be taken to be the rest mass (total baryon number) of the quark core and the energy density ρ_{tr} of the crust at the surface of the quark core. The range of variation of these quantities over which strange dwarfs are stable is determined. This region is referred to as the stability valley for strange dwarfs. The mass and radius obtained from theoretical models of strange dwarfs are compared with observational data obtained through the HIPPARCOS program and the most probable candidate strange dwarfs are identified.

1. Introduction

Strange quark matter may exist in a state that is more bound than the matter in atomic nuclei [1, 2]. If the MIT bag model equation of state [3] is used for the quarks, then for certain values of the phenomenological parameters of the model, a case may arise in which the average energy ϵ_b per baryon at some baryon concentration $n = n_{min}$ has a negative minimum ($\epsilon_b(n_{min}) < 0$) . In this case, the quark matter may be in a self-bound state, so that the existence of self-confining cosmic objects, or so-called "strange stars" (ss), which can exist without gravitation, becomes possible [4]. Gravitation places a limit on the maximum mass of these configurations, which, as in the case of neutron stars, is of the order of $2M_{\odot}$.

At the surface of a strange star the electron density n_e is several orders of magnitude below that of the quarks. Since the electrons are confined only by an electrostatic field, some of them can move away from the quark surface of the strange star by hundreds of fermis ($l \sim n_e^{-1/3}$) to form a thin charged layer where the electric field reaches $10^{17} - 10^{18}$ V/cm [5]. This field isolates the crust, which is made up of atomic nuclei and a degenerate electron gas ("Ae" matter), and is not in thermodynamic equilibrium with the strange quark matter; it is coupled to the quark core only by gravitation. Since free neutrons, with no electrical charge, can pass unimpeded through the electrostatic barrier and be absorbed by the strange quark matter, the maximum density of the crust is limited by the rate of escape of neutrons from nuclei, $\rho_{drip} \approx 4.3 \cdot 10^{11}$ g/cm^3.

Models of strange stars with a crust have been examined [6] over the entire range of variation of the central density of a star. It was found that for strange stars with strange quark core masses $M_{core}/M_{\odot} > 0.1$, the thickness and mass of the crust are negligibly small compared to the star's radius and mass. The situation is different for strange stars with low core masses $M_{core}/M_{\odot} < 0.02$. In configurations of this sort, the crust swells greatly, and their mass and radius are as larger as found for white dwarfs (wd). They differ from wd, however, by the fact that they have a core in the form of a small-radius, low-mass strange star and a crust in which the density can be two orders of magnitude greater than the limiting density reached in white dwarfs. These models are referred to as strange dwarfs (sd) [7, 8].

2. The stability valley for strange dwarfs

Only stable equilibrium configurations of superdense stars are of physical interest. Glendenning, et al. [7, 8], have studied the stability of strange dwarfs (sd) by the method of small radial perturbations developed in the general theory of relativity by Chandrasekhar [9, 10]. They examined the dependence of the mass M on the central energy density of the quark core, ρ_c, for a sequences of sd in which the energy density of the nuclear-electron crust at the surface of the quark core ρ_{tr} was set equal to its maximum value, i.e. on the curve $M(\rho_c, \rho_{tr} = \rho_{drip})$. It was found that, as opposed to ordinary white dwarfs and neutron stars, sd configurations for which $dM/d\rho_c < 0$ have $\omega_0^2 > 0$ (ω_0 is the fundamental mode frequency for radial pulsations), i.e., they are stable, while configurations with $dM/d\rho_c > 0$ have $\omega_0^2 < 0$, i.e., are unstable.

This, however, is nothing surprising, since complete information on the stability of sd can be obtained only by examining the entire range of variation of ρ_c and ρ_{tr}, i.e., we have to consider the functional surface, i.e., the mass of the sd as a function of these variables, $M(\rho_c, \rho_{tr})$. Here it is convenient to take u and ρ_{tr} as the independent variables rather than ρ_c and ρ_{tr}, where u is the total number of baryons in the quark core N_{core} of the sd expressed in solar masses, i.e., $u = mN_{core}/M_{\odot}$, where $m = M(^{56}\text{Fe})/56$ and $N_{core} = 4\pi \int_0^{R_0} n e^{\lambda/2} r^2 dr$, where n is the baryon concentration, R_0 is the radius of the quark core, and $exp(\lambda)$ is the radial component of the metric tensor. In fact, as opposed to neutron stars, the $M(\rho_c)$ curve for bare (without a crust) strange dwarfs is not bounded from below; it approaches zero when the central density ρ_c approaches the density of self-bound quark matter neglecting gravitation. The surface of strange stars without a crust, as usual, is determined by the radius over which the pressure goes to zero: $P(R_{ss} = 0)$. In the case of strange dwarfs, on the other hand, where there is an extended nuclear-electron crust over the surface of a small quark core ($u \leq 0.02M_{\odot}$), the energy density ρ_{tr} and pressure P_{tr} at the surface of a fixed quark core will increase as the mass of the crust increases. In the bag model used for the quark core, the energy density at the center of the quark core will increase extremely slowly. As noted above, the increase in ρ_{tr} is limited by maximum value $\rho_{tr} \leq \rho_{drip}$.

Each small quark core becomes the basis of a family of sd with different crust masses. Thus, in studying the stability of strange dwarfs it is necessary to study the stability of individual families of this sort [11, 12]. Here it becomes possible to use a static criterion for stability developed by Zel'dovich [13] and generalized by Bisnovatyi-Kogan [14] that is free of the cumbersome mathematical calculations characteristic of Chandrasekhar's method. Since the electrostatic field at the surface of the quark core inhibits the penetration of matter from the crust into the quark core, for small radial oscillations the masses of both the crust and the core are unchanged. Thus, the static criterion is applicable to the sequences of this kind. At the maxima of the dependence of the sd mass on ρ_{tr} for a fixed value of u (the $M_u(\rho_{tr})$ curves), the sd loses stability.

Therefore, as opposed to neutron stars and white dwarfs, models of the stability of strange dwarfs in M, u, ρ_{tr} space occupy the part of the $M(u, \rho_{tr})$ surface that is bounded from above by the curve joining the maxima of the $M(u, \rho_{tr})$ curves (see Fig. 1). We refer to this region of the $M(u, \rho_{tr})$ surface as the stability valley for strange dwarfs [15].

For the chosen equations of state of the quark core and crust, the integral parameters of an *sd* (mass, total baryon number, radius) are uniquely determined by the energy density ρ_c at the center of the quark core and the energy density ρ_{tr} of the nuclear-electron matter at the boundary between the quark core. The integral parameters are then obtained by integrating the relativistic Tolman-Oppenheimer-Volkov (TOV) equations of hydrostatic equilibrium [16, 17]. Many sequences of strange dwarfs with fixed values of u within $10^{-4} \geq u \geq 0.1$ were studied.

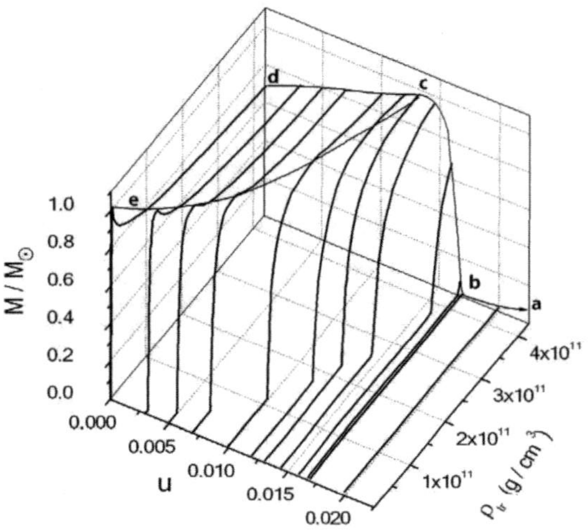

Figure 1. The mass M of strange dwarfs as a function of the parameters u and ρ_{tr} (see the explanations in the text). The leading part of the surface $M(u, \rho_{tr})$ is bounded from above by the curve ecba, which is the stability valley for strange dwarfs.

Some of the results of the calculations for typical sequences are shown in Fig. 1. Here we examine the M, u, ρ_{tr} space (M is the *sd* mass), and plot the $M_u = M(u, \rho_{tr})$ curves (the mass as a function of ρ_{tr}) on the $M = M(u, \rho_{tr})$ surface for $u = const$. The curves are extended up to their intersection with the coordinate plane $\rho_{tr} = \rho_{drip}$. It is seem that the mass increases with increasing ρ_{tr} for each individual sequence, and the stability is lost at M_{max}. For the sequences with $u = 0.013$, the *sd* mass reaches M_{max} for $\rho_{tr} = \rho_{drip}$ (the point c in Fig. 1). For sequences of strange dwarfs with $u > 0.013$, the curves intersect the $\rho_{tr} = \rho_{drip}$ plane when the mass has no longer reached the maximum (the segment bc in Fig. 1). For the sequences that intersect the $\rho_{tr} = \rho_{drip}$ plane in the segment ab, where $u \geq 0.02$, the curves are horizontal, i.e., for these sequences the mass of the crust is negligible compared to the core and $M/M_\odot \approx u$.

The segments of the individual $M(u, \rho_{tr})$ curves for stable configurations of strange dwarfs in M, u, ρ_{tr} space on the saddle surface $M = M(u, \rho_{tr})$, occupy the stability valley for the strange dwarfs. This region is bounded by the curves ec (the maxima of the individual series) and cb (the front portion of the surface in Fig. 1). The part of this saddle surface that applies to unstable configurations and is formed by the segments of the $M_u(\rho_{tr})$ curves after the maximum points, intersects the $\rho_{tr} = \rho_{drip}$ plane along the dc curve, where $\partial M/\partial u$ and, therefore,$\partial M/\partial \rho_c$ are greater than zero. Unstable configurations of this sort naturally have $\omega_0^2 < 0$ [7, 8]. But this does not mean that the quark cores of the individual sequences in the segment dc cannot form stable *sd* configurations.

Strange dwarf configurations for which the masses of the individual sequences $M_u(\rho_{tr})$ have not reached the maximum values, so that $\omega_0^2 > 0$ for them, lie in the $\rho_{tr} = \rho_{drip}$ plane along the curve bc. And there was no justification for the surprise of Glendenning, et al. [7, 8], that this is so, for, on going from one series to another along the bc curve, $\partial M/\partial u$ and, therefore, $\partial M/\partial \rho_c$ are negative. Similar results will be obtained for the intersection of the $M(u, \rho_{tr})$ surface with planes parallel to $\rho_{tr} = \rho_{drip}$ for smaller values of ρ_{tr}, as was shown by Glendenning, et al. [8].

Journal of Physics: Conference Series **496** (2014) 012009 doi:10.1088/1742-6596/496/1/012009

Vartanyan, et al. [12] pointed out that, although $\omega_0^2 > 0$ for *sd* on the segment bc, the fact that the condition $\rho_{tr} = \rho_{drip}$ holds for them makes them analogous to the configurations of series with $u < 0.013$, for which the mass is maximal and $\omega_0^2 = 0$ (these are the configurations lying on the curve *ec* in Fig. 1). Thus, if the crust density is increased even slightly for these configurations (e.g., as a result of radial pulsations), ρ_{tr} will become greater than ρ_{drip} and free neutrons will be born at the surface of the quark core and will move into the quark core, increasing its mass (the total number of baryons). Since $\partial M/\partial u < 0$ on the segment cb in this case, the new quark core with its larger mass cannot form *sd* with the initial baryon number. Thus, this kind of configuration loses equilibrium. It undergoes a transition to a branch of strange stars *ss* in a state with the same number of baryons, but with a thin nuclear-electron crust (an extension of the *ab* branch in Fig. 1). The radius R_{ss} of these configurations is of the order of 10 km, as in typical neutron stars. An energy $\Delta W_G \sim GM^2/R_{ss}$ is released during this transition; this is of the same order of magnitude as the energy released in supernova explosions. The error in the conclusion that strange dwarfs with $\omega_0^2 > 0$ and $\rho_{tr} = \rho_{drip}$ can be stable arises from the fact that Chandrasekhar's method is applicable when no irreversible processes take place involving matter in the star during the pulsations. This condition is violated when neutrons pass into the quark core. These configurations are at the edge of loss of stability. In this case, for each fixed quark core with $u \geq 0.013$, the distance of the *sd* from the critical state (the stability "margin") is greater when the difference $\rho_{drip} - \rho_{tr} > 0$ is larger.

3. Comparison with observation

The European Space Agency's HIPPARCOS satellite has played a major role in determining and refining the radii and masses of white dwarfs. HIPPARCOS data have been analyzed [18, 19] to obtain improved values for the masses and radii of 22 white dwarfs within the mass range of $0.4 \div 1.0 M\odot$. In these papers, data on the radii and masses of the observed white dwarfs with the associated measurement errors were shown in a plot of the theoretical dependence [20] of the radius on *wd* mass, obtained separately for four atomic nuclei, ^{4}He, ^{12}C, ^{24}Mg , and ^{56}Fe. These results are shown in Fig. 2. While the $R(M)$ curves are very similar for the first three nuclei (for which $Z/A = 1/2$), in the case of ^{56}Fe, $Z/A = 0.46$ and this curve is significantly lower than the other and, for a given mass, the corresponding *wd* radius is the smallest.

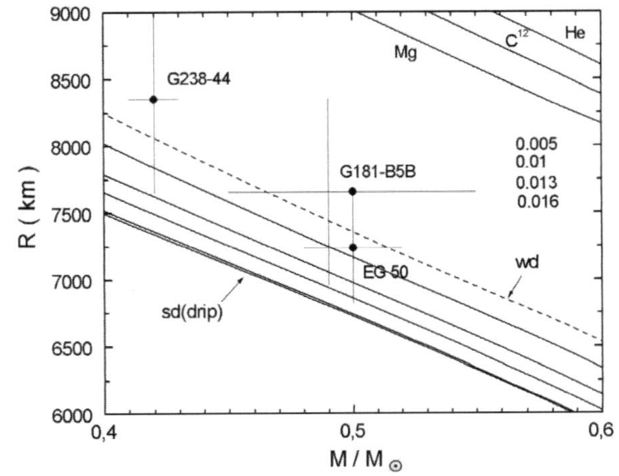

Figure 2. The radius R as a function of mass M for a white dwarf and a strange dwarfs. The curves ^{4}He, ^{12}C, ^{24}Mg and ^{56}Fe are for white dwarfs. The data is taken from from [20]. The other notation are explained in the text.

Journal of Physics: Conference Series **496** (2014) 012009 doi:10.1088/1742-6596/496/1/012009

Vartanyan, et al. [6], have compared the data for observed *wd* with the $R(M)$ curve for strange dwarfs that have $\rho_{tr} = \rho_{drip}$. It was found that the two observed *wd*, EG-50 and G238-44 lie very close to this curve and, therefore, may be candidates for strange dwarfs. The same problem has been examined in more detail in Ref. [21], where sequences of *sd* with $\rho_{tr} = \rho_{drip}$ for a crust containing ^{12}C nuclei were examined.

For completeness of the comparison, it is necessary to examine $R(M)$ not just for one sequence of strange dwarfs with $\rho_{tr} = \rho_{drip}$, but also the curves for characteristic fixed values of u from the stability valley calculated for different elements. This is of special interest for ^{56}Fe nuclei, since only *sd* configurations can lie below the $R(M)$ curve for *wd* of this element.

Under the $R(M)$ curve for white dwarfs containing ^{56}Fe in Fig. 2 we have plotted $R(M)$ for sequences of stable strange dwarfs with $u = 0.005, 0.01, 0.013$, and 0.016 based on data of Ref. [12]. The curves are lower for larger u. The last sequence corresponds to the maximum sized quark core, for which an extended crust can develop, i.e., a strange dwarfs can be formed. The $R(M)$ curve for the *sd* sequences with $\rho_{tr} = \rho_{drip}$ is also shown in this figure.

Thus, under $R(M)$ curve for iron *wd* in Fig. 2, there is a limiting strip on which only *sd* can lie. If the observed white dwarfs include candidates with masses and radii (with small enough measurement errors) that locate them below the $R(M)$ curve for *wd* consisting of ^{56}Fe, then these observed *wd* may be identified with a strange dwarf. Only EG-50 is very close to satisfying this requirement among the observed *wd* mentioned here.

References

[1] Bodmer A R 1971 *Phys. Rev.* D **4** 160
[2] Witten E 1984 *Phys. Rev.* D **30** 272
[3] Chodos A, Jaffe R, Johnson K, Thorne C B, and Weisskopf V F 1974 *Phys. Rev.* D **9** 3471
[4] Farhi E and Jaffe R, 1984 *Phys. Rev.* D **30** 2397
[5] Alcock C, Farhi E and Olinto A, 1986 *Astrophys. J.* **310** 261
[6] Vartanyan Yu L, Grigoryan A K and Sarkisyan T R 2004 *Astrophysics* **47** 189
[7] Glendenning N K, Kettner Ch and Weber F 1995 *Phys. Rev. Lett.* **74** 3519
[8] Glendenning N K, Kettner Ch and Weber F 1995 *Astrophys. J.* **450** 253
[9] Chandrasekhar S 1964 *Phys. Rev. Lett.* **12** 114
[10] Chandrasekhar S 1964 *Astrophys. J.* **140** 417
[11] Vartanyan Yu L, Hajyan G S, Grigoryan A K and Sarkisyan T R, 2009 *Astrophysics* **52** 300
[12] Vartanyan Yu L, Hajyan G S, Grigoryan A K and Sarkisyan T R, 2009 *Astrophysics* **52** 440
[13] Zel'dovich Ya B, 1963 *Voprosy kosmologii, Izd. AN SSSR* **9** 36
[14] Bisnovatiy-Kogan G S, 1974 *Astron. Astrophys.* **31** 3910
[15] Vartanyan Yu L, Hajyan G S, Grigoryan A K and Sarkisyan T R, 2012 *Astrophysics* **55** 98
[16] Tolman R C, 1939 *Phys. Rev.* **55** 354
[17] Oppenheimer J R and Volkoff, 1939 *Phys. Rev.* **55** 374
[18] Provencal J L, Shipman H L, Hog E and Thejll P, 1998 *Astrophys. J.* **494** 759
[19] Provencal J L, Shipman H L, Koester D, Wesemael F and Berg P, 2002 *Astrophys. J.* **568** 324
[20] Hamada T and Salpeter F E, 1961 *Astrophys. J.* **134** 683
[21] Mathews G L, Suh L S, Gorman B O and et al. 2006 *J.Phys.* G **32** 747

Journal of Physics: Conference Series **496** (2014) 012010 doi:10.1088/1742-6596/496/1/012010

Coulomb crystals in neutron star crust

D. A. Baiko
A.F. Ioffe Physical Technical Institute, Saint-Petersburg, Russian Federation
E-mail: baiko@astro.ioffe.ru

Abstract. It is well known that neutron star crust in a wide range of mass densities and temperatures is in a crystal state. At a given density, the crystal is made of fully ionized atomic nuclei of a single species immersed in a nearly incompressible (i.e., constant and uniform) charge compensating background of electrons. This model is known as the Coulomb crystal model. In this talk we analyze thermodynamic and elastic properties of the Coulomb crystals and discuss various deviations from the ideal model. In particular, we study the Coulomb crystal behavior in the presence of a strong magnetic field, consider the effect of the electron gas polarizability, outline the main properties of binary Coulomb crystals, and touch the subject of quasi-free neutrons permeating the Coulomb crystal of ions in deeper layers of neutron star crust.

1. Introduction

Coulomb crystals are expected to form in such astrophysical objects as cores of white dwarfs and crusts of neutron stars. At a given density, the crystal is made of fully ionized atomic nuclei of a single species (with charge Ze, mass M, and number density n) immersed in a nearly incompressible (i.e., constant and uniform) charge compensating background of electrons. In this paper we review the structure, thermodynamics, and elastic properties of Coulomb crystals and discuss a number of modifications of the ideal Coulomb crystal model. In particular, we consider realistic (i.e., polarizable) electron background, analyze Coulomb crystal behavior in a strong magnetic field, and study ordered binary Coulomb crystal mixtures.

Let us recall the structure of the crust of a not too hot neutron star or an old white dwarf (e.g., [1]). On the top, there is a region composed of partially ionized atoms and non-degenerate electrons. As one goes deeper, the electrons form strongly degenerate gas, that to a good approximation can be treated as a uniform charge-neutralizing background, while ions get completely ionized by electron pressure. At densities greater than 10^6 g cm^{-3} electrons become relativistic. Central densities of white dwarfs are $\lesssim 10^{10}$ g cm^{-3}. Yet even deeper, above the neutron drip density of $\sim 4 \cdot 10^{11}$ g cm^{-3}, the atomic nuclei become extremely neutron-rich and, in addition to nuclei and electrons, free neutrons appear. These neutrons are most likely superfluid and they practically play no role in our considerations. Finally, at densities greater than 10^{14} g cm^{-3}, the atomic nuclei acquire non-spherical shapes, start touching each other, and form the so-called region of nuclear pasta. In this work, we shall limit ourselves to the intermediate regions and exclude the partially ionized and nuclear pasta phases.

2. Review of thermodynamic and elastic properties

Ideal Coulomb crystal is expected to have a body-centered cubic (bcc) lattice, which is thermodynamically preferable at all temperatures (below melting). The lattice phonon wave

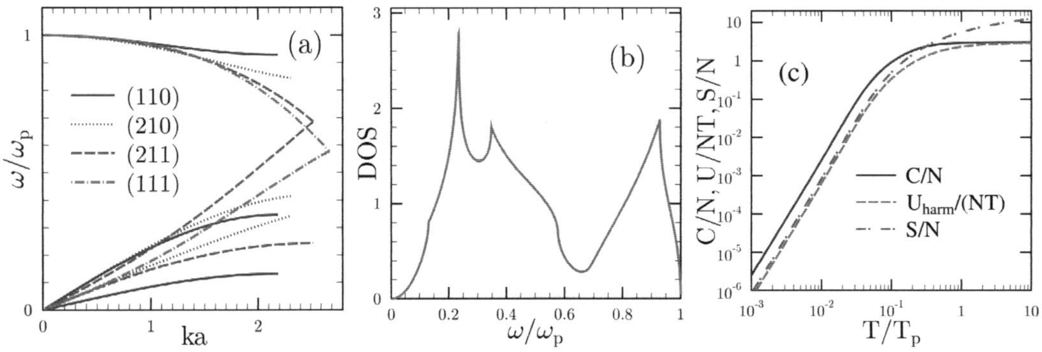

Figure 1. (a) Phonon dispersion curves in 4 directions of the wave vector vs. ka (cf. Fig. 2a for our direction specification convention), where $a = (4\pi n/3)^{-1/3}$ is the ion sphere radius; (b) density of phonon states; (c) specific heat, energy, and entropy of the Coulomb bcc lattice.

vectors $\mathbf{k}$ are restricted to its first Brillouin zone (BZ), which has the shape of a rhombic dodecahedron. To study crystal thermodynamics one needs to find phonon frequencies ω on a dense $\mathbf{k}$-grid in the BZ. These are solutions of a secular equation $\det\{D^{\alpha\beta}(\mathbf{k}) - \omega_i^2(\mathbf{k})\delta^{\alpha\beta}\} = 0$, where $D^{\alpha\beta}(\mathbf{k})$ is the dynamic matrix. The tricky part about Coulomb crystals is that one needs to take into account interactions of a given ion with all other ions (not just with the nearest neighbors) when constructing the dynamic matrix. At any wave vector $\mathbf{k}$ in the BZ there are 3 phonon modes (cf. Fig. 1a). In each direction 2 modes are acoustic ($\omega_{1,2} \propto k$ as $k \to 0$; in certain high symmetry directions these 2 modes coincide), while the third one is optic ($\omega_3 \to \omega_{\rm p}$ as $k \to 0$). At any $\mathbf{k}$ the phonon frequencies satisfy the Kohn's sum rule $\sum_{i=1}^{3} \omega_i^2 = \omega_{\rm p}^2$, where $\omega_{\rm p} = \sqrt{4\pi n Z^2 e^2/M}$ is the ion plasma frequency.

In Fig. 1b we show the phonon density of states (DOS) $\nu(x)$ (where $x \equiv \omega/\omega_{\rm p}$) of the ideal Coulomb bcc lattice [2]. By definition DOS is the fraction of phonon states per unit interval of frequencies. At $x \to 0$, DOS depends quadratically on x, which is a well-known sign of the presence of acoustic phonons. There are points where the derivative of DOS becomes infinite. These are the Van Hove singularities associated with saddle points of the frequency dependence on $\mathbf{k}$. For the Coulomb crystal there is no forbidden gap between acoustic and optic phonons.

Consider the ideal Coulomb crystal energy. Its part related to ions can be subdivided into 4 terms $E = U_{\rm M} + E_{\rm 0ph} + U_{\rm harm} + U_{\rm anharm}$. In this case, $U_{\rm M}$ is the electrostatic (Madelung) energy of static ions and neutralizing background, $E_{\rm 0ph} + U_{\rm harm}$ is the phonon energy, in which one can distinguish the zero-point, $E_{\rm 0ph}$, and the thermal, $U_{\rm harm}$, contributions, and $U_{\rm anharm}$ is the anharmonic energy. The latter can be important, but is neglected for the purpose of this work. Also neglected are the energy of the electron gas alone, which is the same for any crystal structure, and the polarization correction to the electron-ion interaction energy (see Sect. 3). Phonon thermodynamic quantities are given either by three-dimensional BZ integrals or by one-dimensional integrals with the DOS and can be computed numerically [3, 4]. For instance,

$$U_{\rm harm} = \sum_{\mathbf{k}i} \frac{\hbar\omega_i(\mathbf{k})}{e^{\hbar\omega_i(\mathbf{k})/T} - 1} = 3N\hbar\omega_{\rm p} \int_0^\infty \frac{x\nu(x){\rm d}x}{\exp{(x/t)} - 1}, \tag{1}$$

where $t = T/T_{\rm p}$ and $T_{\rm p} = \hbar\omega_{\rm p}$ is the ion plasma temperature. The lattice specific heat $C/N = {\rm d}(U_{\rm harm}/N)/{\rm d}T$, energy $U_{\rm harm}/(NT)$, and entropy S/N are illustrated in Fig. 1c. At high T, $C/N \approx 3$ (Dulong-Petit law) and at low T, $C \propto T^3$ according to the Debye law.

Let us outline the calculation of the Coulomb crystal elastic moduli. Suppose one applies an infinitesimal uniform deformation to the crystal $R_\alpha \to R_\alpha + u_{\alpha\beta}R_\beta$, where $\mathbf{R}$ are lattice vectors and $u_{\alpha\beta}$ are small constant displacement gradients, Fig. 2a (arrows in this plot show our

Journal of Physics: Conference Series **496** (2014) 012010 doi:10.1088/1742-6596/496/1/012010

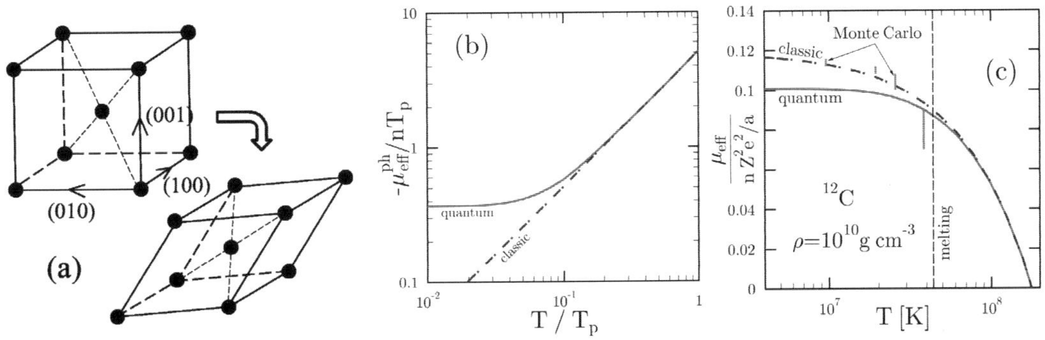

Figure 2. (a) Bcc lattice structure before and after a uniform deformation; (b) phonon contribution to the effective shear modulus; (c) total effective shear modulus.

specification convention for vector directions). The electrostatic energy of the deformed lattice can be expanded in powers of $u_{\alpha\beta}$. The zero-order term is the Madelung energy, the linear term is the electrostatic pressure times the volume change, and the coefficients of the quadratic terms are various static lattice elastic moduli. All these quantities have been found by Fuchs [5].

The potential energy of harmonic lattice vibrations can be also expanded in powers of $u_{\alpha\beta}$, which results in the phonon Hamiltonian $\hat{H}_{\rm ph} = \hat{H}_0 + \delta\hat{H}$. In this case $\hat{H}_0$ is the standard (nondeformed) expression $\hat{H}_0 = \sum_{\mathbf{k}i} \hbar\omega_{\mathbf{k}i}(0.5 + a^\dagger_{\mathbf{k}i}a_{\mathbf{k}i})$ and

$$\delta\hat{H} = 0.5\sum_{\mathbf{k}ii'}\left(\Phi^{\mathbf{k}ii'}_{\alpha\beta}u_{\alpha\beta} + 0.5\,\Phi^{\mathbf{k}ii'}_{\alpha\beta\gamma\lambda}u_{\alpha\beta}u_{\gamma\lambda}\right)(a_{\mathbf{k}i} + a^\dagger_{-\mathbf{k}i})(a_{-\mathbf{k}i'} + a^\dagger_{\mathbf{k}i'})\,, \qquad (2)$$

where $a^\dagger$ and a are the phonon creation and annihilation operators, while the quantities Φ are closely related to derivatives of the dynamic matrix [6]. The term $\delta\hat{H}$ should now be treated as a perturbation and the respective change of the crystal free energy can be found with the aid of the thermodynamic perturbation theory (e.g., [7]). Expanding the free energy in powers of $u_{\alpha\beta}$ one obtains phonon contributions to crystal elastic moduli as the quadratic term coefficients [6]. It is worth mentioning that in this way the quantization of ion motion is naturally included.

In isotropic matter there are only 2 independent elastic moduli, the compressibility modulus K and the shear modulus μ (e.g., [8]). In cubic crystals there are 3 independent elastic coefficient (e.g., [9]). However, we do not expect the neutron star crust to represent a single crystal. Likely, there are small domains with various orientations of the crystal axes and the matter is "effectively" isotropic. To describe this we follow an isotropisation procedure for the elastic coefficients proposed in [10] and obtain an effective shear modulus of the crystal $\mu_{\rm eff}$.

In Fig. 2b we plot the phonon contribution to the effective crystal shear modulus. It is negative and it grows in magnitude with temperature. As expected, excitation of phonons reduces the shear modulus and the crystal resistance to the shear stress. At low T a quantum effect is observed. The phonon contribution does not disappear even at $T = 0$. It saturates at a constant negative value associated with zero-point vibrations. The total crystal shear modulus

$$\mu_{\rm eff} = \mu^{\rm static}_{\rm eff} + \mu^{\rm ph}_{\rm eff} = 0.1194\,nZ^2e^2/a - n\hbar\omega_{\rm p}\left[0.05008 + 136.6(T/T_{\rm p})^3\right]^{1/3}\,. \qquad (3)$$

It has 2 contributions, the static lattice part proportional to the typical Coulomb energy Z^2e^2/a and the phonon part proportional to the typical phonon energy $\hbar\omega_{\rm p}$. In Fig. 2c we show the total effective shear modulus of carbon at 10^{10} g cm^{-3} as a function of temperature. The bars (1σ) represent classic Monte-Carlo (MC) results from [10]. The dot-dashed (blue) line is our analytic calculation without quantum effects (cf. dot-dashed curve in Fig. 2b). It describes the bars reasonably well. The solid (red) line is our more accurate results with quantum effects included.

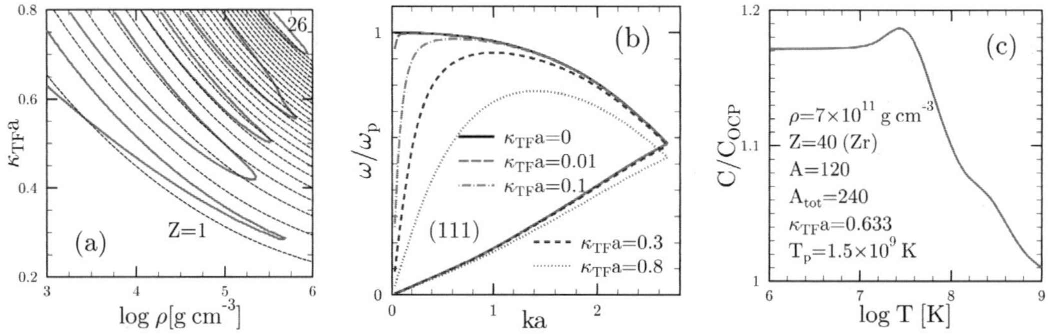

Figure 3. (a) Contours of zero energy difference between bcc and fcc lattices at $T = 0$; (b) phonon dispersion curves for the TF model of polarizable electron background (note that lower acoustic branch curves for $\kappa_{\rm TF}a = 0, 0.01, 0.1$ have merged); (c) ratio of phonon heat capacity with screening to that of the ideal Coulomb crystal (OCP).

3. Electron polarization

In reality the electron gas is not uniform. It is polarizable or, in other words, there is electron screening of the ion charge. This screening adds a term U_ϵ to the lattice energy, which is different for different crystal structures. It can be readily written in the linear response framework as

$$U_\epsilon = \frac{Z^2e^2}{2}\int \frac{d\mathbf{q}}{(2\pi)^3}\frac{4\pi}{q^2}\left[\frac{1}{\epsilon(q)} - 1\right]\left(\sum_{\mathbf{R}} e^{i\mathbf{qR}} - n\int d\mathbf{r}e^{i\mathbf{qr}}\right) , \tag{4}$$

where $\epsilon(q)$ is the electron dielectric function. We study 2 models of the dielectric function, the simple Thomas-Fermi (TF) model $\epsilon_{\rm TF}(q) = 1 + \kappa_{\rm TF}^2/q^2$ and the more realistic RPA dielectric function of the degenerate relativistic electron gas $\epsilon_{\rm J}$ obtained by Jancovici [11]. In this case $\kappa_{\rm TF} = 2k_{\rm F}\sqrt{\alpha/(\pi\beta)}$ is the TF wave number, which measures the screening strength, $\beta = v_{\rm F}/c$, $\alpha = 1/137$, while $v_{\rm F}$ and $k_{\rm F}$ are the electron Fermi velocity and wave number. It is important to realize that the linear response formalism is strictly valid only for $\kappa_{\rm TF}a \ll 1$.

First of all, if one adds the term U_ϵ to the energy, structural transitions appear [12]. Solid (red) lines in Fig. 3a correspond to zero energy difference between the face-centered cubic (fcc) and the bcc structures if the lattice energy is approximated as $U_{\rm M} + U_\epsilon$. Inside the domains the fcc lattice is preferable at $T = 0$. Dashed lines are lines of constant charge from $Z = 1$ to $Z = 26$. The structural transitions appear only for the more realistic Jancovici dielectric function. Calculations with $\epsilon_{\rm TF}$ always predict a lower energy for the bcc lattice.

The phonons in the crystal with electron screening are also different. In Fig. 3b we show dispersion curves calculated with $\epsilon_{\rm TF}$ in the (111) direction of the BZ, where 2 acoustic modes coincide. The main effect is that the optic mode is converted into an acoustic one at wave vectors smaller than the TF wave number [13]. It is remarkable that there is virtually no difference between the dispersion curves calculated with the TF and Jancovici dielectric functions [12] and thus the TF description of the phonon modes is quite adequate (unlike U_ϵ).

Once the phonon frequencies are known, one can analyze various crystal thermodynamic functions. For instance, one can calculate and compare free energies of the fcc and bcc lattices to see whether the structural transitions exist at higher temperatures. It appears that at $T = T_{\rm p}$ the bcc lattice is always preferable [12] and, therefore, the structural transitions may start happening at some point after solidification during the star cooling process.

Another example is the heat capacity. In Fig. 3c we show the ratio of the phonon heat capacities with and without electron screening (TF model) for zirconium in the inner neutron star crust at $\rho = 7\cdot 10^{11}$ g cm^{-3}. At these high densities electrons are ultrarelativistic and the

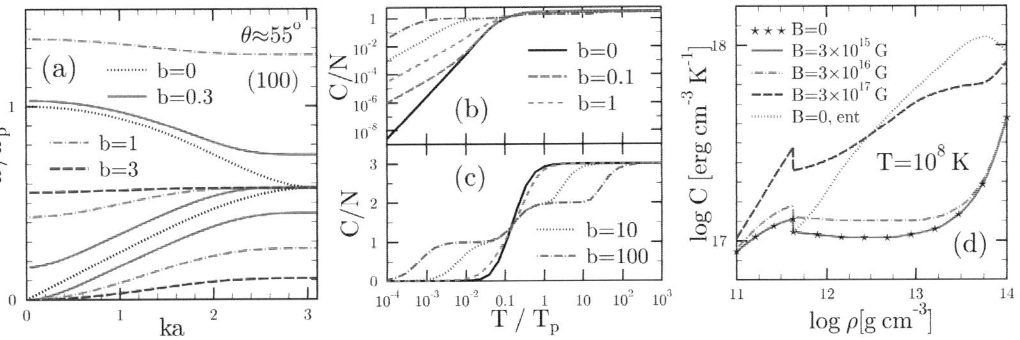

Figure 4. (a) Phonon dispersion curves of a magnetized crystal; (b) and (c) magnetized crystal specific heat in logarithmic and linear scales; (d) ion specific heat of magnetized inner neutron star crust.

screening parameter $\kappa_{\rm TF}a$ is determined only by Z. One sees that the background polarizability enhances the heat capacity in quantum regime but rather moderately (by about 20% at best). The reason for the increase is the conversion of the optic mode into the acoustic one, which can then be excited at arbitrarily low temperatures and contributes to the Debye T^3 asymptote.

4. Magnetized Coulomb crystal

Suppose the ideal Coulomb crystal is placed in an external uniform magnetic field. All results are weakly sensitive to the field direction. We assume that the field is directed towards one of the nearest neighbors, i.e. direction (111). The secular equation takes the form $\det\{D^{\alpha\beta}(\mathbf{k}) - \omega_i^2(\mathbf{k})\delta^{\alpha\beta} - i\omega_i(\mathbf{k})\omega_{\rm B}\epsilon^{\alpha\gamma\beta}n^\gamma\} = 0$, where $\mathbf{n} = \mathbf{B}/B$ and $\omega_{\rm B} = ZeB/(Mc)$ is the ion cyclotron frequency. The most important parameter turns out to be $b = \omega_{\rm B}/\omega_{\rm p}$. When $b \gg 1$ the magnetic field is really important and the crystal is said to be strongly magnetized.

It is very useful to recall a simple problem of a charged harmonic oscillator in a magnetic field (e.g., [14]). We have the obvious equation of motion $\ddot{\mathbf{r}} + \omega_0^2\mathbf{r} = \omega_{\rm B}\mathbf{v}\times\mathbf{n}$ with the Lorenz force on the right-hand side, ω_0 being the oscillator frequency in the absence of the field. Assuming the field is along the z-axis and introducing the new variable $\xi = x + iy$, we arrive at 2 equations: $\ddot{z} + \omega_0^2 z = 0$ and $\ddot{\xi} + \omega_0^2\xi = -i\omega_{\rm B}\dot{\xi}$, from which we immediately get the eigenfrequencies of the magnetized system as $\omega_{1,3} = \sqrt{\omega_0^2 + \omega_{\rm B}^2/4} \mp \omega_{\rm B}/2$ and $\omega_2 = \omega_0$. If we assume now that the frequency ω_0 is that of an acoustic phonon, $\omega_0 = ck$, and $\omega_{\rm B} \gg \omega_0$, we get 3 solutions, one of which (ω_1) depends on k quadratically and is $\propto 1/B$, another (ω_3) is at cyclotron frequency, so that it is $\propto B$, and the last one (ω_2) is independent of B. This simple picture is quite adequate for qualitative understanding of the magnetized crystal phonon spectrum.

Exact solution of the crystal secular equation [15] shows that the behavior of its phonon modes is qualitatively different for $\mathbf{kB} \neq 0$ and $\mathbf{kB} = 0$. In the former case (see Fig. 4a, where the angle between $\mathbf{k}$ and $\mathbf{B}$ is $\approx 55^{\rm o}$) branch 1 has quadratic dependence on k at $k \to 0$, while branch 3 to a good accuracy is at the ion cyclotron frequency. Branch 2 is optic and sweeps the space between the upper acoustic field-free branch and the ion plasma frequency becoming B-independent at $b \gg 1$. The situation is different for $\mathbf{k}$ perpendicular to $\mathbf{B}$. In this case branches 1 and 2 remain acoustic, while branch 3 is optic. At any $\mathbf{k}$ the frequencies satisfy a modified Kohn's sum rule $\sum_{i=1}^3 \omega_i^2 = \omega_{\rm p}^2 + \omega_{\rm B}^2$.

Once the phonon frequencies are known the magnetized crystal thermodynamic functions can be calculated via BZ integration using the same general formulas as in the field-free case [15, 16]. The most important feature of the magnetized crystal heat capacity is that the branch 1 contribution (and the total heat capacity) behaves as $T^{3/2}$ at low temperatures instead of the Debye T^3 law [17]. This is a direct consequence of the quadratic dependence of the branch

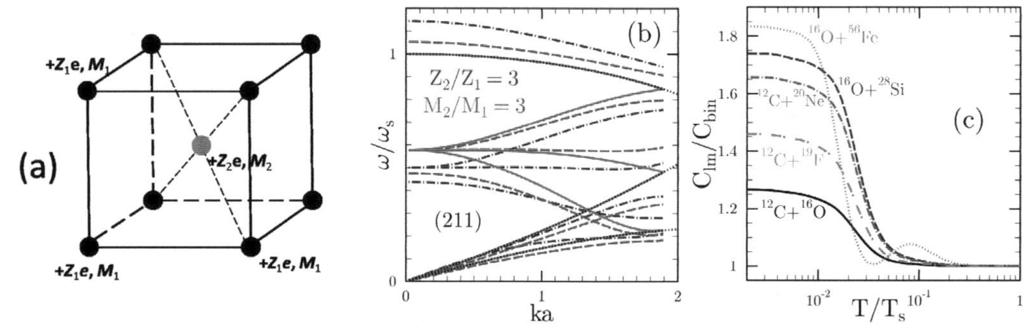

Figure 5. (a) Binary crystal structure; (b) binary crystal dispersion curves; (c) the ratio of binary crystal heat capacities obtained from the linear mixing theory and from the exact phonon spectrum.

frequency on k. Branch 2 contribution behaves as T^4 as a result of these phonons being acoustic only in the plane perpendicular to $\mathbf{B}$ and optic otherwise. Branch 3 heat capacity is exponentially suppressed at low T as a result of its optic nature. Consequently at low T the total specific heat of a strongly magnetized Coulomb crystal is bigger than that of the non-magnetized one by many orders of magnitude, Fig. 4b. At high T the specific heat has a peculiar staircase structure for strong magnetization, Fig. 4c. The reason is that the frequencies of the branches are very different ($\propto 1/B$, independent of B, and $\propto B$) and get saturated at very different temperatures.

Let us see if magnetization can modify the heat capacity of the inner neutron star crust [16], where practically all the crust mass is located. Curves in Fig. 4d show the total ion heat capacity as a function of mass density assuming the smooth composition model of ref. [1]. We see that the non-magnetized curve (stars) merges with the $B = 3 \cdot 10^{15}$ G curve (solid, red), while the $B = 3 \cdot 10^{16}$ G curve (dot-dashed, magenta) is somewhat above. In order to have a meaningful enhancement of the heat capacity at these high densities it is necessary to have a really huge field on the order of 10^{17} G (dashed, blue curve). If such fields exist in magnetars, this will be very important for their cooling models. At lower densities and temperatures in the external crust a smaller field will have a similar substantial effect, which should be important for the dynamics of bursts and superbursts as well as for afterburst relaxation modeling.

Recently, it has been argued by Chamel (e.g., [18]) that free neutrons in the inner neutron star crust are not actually free but are entrained by nuclei. In this case they increase the effective mass of the nuclei and decrease the ion plasma and cyclotron frequencies. By doing this at given T they make the crystal less quantum and thus increase its specific heat. We have included this effect by adopting the fraction of entrained neutrons from Tab. I in [18]. The result (dotted, green curve "ent" in Fig. 4d) is the dramatic increase of the specific heat comparable to that of the superstrong magnetic field in the standard model without neutron entrainment.

5. Binary Coulomb crystals

The last topic of this work is binary Coulomb crystals. This subject may be important for the white dwarf cores, where one has CO mixture, and also for the neutron star crusts, where one can expect various mixtures at various densities. In this work the one-to-one ratio of two constituents is assumed as well as their ordered distribution over nodes of a bcc lattice (Fig. 5a). This structure has to be treated as a simple cubic lattice with 2 different ions in the elementary cell. The BZ of this lattice is a cube. According to [19] such a crystal is unstable if the charge ratio is ≥ 3.6 irrespective of the mass ratio.

As in Sect. 2 one can construct the dynamic matrix and solve the respective secular equation at every point $\mathbf{k}$ of the BZ to find phonon frequencies [19]. In this case there are

6 frequencies at each **k**, which satisfy a modified Kohn's sum rule $\sum_{i=1}^{6} \omega_i^2 = 2\omega_s^2$, where $\omega_s = T_s/\hbar = \sqrt{\pi n e^2 (Z_1+Z_2)(Z_1/M_1+Z_2/M_2)}$ is the binary mixture plasma frequency (n is the number density of all ions). In Fig. 5b, 3 sets of dispersion curves are plotted in the direction (211) up to the BZ boundary in this direction. The solid (red) curves correspond to identical ions while dotted (black) curves show dispersion relations of a bcc lattice of identical ions in the same direction. Naturally, solid and dotted curves coincide but the dotted ones can be extended further (cf. Fig. 1a), up to the boundary of the bcc lattice BZ. The dot-dashed (blue) and dashed (green) curves correspond to charge and mass ratios of 3, respectively.

It is customary to obtain thermodynamic functions of any mixture from the so-called linear mixing theory, in which, say, the binary mixture total heat capacity is $C_{\rm lm} = N_1 c_1 + N_2 c_2$, where N_i is the total number of particles of the i-th type and c_i is the specific heat of the system composed only of the i-th ion species at the same temperature and electron number density as in the mixture. But now we know the exact frequencies of the ordered binary crystal mixture and thus are in a position to calculate its exact specific heat $C_{\rm bin}$ [19]. This allows one to assess the validity of the linear mixing theory for application to such mixtures.

In Fig. 5c we plot the ratio of the linear mixing specific heat to the exact binary crystal specific heat. At high temperatures the linear mixing theory reproduces the Dulong-Petit law and consequently is very accurate. But at low temperatures it overestimates the specific heat substantially. For instance, by 80% for iron/oxygen mixture and by almost 30% for CO mixture [19, 2]. This will affect quantitatively the cooling curves of white dwarfs and may be important for modeling neutron star crust thermal relaxation if this structure is realized.

6. Conclusion

The work on Coulomb crystal theory continues. For the sake of white dwarf cooling theory it is desirable to calculate the crystal anharmonic energy in quantum regime (only classic crystal anharmonic energy is currently known). It is also worthwhile to study Coulomb crystal mixtures with substitutional disorder. Both topics are currently work in progress.

Acknowledgments

The author is grateful to D.G. Yakovlev and A.A. Kozhberov for discussions. This work was partially supported by RFBR (grant 14-02-00868) and by Rosnauka (grant NSh 294.2014.2).

References

[1] Haensel P, Potekhin A Y and Yakovlev D G 2007 *Neutron Stars 1: Equation of State and Structure* (New York: Springer)
[2] Kozhberov A A and Baiko D A *in preparation*
[3] Baiko D A 2000 Ph.D. thesis, A.F. Ioffe Physical-Technical Institute
[4] Baiko D A, Potekhin A Y and Yakovlev D G 2001 *Phys. Rev.* E **64** 057402
[5] Fuchs K 1936 *Proc. Roy. Soc. London* **153** 622
[6] Baiko D A 2011 *Mon. Not. Roy. Astron. Soc.* **416** 22
[7] Landau L D and Lifshitz E M 1980 *Statistical Physics. Part I* (Oxford: Pergamon Press)
[8] Landau L D and Lifshitz E M 1986 *Theory of Elasticity* (Oxford: Pergamon Press)
[9] Wallace D C 1967 *Phys. Rev.* **162** 776
[10] Ogata S and Ichimaru S 1990 *Phys. Rev.* A **42** 4867
[11] Jancovici B 1962 *Nuovo Cimento* **25** 428
[12] Baiko D A 2002 *Phys. Rev.* E **66** 056405
[13] Pollock E L and Hansen J P 1973 *Phys. Rev.* A **8** 3110
[14] Landau L D and Lifshitz E M 1975 *The Classical Theory of Fields* (Oxford: Pergamon Press)
[15] Baiko D A 2009 *Phys. Rev.* E **80** 046405
[16] Baiko D A and Yakovlev D G 2013 *Mon. Not. Roy. Astron. Soc.* **433** 2018
[17] Usov N A, Grebenschikov Yu B and Ulinich F R 1980 *J. Exp. Theor. Phys.* **78** 296
[18] Chamel N 2012 *Phys. Rev.* C **85** 035801
[19] Kozhberov A A and Baiko D A 2012 *Contrib. Plasma Phys.* **52** 153

Journal of Physics: Conference Series **496** (2014) 012011 doi:10.1088/1742-6596/496/1/012011

Dense matter in strong magnetic fields

Monika Sinha

Institute for Theoretical Physics, J. W. Goethe-University, D-60438 Frankfurt-Main, Germany
Indian Institute of Technology Rajasthan, Jodhpur 342011, Rajasthan, India

Abstract. Compact stars having strong magnetic fields (magnetars) have been observationally determined to have surface magnetic fields of order of $10^{14} - 10^{15}$ G, the implied internal field strength being several orders larger. We study the equation of state and composition of hypernuclear matter and quark matter - two forms of dense matter in strong magnetic fields. We find that the magnetic field has substantial influence on the properties of hypernuclear matter and quark matter for magnetic field $B \geq 10^{17}$ G and $B \geq 10^{18}$ G respectively. In particular the matter properties become anisotropic. Moreover, above a critical field B_{cr}, both hypernuclear and quark matter show instability, although the values of B_{cr} are different for two kinds of matter.

1. Introduction

Soft γ-ray repeaters and anomalous X-ray pulsars are commonly believed to be magnetars - the compact stars with surface magnetic fields $B_s \sim 10^{14} - 10^{15}$ G. The interpretation of astrophysical observations of magnetars requires good knowledge of the properties of dense matter in the presence of large magnetic fields. The properties of dense matter inside the compact objects is poorly known due to lack of the proper knowledge of strong interactions that are relevant at densities of interest. As a consequence, many phenomenological models of dense matter have been proposed over the years. The models of dense matter can tentatively be divided into two broad classes: one class includes matter made of ordinary hadronic matter; the second class is deconfined quark-gluon state with about equal number of up, down and strange quarks, known as the strange quark matter (SQM). In both cases the underlying constituent particles are fermions which are interacting via exchange of bosons. Fermions in strong magnetic field experience two well-known quantum mechanical effects: the Pauli paramagnetism and the Landau diamagnetism. The first is due to the interaction of the spin of the fermion with the magnetic field and therefore, is relevant for both charged and uncharged fermions. The second effect is relevant only for charged fermions, and is particularly strong for light particles, which in the case of compact stars are the leptons.

Since one can not exclude the possibility of a density dependent field profile which is favorable for local magnetostatic equilibrium inside a star, we consider the hyperonic matter under the influence of density dependent magnetic field. Among many effective phenomenological models to study the quark matter, the MIT bag model [4] and the Nambu-Jona-Lasinio (NJL) model [5, 6] are the most popular ones. However, the NJL model does not account for confinement property of QCD, while the bag model cannot account for the chiral symmetry breaking, although it is built in a manner as to confine through an *ad hoc* bag pressure. We will base our discussion of quark matter in strong magnetic field and at nonzero temperature on a model originally introduced by Dey *et. al.* [7]. In this model quark masses are density dependent ensuring chiral symmetry restoration at high density and the quarks interact among themselves via the Richardson potential (RP) [8] in which asymptotic freedom and confinement are built in. In the next section we discuss the general effect of magnetic field on fermionic matter. In Sec. 3 we discuss the results concerning the equation of state (EoS) of hypernuclear and quark matter in strong magnetic fields. Our conclusions are presented in Sec. 4.

2. Magnetized matter

In the presence of a magnetic field the motion of charged fermions is Landau quantized in the plane perpendicular to the direction of the magnetic field. If the field direction is assumed to be the direction of z-axis the single particle energy of a charged particle with mass m in the n-th Landau level is $\epsilon_n = \sqrt{p_z^2 + m^2 + 2ne|Q|B}$, p_z being the momentum parallel to the field (which is in the z direction of Cartesian system of coordinates) and Q being the charge of the particle in units of proton charge. Evidently, the momentum of a particle in the x-y plane is quantized. The phase space sampling for such a particle is therefore modified according to the rule (the spin degeneracy is included)

$$2\int_0^{p_F} d^3p \longrightarrow e|Q|B \sum_{n=0}^{n_{max}} (2-\delta_{n,0}) \int_{-p_{F,n}}^{p_{F,n}} dp_z \int_0^{2\pi} d\phi, \tag{1}$$

where ϕ is the azimuthal angle and $p_{F,n} = \sqrt{p_F^2 - 2ne|Q|B}$ with p_F being the Fermi momentum.

In the presence of electromagnetic field the energy momentum tensor of a fermionic system in the rest frame of matter is $T^{\mu\nu} = T_m^{\mu\nu} + T_f^{\mu\nu}$, where the matter part of energy momentum tensor is

$$T_m^{\mu\nu} = \begin{bmatrix} \varepsilon_m & 0 & 0 & 0 \\ 0 & P_m - MB & 0 & 0 \\ 0 & 0 & P_m - MB & 0 \\ 0 & 0 & 0 & P_m \end{bmatrix}, \tag{2}$$

and in the absence of electric field the field part of the energy momentum tensor is

$$T_f^{\mu\nu} = \frac{B^2}{8\pi} \begin{bmatrix} 1 & 0 & 0 & 0 \\ 0 & 1 & 0 & 0 \\ 0 & 0 & 1 & 0 \\ 0 & 0 & 0 & -1 \end{bmatrix}. \tag{3}$$

The matter energy density ε_m is obtained by integrating the single particle energies of each species over appropriate phase space volume and then adding their contributions. The thermodynamic pressure at zero temperature is $P_m = \sum_i \mu_i n_i - \varepsilon_m$, where μ and n are chemical potential and number density of corresponding species. The total energy density of the system is given by the sum of the matter and field contributions $\varepsilon = \varepsilon_m + B^2/8\pi$. It is clear from Eqs. (2) and (3) that the presence of magnetic field makes the pressure anisotropic: the pressure in the perpendicular to the magnetic field direction is $P_\perp = P_m - MB + B^2/8\pi$, whereas the pressure in the direction parallel to the magnetic field is $P_\parallel = P_m - B^2/8\pi$.

We study the hyperonic matter in the presence of strong magnetic field within the nonlinear Boguta-Bodmer-Walecka model [9, 10]. The model is described in detail in Ref. [11]. The field profile has been adopted according to [12]

$$B\left(\frac{n_b}{n_0}\right) = B_s + B_c\left\{1 - \exp\left[-\beta\left(\frac{n_b}{n_0}\right)^\gamma\right]\right\}. \tag{4}$$

The parameters β and γ control the relaxation from the central value B_c to the asymptotic value at the suface B_s. Here n_b is the baryon number density, n_0 is the normal nuclear matter density.

In the case of quark matter at finite temperature, we consider matter composed of u, d and s quarks which interact via the Richardson potential [8]

$$V(q^2) = -\frac{4}{9}\frac{\pi}{\ln[1+(q^2+m_g^2)/\Lambda^2]}\frac{1}{(q^2+m_g^2)}, \tag{5}$$

where m_g is gluon mass and Λ is a scale parameter. The finite gluon mass is responsible for screening in medium and is related to the screening length D via

$$m_g^2 = D^{-2} = \frac{2\alpha_0}{\pi}\sum_{i=u,d,s} k_F^i \mu_i^*, \tag{6}$$

where α_0 is the perturbative quark gluon coupling, $\mu_i^* = \sqrt{(k_F^i)^2 + m_i^2}$ is the Fermi energy (the chemical potential at zero temperature), k_F^i is the Fermi momentum and m_i the quark mass. In this model the quark masses vary with density as [7]

$$m_i = M_i + M_q \operatorname{sech}\left(\nu\frac{n_b}{n_0}\right), \qquad i = u, d, s, \tag{7}$$

where ν is a parameter. At large n_b the quark mass m_i falls off from its constituent value M_q to its current value M_i.

The kinetic part of the energy density for a particular quark flavor in the presence of magnetic field and at non-zero temperature is given by

$$\varepsilon_{kin} = \frac{3}{(2\pi)^3}e|Q|B\sum_{n=0}^{\infty}(2-\delta_{n,0})\int_0^{2\pi}d\phi\int_{-\infty}^{\infty}f(\epsilon)\epsilon\ dk_z, \tag{8}$$

where $f(\epsilon)$ is Fermi distribution function. The potential part of the energy density due to interaction between the flavors i and j is given by

$$\begin{aligned}\varepsilon_{pot}^{ij} &= \frac{e^2|Q_i||Q_j|}{(2\pi)^5}B^2\sum_{n_i}\sum_{n_j}(2-\delta_{n_i,0})(2-\delta_{n_j,0})\\ &\int_0^{2\pi}d\phi_i\int_0^{2\pi}d\phi_j\int_{-\infty}^{\infty}dk_z^i\int_{-\infty}^{\infty}dk_z^j f(\epsilon_i)f(\epsilon_j)NV(q^2)S,\end{aligned} \tag{9}$$

where

$$N = \frac{(\epsilon_i+m_i)(\epsilon_j+m_j)}{4\epsilon_i\epsilon_j}, \quad S = 1 + \frac{k_i^2k_j^2}{(\epsilon_i+m_i)^2(\epsilon_j+m_j)^2} + \frac{2\mathbf{k}_i\ cdot\mathbf{k}_j}{(\epsilon_i+m_i)(\epsilon_j+m_j)}.$$

The matter pressure is then given by

$$P_m = \sum_i \mu_i n_i + Ts - \varepsilon, \tag{10}$$

where T is the temperature, s is the entropy density and ε is the sum of the kinetic and potential energies. For further details see Ref. [13].

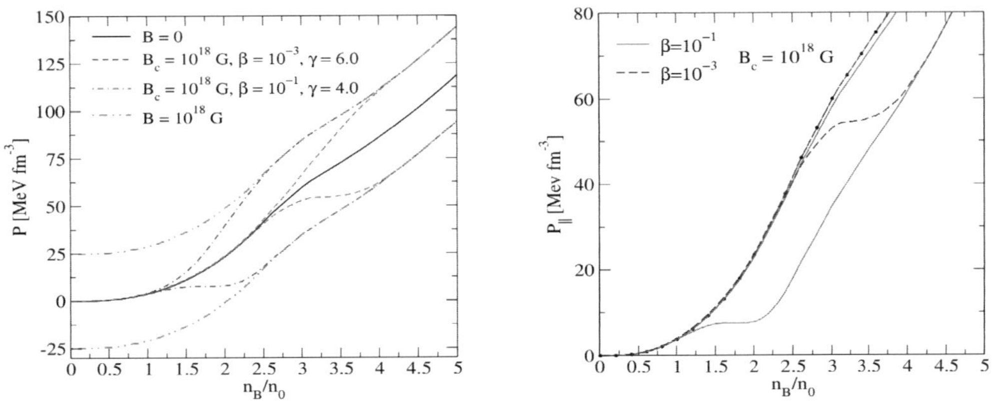

Figure 1. *Left panel* (a): Variation of total pressure as a function of normalized baryon number density for fixed magnetic fields $B_c = 0$ (solid line) and $B_c = 10^{18}$ G with several field profiles, $\beta = 10^{-3}$, $\gamma = 6$ (dashed lines), $\beta = 10^{-1}$, $\gamma = 4$ (dashed-dotted lines), and $\beta \to \infty$, *i.e.*, $B_c =$ constant (dashed-double-dotted lines). For each pair of curves the upper branch is for $P_\perp$ and the lower branch for $P_\parallel$. *Right panel* (b): Dependence of $P_\parallel$ on the normalized baryon number density for different magnetic field profiles and $B_c = 10^{18}$ G. The dots show the reference case $B_c = 0$. The solid and dashed lines correspond to $\beta = 0.1$ and 0.001, respectively. For each β we choose a pair of γ's; in the first case we have $\gamma = 1$ and $\gamma = 4$, whereas in the second case $\gamma = 1$ and $\gamma = 6$.

3. Results

Fig. 1a shows the EoS of hyperonic matter in the cases of no magnetic field, constant field as well as for fields with various density profiles. For nonzero magnetic fields the pressure splits into parallel and perpendicular components showing anisotropy which arises from the magnetic field contribution. From Fig. 1a it is also evident that there is an onset of instability for some field profiles due to $P_\parallel$ component of the pressure. This is shown in Fig. 1b in a more systematic manner. It is clear that for a given value of β the corresponding EoS becomes softer as γ is increased. Consequently, beyond a certain critical value of γ and in a certain density regime $P_\parallel$ ceases to increase and subsequently decreases with further increase in n_b. This implies that matter becomes unstable above that value of density for that particular B_c and magnetic field profile. For comparison we also show results for each β with the minimum value of γ taken to be 1. Note that the maximum value of γ is taken such that $P_\parallel$ forms a plateau as a function of n_b.

The instability arises due to the negative contribution from the field energy density (pressure) to the net pressure of magnetized matter in the direction of the magnetic field. Since for any particular B_c and magnetic profile the field strength increases with the increase of n_b more negative contribution is added to $P_\parallel$ with the increase of n_b. Consequently, at a certain density, $P_\parallel$ ceases to increase and then decreases with the increase of n_b.

We turn now to the problem of quark matter in strong magnetic fields and we focus below on the effect of the RP model on the magnetized SQM at finite temperature. Our investigation shows that the results are not sensitive to the temperature in the range relevant to the physics of compact stars. Hence, we report here results with a fixed temperature $T = 20$ MeV. In Fig. 2a we show the EoS of SQM for the RP model and the bag model. Magnetic field introduces some oscillations in the pressure with density; in each case the increase of pressure after a

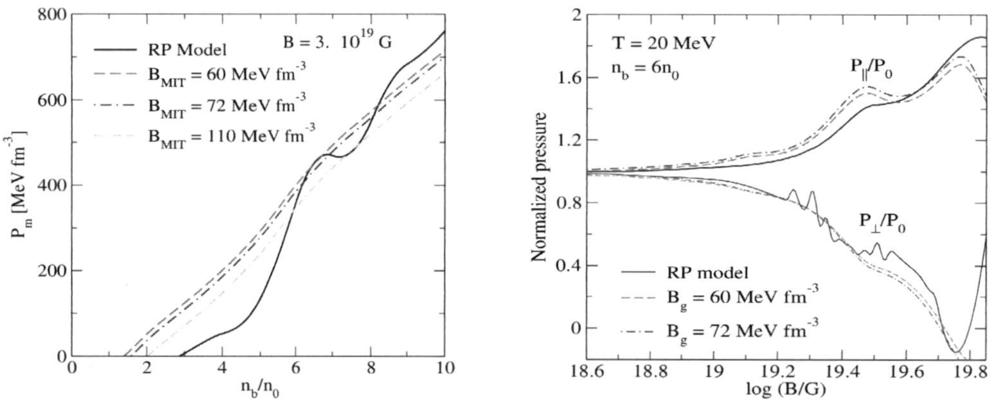

Figure 2. *Left panel* (a): Dependence of the thermodynamic pressure P_m on the normalized baryon number density at $T = 20$ MeV for $B = 3 \times 10^{19}$ G. The pressure is shown for the RP model (solid), for the MIT bag model with $B_{\text{MIT}} = 60$ MeV fm^{-3} (dashed), $B_{\text{MIT}} = 72$ MeV fm^{-3} (dash-dotted) and $B_{\text{MIT}} = 110$ MeV fm^{-3} (double-dash-dotted). *Right panel* (b): Dependence of the normalized pressure components that are parallel and perpendicular to the magnetic field on the strength of the magnetic field at $n_b = 6n_0$ and $T = 20$ MeV for various models. The upper three curves correspond to the parallel pressure, the lower three - to the perpendicular pressure.

plateau is caused by the opening of a new Landau level. The oscillations are much stronger in the RP model and this can be traced back to the momentum dependence of the potential. The major contribution comes from the static gluon propagator part of the potential, while the logarithmic factor in the potential depends weakly on momentum. Note that at a certain density the pressure has a plateau and slight negative downturn, which can be interpreted as an instability of homogenous magnetized matter towards phase separation.

For large magnetic fields, the effects of anisotropy become important, in particular the parallel and perpendicular components of pressure differ substantially. We show the variations of $P_{||}$ and $P_{\perp}$ with B within the RP and bag models at $n = 6n_0$ in Fig. 2b. We note that below $B = 3 \times 10^{18}$ G, both $P_{||}$ and $P_{\perp}$ are practically equal to the pressure of matter in absence of magnetic field. Hence, we conclude that for the SQM the effect of magnetic field is not significant below $B \sim 10^{18}$ G in the framework of our current modes. For larger fields $P_{||}$ increases whereas $P_{\perp}$ decreases in both models of SQM. For large enough fields $P_{\perp}$ becomes negative starting from some critical value of B; this critical value is almost the same in both models. We now recall that for very large magnetic field $P_{\perp} \to 0$ in the models without confinement [14, 15]. We see that the addition of the confining potential provides additional "attraction" inside the SQM and its effect becomes more transparent at larger B.

4. Conclusions

We have found that above a certain critical field value $B_{cr} \sim 10^{18}$ G the hyperonic matter may become unstable. The instability arises due to negative contribution of the field pressure to the pressure of matter. The details of the onset of instability depend on the assumed central field value B_c as well as the parameterization of the field profile. The instability puts a natural upper bound on the possible central magnetic field of a neutron star.

We find significant differences between the equation of state of the strange quark matter predicted by the RP and the bag models. This is due to the intrinsic momentum-dependent interaction between quarks in the RP model, which mimics the one-gluon-exchange interaction of the QCD. Specifically, we find that the thermodynamic pressure in the RP model is more sensitive to baryon density when the magnetic field is strong. The de Haas-van Alfven type oscillations in the transverse pressure $P_\perp$ are much more pronounced in the RP model than in the MIT bag model.

Furthermore, we find that the presence of a confining potential, modeled either in terms of the Richardson potential or the MIT bag, suppresses the pressure components $P_{||}$ and $P_\perp$ and, at large B, the anisotropy in the equation of state. The splitting between the longitudinal pressure $P_{||}$ and the transverse pressure $P_\perp$ was found to be weaker than that in free (non-interacting) strange quark matter. This underlines the importance of taking into account the confining potential in studies of strongly magnetic strange quark matter in cores of neutron stars and in strange stars. It remains an interesting task to explore the effects of the confining potential on the structure and geometry of strongly magnetized stars.

Acknowledgments

The author acknowledges the support of the Alexander von Humboldt Foundation.

References

[1] Usov V V 1992 *Nature* **357** 472
[2] Thompson C and Duncan R C 1995 *Mon. Not. Roy. Astron. Soc.* **275** 255
[3] Thompson C and Duncan R C 1996 *Astro. Phys. J.* **473** 322
[4] Chodos A, Jaffe R L, Johnson K, Thorn C B and Weisskopf V F 1974 *Phys. Rev. D* **9** 3471
[5] Nambu Y and Jona-Lasinio G 1961 *Phys. Rev.* **124** 246
[6] Nambu Y and Jona-Lasinio G 1961 *Phys. Rev.* **122** 345
[7] Dey M, Bombaci I, Dey J, Ray S and Samanta B C 1998 *Phys. Lett. B* **438** 123
[8] Richardson J L 1979 *Phys. Lett. B* **82** 272
[9] Walecka J D 1974 *Ann. Phys.* **83** 491
[10] Glendenning N K 1987 *Zeitschrift fur Physik A Hadrons and Nuclei* **327** 295
[11] Sinha M, Mukhopadhyay B and Sedrakian A 2013 *Nucl. Phys. A* **898** 43
[12] Bandyopadhyay D, Chakrabarty S and Pal S 1997 *Phys. Rev. Lett.* **79** 2176–2179
[13] Sinha M, Huang X G and Sedrakian A 2013 *Phys. Rev. D* **88** 025008
[14] Huang X G, Huang M, Rischke D H and Sedrakian A 2010 *Phys. Rev. D* **81** 045015
[15] Dexheimer V, Negreiros R and Schramm S 2012 *Euro. Phys. J. A* **48** 189

Journal of Physics: Conference Series **496** (2014) 012012 doi:10.1088/1742-6596/496/1/012012

The influence of small scale magnetic field on the polar cap X-ray luminosity of old radio pulsars

D. P. Barsukov

Ioffe Physical-Technical Institute of the Russian Academy of Sciences, Saint-Petersburg, Russian Federation
Saint-Petersburg State Polytechnical University, Saint-Petersburg, Russian Federation

O. A. Goglichidze and A. I. Tsygan

Ioffe Physical-Technical Institute of the Russian Academy of Sciences, Saint-Petersburg, Russian Federation

E-mail: tsygan.astro@mail.ioffe.ru

Abstract. The influence of small-scale magnetic field on the polar cap heating by returning positrons is considered. The returning positron current is calculated in the framework of two models of rapid and gradual screening. To calculate the electron-positron pair production rate we take into account only the curvature radiation of primary electrons and its absorption in magnetic field. We use the polar cap model with steady space charge limited electron flow. It is shown that the rapid screening model is in the better agreement with observations of old (age $> 10^6$ years) radio pulsars. The gradual screening model usually leads to too strong heating and too large X-ray luminosities.

1. Introduction

The problem of magnitude of the returning positron current which flows through inner gaps and causes the polar cap heating has not been yet definitely solved. Here we present the comparison of two models of returning current calculation for old isolated normal radio pulsars with characteristic age $\tau = P/(2\dot{P}) \geq 10^6$ years. We assume that pulsar magnetosphere can be described by Goldreich-Julian model [1]. We assume that the inner gaps occupy entire pulsar tube cross-section and neglect the input of outer gaps. Also we assume that inner gaps are stationary and operate in regime of space charge limited flow with electron emission from star surface.

2. The small-scale magnetic field

Let the neutron star have a radius r_{ns} and dipolar magnetic moment $\vec{m}$ (its field at magnetic pole is $B_{dip} = 2m/r_{ns}^3$). We assume also that a small-scale magnetic field with strength B_{sc} and characteristic scale $\ell \approx r_{ns}/10$ presents nearby the polar cap. For simplicity we suppose that $\vec{B}_{sc}$ is parallel to surface (and $\vec{B}_{sc} \cdot \vec{m} = 0$) and denote by ψ $(0 \leq \psi \leq \pi)$ the angle between $\vec{B}_{sc}$ and plane containing $\vec{\Omega}$ and $\vec{m}$ (see Fig. 1). We describe this component by additional magnetic

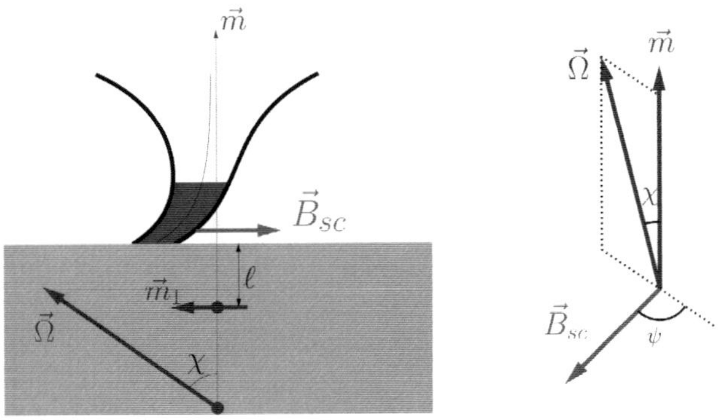

Figure 1. The orientation of small-scale magnetic field $\vec{B}_{sc}$.

moment $\vec{m}_1$ located in the polar region of the neutron star at depth ℓ [2]:

$$\vec{B} = \frac{3\vec{r}(\vec{r}\cdot\vec{m}) - \vec{m}r^2}{r^5} + \frac{3\vec{\rho}(\vec{\rho}\cdot\vec{m}_1) - \vec{m}_1\rho^2}{\rho^5}, \quad (1)$$

where $\vec{\rho} = \vec{r} - (r_{ns} - \ell)\vec{e}_z$, $\vec{m} = m\vec{e}_z$.
In the reference frame rotating with the star all quantities do not depend on time, so we can write

$$\Delta\Phi = -4\pi(\rho - \rho_{GJ}), \quad \vec{E} = -\vec{\nabla}\Phi, \quad (2)$$

where $\rho_{GJ} = \frac{\Omega B}{2\pi c}\tilde{\rho}_{GJ}$ is Goldreich-Julian density [1], $\rho = \frac{\Omega B}{2\pi c}\tilde{\rho}$ and B is magnetic field strength. It is worth to note that without frame dragging $\tilde{\rho}_{GJ}(z) \approx -\cos\tilde{\chi}$, where $\tilde{\chi}$ is the angle between $\vec{B}$ and $\vec{\Omega}$. We choose $\Phi = 0$ at surface and pulsar tube boundaries and take into account free electron emission condition $E_{||}\big|_{z=0} = 0$ [3], where $E_{||} = (\vec{B}\cdot\vec{E})/B$ and z is the altitude above star surface. We assume that the particles inside the pulsar tube move along field lines $\vec{v} \parallel \vec{B}$ with relativistic velocity $v \approx c$, so continuity equation div $(\rho\vec{v}) = 0$ may be rewritten as $(\vec{B}\cdot\vec{\nabla})\tilde{\rho} = 0$ and, hence, $\tilde{\rho}$ is constant along field lines.

3. Rapid screening model

The rapid screening model was proposed in [3]. According to this model pulsar tube is divided into three parts (see Fig. 2):

1) $0 < z < z_c$ acceleration region (pulsar diode)
There is no pair production and no pair-plasma to screen the electric field. Thus, a large parallel electric field $E_{||}$ is present here which accelerates the electrons. The accelerated electrons, moving along curved field lines, radiate curvature gamma photons. The latter photons after some propagation are converted into electron-positron pairs. This determines the upper boundary $z = z_c$ of this region. The altitude $z = z_c$ is practically equal to the altitude of pair formation front [4].

2) $z_c < z < z_r$ partial screening region
There is pair production but the pairs do not have time to screen electric field completely. So, small electric field $E_{||}$ is present here which decelerates positrons. Some of them change the velocity sign and begin to move downwards, forming the returning positron current.

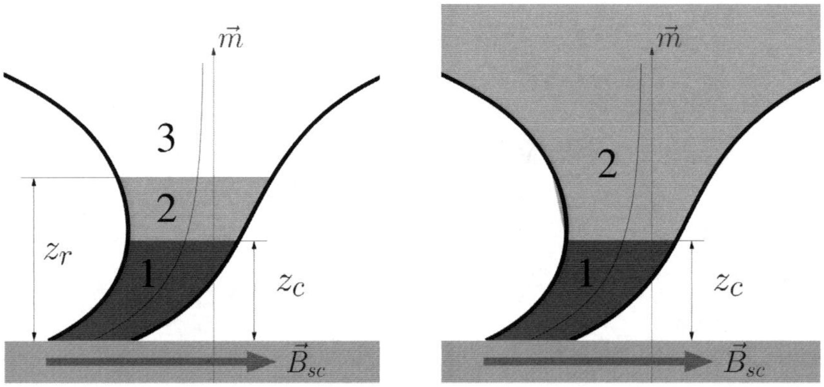

Figure 2. A schematic view of pulsar tube in the framework of rapid screening model (left panel) and gradual screening model (right panel). Different regions are marked as follows: 1 – the acceleration region (diode), 2 – the partial screening zone, 3 – the full screening zone.

These positrons return back into pulsar diode, become accelerated by the large electric field and hit the star's surface.
This leads to polar cap heating.

3) $z > z_r$ full screening region
This region is filled by large amount of electron-positron pairs, which totally screen the electric field. In [3] it was assumed that $E_{||} = 0$ in this region. And, hence, the positrons do not decelerate and change the velocity sign here. Consequently, this region does not give an input into returning positron current.

To determine the altitude z_r the following conditions were proposed in [3]

$$E_{||}\Big|_{z=z_r} = 0 \qquad \text{and} \qquad (\vec{B}\cdot\vec{\nabla})E_{||}\Big|_{z=z_r} = 0. \tag{3}$$

The first condition means that the electric field is continuous at $z = z_r$, the second means that at $z = z_r$ the charge density is continuous too. If electron-positron pairs are mainly generated by curvature radiation photons absorption in magnetic field, the energy of positrons is relatively small $\sim 10^2 - 10^4$ MeV and its number increases very rapidly with the altitude. Hence, the length of partial-screening region is small $z_r - z_c \ll z_c$, ℓ [4]. So, assuming that the pulsar tube is thin $r_t \ll \ell$, where r_t is the pulsar tube radius, it is possible to estimate returning positron density flowing along central open field line as

$$\tilde{\rho}_{ret} \approx -r_t \left.\frac{\partial\tilde{\rho}_{GJ}}{\partial z}\right|_{z=z_c} F\left(\frac{z_c}{r_t}\right), \tag{4}$$

where function $F(x)$ may be approximated as $F(x) \approx \frac{4x}{16+15x}\left(1 + 1.19\frac{x}{1+x^2}\right)$ and $F(x) \approx \frac{x}{4}$ for $x \ll 1$, $F(x) \approx \frac{4}{15}$ for $x \gg 1$.

4. Gradually screening model

Harding and Muslimov have shown [4] that if we accept the assumptions that

1) all quantities do not depend on time t (stationary case)

2) pairs are affected only by the mean electric field

3) $\tilde{\rho}_{GJ}$ monotonically decreases with altitude z

then the conditions (3) can not be satisfied at the same point. So, there is no stationary solution. To solve this problem they proposed that the full screening region does not exist, and pulsar tube consists of two parts: 1) acceleration region and 2) partially screening region (see Fig. 2). Hence, the electric field which decelerates positrons exists everywhere. In [4] it is assumed that for $z > z_c$ electric field falls exponentially and very quickly becomes negligibly small and, consequently, it does not cause large positron deceleration. Therefore, Ref. [4] proposed to replace (3) by the condition

$$\Phi \to \text{const.} \quad \text{at} \quad z \to +\infty. \tag{5}$$

But if the electric field (very small, of course) penetrates deep into plasma, all altitudes above z_c give an input into returning current [5]. Potential Φ does not change significantly along the field lines at $z \geq z_c$ and positrons are reversed to support $\tilde{\rho} - \tilde{\rho}_{GJ} \approx$ const. along the field lines at $z \geq z_c$. Hence, the value of the returning positron current formed at $z_c \leq z \leq z_f$ may be estimated as

$$\tilde{\rho}_{ret} \approx -\frac{1}{2}\left(\tilde{\rho}_{GJ}(z_f) - \tilde{\rho}_{GJ}(z_c)\right)$$

We suppose that $z_f \approx (3 \div 15) r_{ns}$ is a good choice. This choice satisfies two conditions:

(i) $z_f < z_{rad}$, where $z_{rad} \approx (5 \div 50) r_{ns}$ is the altitude at which plasma waves begin to affect the pair-dynamics, so that the assumption 2) becomes invalid. Hence, at $z \geq z_{rad}$ it is possible to satisfy both conditions (3).

(ii) $z_f < z_{max}$, where $z_{max} \approx (1 \div 5) r_{ns}$ is the altitude at which the function $\tilde{\rho}_{GJ}$ has a minimum. It is easy to show that there exists a solution satisfying the both conditions (3) at $z_r \approx z_{max}$.

Also it is worth noting that $\tilde{\rho}_{GJ}$ does not change significantly at those altitudes, so any z_f value from the selected range may be appropriate.

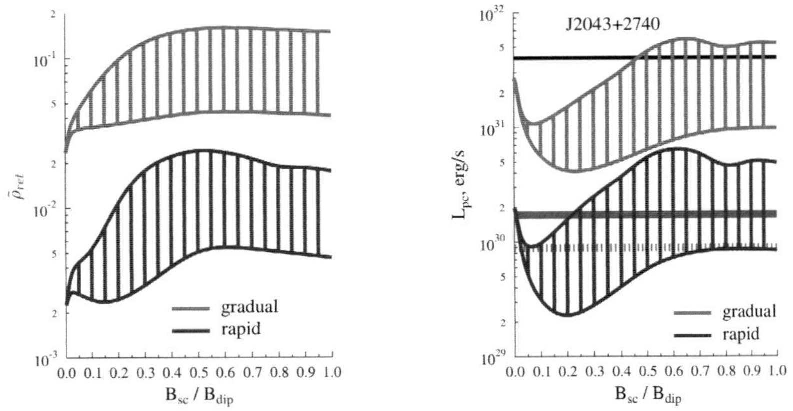

Figure 3. The left panel shows the returning positron current calculated in the framework of two models for pulsar J2043+2740, $B_{dip} = 7.1 \cdot 10^{11}$ G, $P = 96$ ms, $\tau = 1.2 \cdot 10^6$ years, $\chi = 55°$. The corresponding polar cap luminosity is shown in right panel. Upper limits of polar cap emission from [9] are shown by green lines, solid when we see one cap, dashed when we see both caps. The emission from the stellar surface is taken from [10] and is shown by black line.

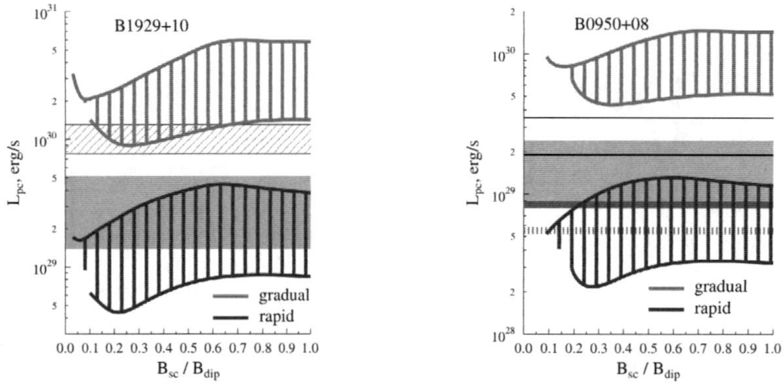

Figure 4. The polar cap luminosity L_{pc} for B1929+10 is shown in the left panel, $B_{dip} = 1.0 \cdot 10^{12}$ G, $P = 0.23$ s, $\tau = 3 \cdot 10^6$ years, $\chi = 45°$. The luminosity L_{pc}, which is taken from [12], is shown by orange area. The L_{pc} luminosity range from [13] is shown by black dashed area. The polar cap luminosity L_{pc} for B0950+08 is shown in right panel, $B_{dip} = 4.9 \cdot 10^{11}$ G, $P = 0.25$ s $\tau = 17.5 \cdot 10^6$ years, $\chi = 30°$. The luminosities L_{pc} from [10] are shown by orange area and black lines. The upper limits from [9] are shown by green lines, solid when we see one cap, dashed when we see both caps.

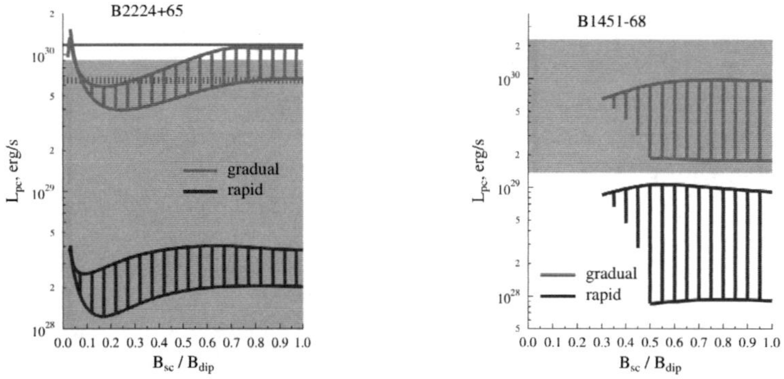

Figure 5. The polar cap luminosity L_{pc} for B2224+65 is shown in left panel, $B_{dip} = 5.2 \cdot 10^{12}$ G, $P = 0.68$ s $\tau = 1.1 \cdot 10^6$ years, $\chi = 16°$. L_{pc} from [14] is shown by solid green line. Upper limit from [15] is shown by dashed green line, upper limit from [10] is shown by orange area. The polar cap luminosity L_{pc} for B1451-68 is shown in right panel, $B_{dip} = 3.2 \cdot 10^{11}$ G, $P = 0.26$ s $\tau = 42.5 \cdot 10^6$ years, $\chi = 50°$. The luminosities L_{pc} taken from [16] are shown by orange area.

5. Results

The polar cap luminosity L_{pc} is estimated as

$$L_{pc} = \int e\Phi|_{z=z_c} \left.\frac{\Omega B}{2\pi c e}\right|_{z=0} \tilde{\rho}_{ret} dS.$$

The results of calculation for J2043+2740, B1929+10, B0950+08, B2224+65 and B1451-68 pulsars for $B_{sc} \leq B_{dip}$ and $0 \leq \psi \leq \frac{\pi}{2}$ are shown in Figs. 3–5. The pulsars parameters are taken from [6] and the inclination angles χ from [7, 8]. The possible values L_{pc} calculated in the framework of rapid screening model are shown by blue region and calculated in the framework of gradual screening model is shown by red region. In both cases, lower boundaries of the regions correspond to $\psi = \frac{\pi}{2}$ and upper boundaries to $\psi = 0$. For J2043+2740, B1929+10 and B0950+08 gradual screening model predicts luminosities exceeding the observed values. However, for B2224+65 and B1451-68 pulsars gradual screening model may be more appropriate.

6. Discussion

The disagreement between the prediction of gradual screening model and observations may be explained in several ways. The most obvious explanation is the existence of large surface magnetic field $B_{surf} > 10^{14}$ G. In this case, there is no free charge emission and instead of steady charge limited flow model the model of vacuum gap must be used [17]. The disagreement may be explained by large redshift which we do not take into account at all. However this means that $r_{ns} < 2r_g$, where r_g is gravitational radius of star.

The disagreement can also arise due to some viscous force acting on secondary particles [18]. This force may be related to backflowing radiation [19, 20, 21], the radiation closed inside the diode [22, 23] or the radiation coming from deep layers of neutron star [24].

Acknowledgments

The work has been supported by the RFBR (project 13-02-00112), by the State Program "Leading Scientific Schools of the Russian Federation" (grant NSh-4035.2012.2) and by the Ministry of Education and Science of Russian Federation (agreement No.8409).

References

[1] Beskin V S, Gurevich A V and Istomin Ya N *Physics of The Pulsar Magnetosphere* (Cambridge University Press, 1993)
[2] Gil J A, Melikidze G I and Mitra D 2002 *A&A* **388** 235
[3] Arons J, Fawley W M and Scharlemann E T 1979 *ApJ* **231** 854
[4] Harding A K and Muslimov A G 2001 *ApJ* **556** 987
[5] Lyubarskii Yu E 1992 *A&A* **261** 544
[6] Manchester R N, Hobbs G B, Teoh A and Hobbs M 2005 *Astron. J.* **129** 1993 (*Preprint astro-ph/0412641*) http://www.atnf.csiro.au/research/pulsar/psrcat
[7] Rankin J M 1993 *ApJSS* **85** 146
[8] Malov I F and Nikitina E B 2011 *Astron. Rep.* **55** 19
[9] Becker W, Weisskopf M C, Tennant A F, Jessner A, Dyks J, Harding A K and Zhang Shuang N 2004 *ApJ* **615** 908
[10] Zavlin V E and Pavlov G G 2004 *ApJ* **616** 452
[11] Noutsos A et al. 2011 *ApJ* **728** 77
[12] Misanovic Z, Pavlov G G and Garmire G P 2008 *ApJ* **685** 1129
[13] Gil J, Haberl F, Melikidze G, Geppert U, Zhang B and Melikidze G Jr 2008 *ApJ* **686** 497
[14] Hui C Y, Huang R H H, Trepl L, Tetzlaff N, Takata J, Wu E M H and Cheng K S 2012 *ApJ* **747** 74
[15] Hui C Y and Becker W 2007 *A&A* **467** 1209
[16] Posselt B, Pavlov G G, Manchester R N, Kargaltsev O and Garmire G P 2012 *ApJ* **749** id 146
[17] Gil J, Melikidze G I and Geppert U 2003 *A&A* **407** 315
[18] Shibata S, Miyazaki J and Takahara F 1998 *MNRAS* **295** L53
[19] Melikidze G and Gil J 2006 *Chin. J. Astron. Astrophys.* **6** Suppl. 2, 81
[20] Dyks J, Frackowiak M, Slowikowska A, Rudak B and Zhang Bing 2006 *Chin. J. Astron. Astrophys.* **6** Suppl. 2, 85
[21] Lomiashvili D, Machabeli G and Malov I 2007 Drift Wave Model of Rotating Radio Transients *Preprint astro-ph/0709.2019*
[22] Kontorovich V M and Flanchik A B 2007 *JETP Letters* **85** 267
[23] Kontorovich V M and Flanchik A B 2013 *Astrophysics and Space Science* **345** 169
[24] Sedrakian D M, Harutyunyan A S and Hayrapetyan M V 2013 *Astrophysics* **56** 229

The spin evolution of the pulsars with non-rigid core

D. P. Barsukov[1,2], **O. A. Goglichidze**[1,*], **A. I. Tsygan**[1]

[1]Ioffe Physical-Technical Institute of the Russian Academy of Sciences, Saint-Petersburg, Russian Federation

[2]Saint-Petersburg State Polytechnical University, Saint-Petersburg, Russian Federation

E-mail: *goglichidze@gmail.com

Abstract. We formulate a model of pulsar spin evolution (braking, inclination angle evolution and radiative precession) taking into account the non-rigidity of neutron star rotation. We discuss two simple limiting cases of this model and show that the evolution of the inclination angle substantially depends on the model of crust-core interaction. The non-rigidity of core rotation accelerates the inclination angle evolution and makes all pulsars evolve to the orthogonal state. The size of the effect depends on the amount of differentially rotating matter and mechanism of its interaction with the rest of the star. Since the rapid evolution of the inclination angle apparently contradicts the observational data, our results may be used as an additional test for the theories of the cores of neutron stars.

1. Introduction

A rotating magnetized neutron star, if it has perfectly spherical shape, can be characterized by two vectors: angular velocity vector $\vec{\Omega}$ and magnetic moment vector $\vec{m}$. During the life of a neutron star the magnitudes of these vectors as well as the inclination angle χ formed by these vectors evolve in time. This evolution is caused by electromagnetic torque acting on rotating magnetized star. A neutron star is surrounded by a large magnetosphere. Strictly speaking, the electromagnetic torque can be calculated only by using a self-consistent theory of this magnetosphere which despite the achieved progress is far from complete at present [1].

The problem becomes even more complicated if one wishes to take into account the internal structure of neutron star. Neutron stars are not perfectly rigid. They contain a liquid core. There are two main mechanisms of crust-core interaction: coupling through the magnetic field and viscosity [2]. The first mechanism is much more effective and, if it works, then the protons, electrons and normal neutrons in the core can be considered as rigidly rotating with the crust. In contrast, the superfluid neutrons which are believed to be present in neutron star's core [3] are decoupled from the rest of the matter and interact with it only by weak mutual friction force (vortex-mediated interaction) [4].

The configurations when the magnetic field does not penetrate the core have been discussed in the literature [5, 6]. The expulsion of magnetic field can be caused, for example, by type-I superconductivity of core protons. Despite the fact that the type-II proton superconductivity is more likely from the point of view of microscopic calculations (see, however, [7]), its implications for the neutron star rotation dynamics are not so clear. The coexistence of proton magnetic fluxoids with neutron superfluidity seems to be inconsistent with the observed long periods of precession in some neutron stars [8].

In the present paper we consider in detail the two limiting cases of strong and weak coupling between the crust and the core of a neutron star, which correspond to the two physical cases discussed above.

2. Basic Assumptions

We will treat a neutron star as a spherical rigid shell containing a liquid core in a spherical cavity. We will denote the inner radius of the crust by R_c and the outer one by R_{ns}. The crust rotates with angular velocity $\vec{\Omega}$. It feels the action of an external (electromagnetic) torque $\vec{K}$ which is assumed to be slowly varying in the reference frame co-rotating with the crust. The core acts on the crust with a torque $\vec{N}$. Thus, the motion of the crust can be described as

$$I_{crust}\dot{\vec{\Omega}} = \vec{K} + \vec{N}, \tag{1}$$

where I_{crust} is the crust moment of inertia. Here and below we use the notation $\dot{\vec{\Omega}} = d_t\vec{\Omega}$.

Let us for clarity consider the simplest possible system of hydrodynamical equations written down in the frame of reference co-rotating with the crust:

$$\partial_t\vec{u} + 2[\vec{\Omega}\times\vec{u}] + (\vec{u}\cdot\vec{\nabla})\vec{u} + \vec{\nabla}P/\rho + \vec{\nabla}\Phi = -[\dot{\vec{\Omega}}\times\vec{r}] + \vec{f}, \tag{2}$$

$$\partial_t\rho + \mathrm{div}(\rho\vec{u}) = 0, \tag{3}$$

$$\Phi(\vec{r}) = -G\int\frac{\rho(\vec{r}')}{|\vec{r}-\vec{r}'|}d^3r' - \frac{1}{2}[\Omega\times\vec{r}]^2. \tag{4}$$

Here, $\vec{f}$ is the sum of forces (per unit mass) acting on the fluid, $2[\vec{\Omega}\times\vec{u}]$ is the Coriolis force, $-[\dot{\vec{\Omega}}\times\vec{r}]$ is the effective force caused by non-uniformity of neutron star rotation.

For a neutron star it is a good approximation to assume that the temperature is constant inside the star core [9]. It allows us to write the pressure gradient as

$$\vec{\nabla}P = \sum_\alpha \rho_\alpha\vec{\nabla}\mu_\alpha \tag{5}$$

where μ_α is the chemical potential per unit mass of α-th constituent, ρ_α is the α-th constituent's mass density.

Let us suppose that the flows in the core are produced only by $\dot{\vec{\Omega}}$. In the case of uniform rotation the star is in hydrostatical equilibrium:

$$\vec{\nabla}P^{(0)} + \rho^{(0)}\vec{\nabla}\Phi^{(0)} = 0. \tag{6}$$

Here and below the superscript (0) denotes the hydrostatical values of variables.

Since $\epsilon = \dot{\Omega}/\Omega^2 \ll 1$ one can consider $\vec{u}$ as a small perturbation to the rigid-body rotation $[\vec{\Omega}\times\vec{r}]$ produced by $\dot{\vec{\Omega}}$. All variables can be expanded in a power series in ϵ:

$$\rho_\alpha = \rho_\alpha^{(0)} + \rho_\alpha^{(1)} + O(\epsilon^2),\quad \Phi = \Phi^{(0)} + \Phi^{(1)} + O(\epsilon^2),\quad \mu_\alpha = \mu_\alpha^{(0)} + \mu_\alpha^{(1)} + O(\epsilon^2) \tag{7}$$

and the terms quadratic in ϵ can be neglected. We also assume that there is some dissipation mechanism in the system that is characterized by a time-scale τ_{dis} which is much smaller than the time-scale of rotational evolution $\tau_{re} = \Omega/\dot{\Omega}$ of the star. If so, the time-derivatives in Eqs. (2) and (3) become negligibly small for the neutron stars with ages much larger than τ_{dis}. Then our equations take the form

$$2[\vec{\Omega}\times\vec{u}] + \sum_\alpha\frac{\rho_\alpha^{(0)}}{\rho^{(0)}}\vec{\nabla}\tilde{\mu}_\alpha^{(1)} = -[\dot{\vec{\Omega}}\times\vec{r}] + \vec{f}^{(1)}, \tag{8}$$

$$\mathrm{div}\left(\rho^{(0)}\vec{u}\right) = 0, \tag{9}$$

where we introduced $\tilde{\mu}_\alpha = \mu_\alpha + \Phi$. Two observations can be made. First, neglecting the temperature term in $\vec{\nabla}P$, one is able to exclude Eq. (4) from consideration (if there is no need to calculate $\rho^{(1)}$, $\mu_\alpha^{(1)}$ or $\Phi^{(1)}$). Second, for neutron stars with ages much larger than τ_{dis} the internal flow "forgets" the initial conditions. At each time instance the velocity field is determined only by instantaneous values of $\vec{\Omega}$ and $\dot{\vec{\Omega}}$ and depends on time only through these two vectors (quasistationary approximation).

3. Rotational dynamics

After the quasistationary approximation becomes valid the interaction torque can be written in the following form

$$\vec{N} = -S_1 I_{core}\vec{e}_\Omega(\vec{e}_\Omega\cdot\dot{\vec{\Omega}}) - S_2 I_{core}\vec{e}_\Omega\times[\vec{e}_\Omega\times\dot{\vec{\Omega}}] + S_3 I_{core}[\vec{e}_\Omega\times\dot{\vec{\Omega}}], \tag{10}$$

where I_{core} is the core moment of inertia, S_1, S_2, S_3 are some dimensionless coefficients and $\vec{e}_\Omega = \vec{\Omega}/\Omega$. It can be shown that $S_1 \approx 1$. The coefficients S_2 and S_3 are determined by a specific crust-core interaction mechanism.

The external torque can be written in the following form

$$\vec{K} = K_0\left(\tilde{k}_\Omega\vec{e}_\Omega + \tilde{k}_m\vec{e}_m + \tilde{k}_\perp[\vec{e}_\Omega\times\vec{e}_m]\right), \tag{11}$$

where $\vec{e}_m = \vec{m}/m$, $\tilde{k}_\Omega$, $\tilde{k}_m$, $\tilde{k}_\perp$ are some dimensionless functions which depend on relative orientation between $\vec{\Omega}$ and $\vec{m}$ (and other possible variables specific for particular magnetosphere model). They are defined in such a manner that K_0 equals the value of the torque acting on aligned ($\chi = 0$) pulsar.

Substituting (11) and (10) into (1) and solving the equation for $\dot{\vec{\Omega}}$ we obtain three scalar equations describing pulsar braking, inclination angle evolution and torque-driven precession

$$\dot{\Omega} = \frac{K_0}{I_{tot}}\left(\tilde{k}_\Omega + \tilde{k}_m\cos\chi\right), \tag{12}$$

$$\dot{\chi} = -\frac{K_0}{\Omega}\frac{(I_{crust} - S_2 I_{core})\tilde{k}_m - S_3 I_{core}\tilde{k}_\perp}{(I_{crust} - S_2 I_{core})^2 + S_3^2 I_{core}^2}\sin\chi, \tag{13}$$

$$\dot{\varphi}_\Omega = -\frac{K_0}{\Omega}\frac{(I_{crust} - S_2 I_{core})\tilde{k}_\perp + S_3 I_{core}\tilde{k}_m}{(I_{crust} - S_2 I_{core})^2 + S_3^2 I_{core}^2}, \tag{14}$$

where $I_{tot} = I_{crust} + I_{core}$. Note that Eq. (12) has the same form as for perfectly rigid star. Roughly speaking, it is a consequence of the energy conservation law.

Eqs. (13) and (14) differ from analogous equations used for rigidly rotating star in two ways. First, the full moment of inertia I_{tot} in the denominator is replaced by the combinations of $|I_{crust} - S_2 I_{core}|$ and $S_3 I_{core}$ which are much less than I_{tot}. It leads to more rapid evolution of the angles. Second, one can see that Eq. (13) contains $\tilde{k}_\perp$ (multiplied by S_3) which does not influence the evolution of the inclination angle in the case of a rigid star (cf. [10]). The precession is affected by $\tilde{k}_m$ through non-zero S_3 as well.

The perpendicular part of $\vec{K}$ is associated with the so-called anomalous torque caused by inertia of the near-zone electromagnetic field [11]. The anomalous torque is larger than the other components of $\vec{K}$ approximately by a factor

$$\frac{c}{\Omega R_{ns}} = 4.7\times10^3\left(\frac{P}{1\mathrm{sec}}\right)\left(\frac{R_{ns}}{10\mathrm{km}}\right)^{-1}. \tag{15}$$

which grows as the star slows down under the action of the braking torque. As we show below, S_2 and S_3 also grow as the star evolves. Therefore, the influence of anomalous torque on the inclination angle evolution increases with time. The effect becomes important when $S_3\tilde{k}_\perp \approx (I_{crust}/I_{core} - S_2)\,\tilde{k}_m$.

4. Weak coupling limit

Let us first consider the case of magnetic field confined to the crust. In this case, crust-core interaction occurs through the viscosity. If neutrons are not superfluid, all constituents can be regarded as moving with the same velocity $\vec{u}$. It can be shown that the composition gradient $\vec{\nabla}(\rho_p/\rho_n)$ causes a strong damping of any stationary radial flow. Thus, the angular momentum can be transfered from the core to the crust only by viscous tensions. In this case, one finds that (detailed calculations can be found in [12])

$$S_2 = -\frac{8\pi\rho^{(b)}R_c^5}{3\sqrt{2}I_{core}}E^{1/2}, \quad S_3 = \frac{8\pi\rho^{(b)}R_c^5}{3\sqrt{2}I_{core}}E^{1/2}, \tag{16}$$

where $\rho^{(b)}$ is the mass density at the crust-core interface, $E = \eta^{(b)}/\Omega R_c^2$ is the so-called Ekman number, $\eta^{(b)}$ is the value of the shear viscosity coefficient at the crust-core interface. The expressions for these coefficients were obtained under the assumption that $E \ll 1$. It is satisfied with a good accuracy for neutron stars. However, η essentially depends on temperature. Therefore, in order to close the system of equations one needs to take into account the thermal evolution of the star, which practically does not depend on the spin evolution. It is determined mostly by the mass of the star, which implies that the viscosity is a known function of star's age.

If neutrons are superfluid, the neutron component decouples from the rest of the matter and an additional dynamical degree of freedom - the velocity of superfluid neutrons $\vec{u}_s$ - appears. The neutron fluid and the charged components interact only by weak mutual friction force. In this situation, the damping of radial flows may not take place. Besides the viscous tensions, the angular momentum can be transfered directly by flow (the so-called Ekman pumping mechanism [13]). However, neutron superfluidity starts to affect the crust-core interaction only when the distance between the crust-core interface and phase transition surface is of the order of $E^{1/2}R_c$ or smaller. The realization of such a configuration depends on the model of neutron superfluidity.

5. Strong coupling limit

The neutron superfluidity becomes more important in the case of strong coupling. In this limit we assume that the charged component and the normal neutrons rotate rigidly with the crust. The second component which rotates differentially consists of superfluid neutrons. The components interact through the mutual friction. These crust-core interaction coefficients are given by (see [14] for the detail)

$$S_2 = -\frac{I_c}{I_{core}} - \frac{8\pi}{3I_{core}}\int_0^{R_s}\frac{(1-\beta')\beta' - \beta^2}{(1-\beta')^2+\beta^2}\rho_s r^4 dr, \quad S_3 = \frac{8\pi}{3I_{core}}\int_0^{R_s}\frac{\beta}{(1-\beta')^2+\beta^2}\rho_s r^4 dr, \tag{17}$$

where β and β' are the mutual friction coefficients, ρ_s is the mass density of superfluid neutrons, R_s is the radius of the spherical surface of the phase transition.

Note that during star's cooling more and more neutrons become superfluid, i.e., the normal neutron component, which co-rotates with the charge component, is converted into the superfluid component. It is not obvious that Eqs. (12)-(14) remain valid in this situation. The left-hand side of Eq. (1) should also contain $\dot{I}_{crust}\vec{\Omega}$. Now the angular momentum exchange between

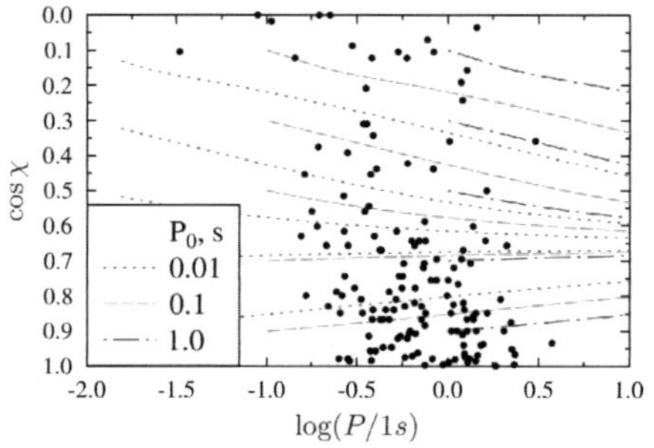

Figure 1. The $\cos\chi - P$ trajectories for rigidly rotating neutron stars with different initial periods and inclination angles.

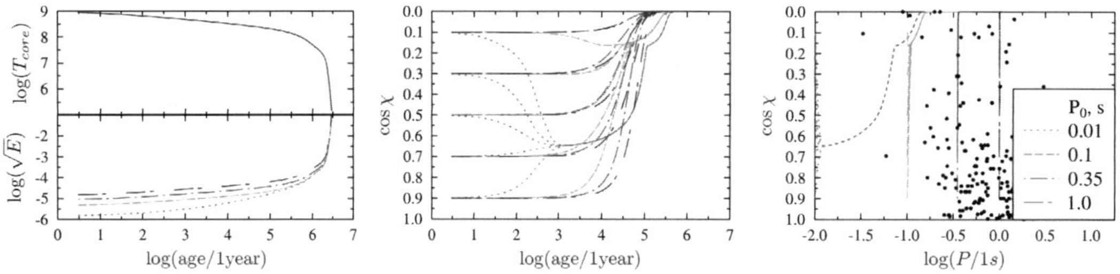

Figure 2. The evolution of neutron star core temperature and $E^{1/2}$ (left panel), the inclination angle evolution trajectories (middle panel) and the same trajectories in $\cos\chi - P$ plane (right panel).

components is produced not only by $\dot{\vec{\Omega}}$ but also directly by matter conversion. It can be shown that in the linear in ϵ approximation two terms should be added to the interaction torque (10):

$$\vec{N}_{new} = \vec{N}_{old} - \dot{I}_{core}\vec{\Omega} - \int \dot{\rho}_s^{(0)}[\vec{r} \times \vec{u}_s]dV. \tag{18}$$

The first additional term cancels with $\dot{I}_{crust}\vec{\Omega}$. As for the second term, we assume that the thermal evolution occurs on approximately the same time-scale as the rotational evolution. In that case, the second term is quadratically small and can be neglected.

6. Inclination angle evolution

In order to plot some exemplary inclination angle trajectories we use the model expression for external torque proposed in [15] with non-dipolarity parameter $\nu_{nd} = 0.5$. The trajectories for a rigidly rotating neutron star are shown in Fig. 1.

The trajectories calculated with the coefficients (16) are shown in Fig. 2. Here, in order to take into account the thermal evolution of the star we have used the cooling code developed by Gnedin, Yakovlev and Potekhin [9]. The cooling curve has been calculated for a light star with a mass $\approx 1M_{\odot}$ and I_{crust}/I_{tot}=0.05.

The trajectories calculated with coefficients (17) for two different neutron superfluid models are shown in Figs. 3 and 4. The superfluidity models used in Fig. (3) are taken from [16]. Fig. 4 is plotted for a model in which the neutron superfluidity is strongly damped in the region of

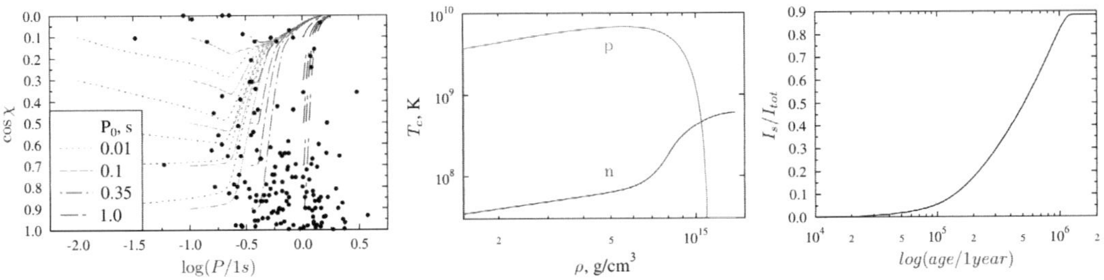

Figure 3. The $P - \cos\chi$ trajectories (left panel), the models of proton and neutron superfluidities (middle panel) and the evolution of relation I_s/I_{tot} with time (right panel).

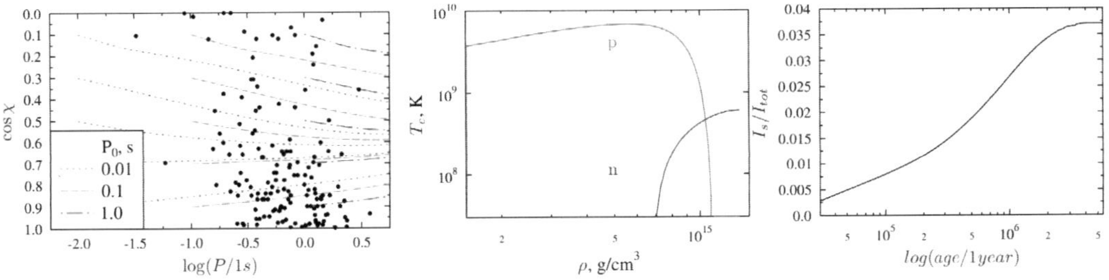

Figure 4. The same as in Fig. 3 for another model of neutron superfluidity.

small densities. The mutual friction coefficients have been calculated using the results of Alpar, Langer and Sauls [17]. All figures include 149 pulsars whose observable values of the inclination angle are taken from [18].

7. Discussion

One can see that the evolution of the inclination angle dramatically differs for different crust-core interaction models. The rate of the evolution depends on the amount of the matter which rotates differentially. Moreover, all trajectories sharply turn upwards when the action of anomalous torque becomes comparable with the action of normal torque. These features should be included in the studies of population synthesis of pulsars. They can also be used as additional tests for the theories of neutron star interiors.

Acknowledgements

This work was supported by the Russian Foundation for the Basic Research (project 13-02-00112), the Programme of the State Support for Leading Scientific Schools of the Russian Federation (grant NSh-4035.2012.2) and Ministry of Education and Science of Russian Federation (Agreement No. 8409).

References

[1] Spitkovsky A 2008 *40 Years of Pulsars: Millisecond Pulsars, Magnetars and More* (*American Institute of Physics Conference Series* vol 983) pp 20–28
[2] Easson I 1979 *ApJ* **233** 711–716
[3] Yakovlev D G, Levenfish K P and Shibanov Y A 1999 *Physics-Uspekhi* **42** 737
[4] Hall H E and Vinen W F 1956 *Royal Society of London Proceedings Series A* **238** 215–234
[5] Pons J A, Miralles J A and Geppert U 2009 *A&A* **496** 207–216

[6] Gourgouliatos K N *et al.* 2013 *MNRAS* **434** 2480–2490
[7] Buckley K B, Metlitski M A and Zhitnitsky A R 2004 *Physical Review Letters* **92** 151102
[8] Link B 2006 *A&A* **458** 881–884
[9] Gnedin O Y, Yakovlev D G and Potekhin A Y 2001 *MNRAS* **324** 725–736
[10] Casini H and Montemayor R 1998 *ApJ* **503** 374
[11] Melatos A 2000 *MNRAS* **313** 217–228
[12] Barsukov D P, Goglichidze O A and Tsygan A I 2014 in preparation
[13] Greenspan H 1990 *The theory of rotating fluids* Cambridge monographs on mechanics and applied mathematics (At the University Press) ISBN 9780962699801
[14] Barsukov D P, Goglichidze O A and Tsygan A I 2013 *MNRAS* **432** 520–529
[15] Barsukov D P, Polyakova P I and Tsygan A I 2009 *Astronomy Reports* **53** 1146–1154
[16] Gusakov M E, Kaminker A D, Yakovlev D G and Gnedin O Y 2005 *MNRAS* **363** 555–562
[17] Alpar M A, Langer S A and Sauls J A 1984 *ApJ* **282** 533–541
[18] Rankin J M 1993 *ApJS* **85** 145–161

Journal of Physics: Conference Series **496** (2014) 012014 doi:10.1088/1742-6596/496/1/012014

Nuclear medium cooling scenario in the era of Cas A cooling data and 2 $M_\odot$ pulsar mass measurements

H. A. Grigorian

Laboratory of Information Technologies, Joint Institute for Nuclear Research, 141980 Dubna, Russia
Yerevan State University, 0025 Yerevan, Armenia

D. B. Blaschke

Instytut Fizyki Teoretycznej, Uniwersytet Wrocławski, 50-204 Wrocław, Poland
Laboratory for Theoretical Physics, Joint Institute for Nuclear Research, 141980 Dubna, Russia

D. N. Voskresensky

National Research Nuclear University (MEPhI), 115409 Moscow, Russia

E-mail: blaschke@ift.uni.wroc.pl

Abstract. We show that all reliably known temperature data of neutron stars including those belonging to Cassiopea A can be comfortably explained in our "nuclear medium cooling" scenario of neutron stars. The cooling rates account for medium-modified one-pion exchange in dense matter, polarization effects in the pair-breaking-formation processes operating on superfluid neutrons and protons paired in the $1S_0$ state, and other relevant processes. The emissivity of the pair-breaking-formation process in the $3P_2$ state is a tiny quantity within our scenario. Crucial for a successful description of the Cassiopeia A cooling proves to be the thermal conductivity from both, the electrons and nucleons, being reduced by medium effects. Moreover, we exploit an EoS which stiffens at high densities due to an excluded volume effect and is capable of describing a maximum mass of 2.1 $M_\odot$, thus including the recent measurements of PSR J1614-2230 and PSR J0348+0432.

1. Introduction

The isolated neutron star in Cassiopeia A (Cas A) was discovered in 1999 by the *Chandra* satellite [1]. Its association with the historical supernova SN 1680 [2] gives Cas A an age of 333 years, in agreement with the nebula's kinematic age [3]. The thermal soft X-ray spectrum of Cas A can be fitted with a non-magnetized carbon atmosphere model, a surface temperature of 2×10^6 K, and an emitting radius of 8 to 17 km [4]. Analyzing the data from 2000 to 2009, Heinke & Ho [5] reported a rapid decrease of Cas A's surface temperature over a 10-year period, from 2.12×10^6 to 2.04×10^6 K. Such a rapid drop in temperature conflicts with standard cooling scenarios based on the efficient modified Urca (MU) process [6, 7]. Interpretations of Cas A's temperature data based on hadronic matter cooling scenarios were provided by Page *et al.* [8], Yakovlev *et al.* [9, 10], Blaschke *et al.* [11, 12]. The new analysis of the *Chandra* data performed by Elshamounty *et al.* [13] including a new measured data point allowed to precisely extract the decade temperature decline. The drop in the temperature lies in the range $2 \ldots 5.5\%$, and the most recent results from ACIS-S detector yield a $3 \ldots 4\%$ decline.

The interpretation of Cas A data by Page *et al.* [8] is based on the "minimal cooling" paradigm [14], where a minimal number of cooling processes is taken into account.

The works of Yakovlev *et al.* [9, 10] include all emission processes which are part of the minimal cooling paradigm including the FOPE model for MU reaction rate. They assume that the proton gap is so large that charged current processes are strongly suppressed in the entire stellar core. The value and the density dependence of the $3P_2$ neutron gap are fitted to the Cas A data, leading to a critical temperature of $0.7 \ldots 0.9 \times 10^9$ K for the neutron pairing gap and a neutron star mass $M = 1.65\ M_\odot$. Both groups therefore came to the striking conclusion that the temperature data of Cas A allow one to extract the value of the $3P_2$ neutron pairing gap. Continuing this approach Elshamouty *et al.* [13] arrive at the same conclusion.

The work of Blaschke *et al.* [11, 12] presents the "nuclear medium cooling scenario" as a model for the successful description of all the known temperature data including those of Cas A. This scenario includes efficient medium modified Urca (MMU) and medium nucleon bremstrahlung (MnB, MpB) processes, as motivated by a softening of the virtual pion mode in dense matter [15, 16], a very low (almost zero) value of the $3P_2$ neutron gap, as motivated by the result of Schwenk and Friman [17], and a small thermal conductivity of neutron star matter caused by in-medium effects, as motivated by calculations of the lepton thermal conductivity of Shternin and Yakovlev [18] and evaluations of the effect of pion softening on the nucleon thermal conductivity.

A strong suppression of the thermal conductivity is justified by results of Shternin and Yakovlev [18], who included the in-medium effect of Landau damping of electromagnetic interactions owing to the exchange of transverse plasmons in the partial electron (and muon) contribution to the thermal conductivity. Earlier, this effect has been studied by Heiselberg and Pethick for a degenerate quark plasma [19] and by Jaikumar *et al.* [20] for neutrino bremsstrahlung radiation via electron-electron collisions in neutron star crusts and cores. Now, we incorporate the in-medium modifications of the electron-electron interaction into our scenario, precisely as it has been done in [18]. Moreover, the partial NN thermal conductivity should be suppressed within our scenario owing to the increase of the squared NN interaction matrix element with density caused by the medium modification of the FOPE. Thereby, we additionally suppress the NN thermal conductivity term calculated in [21] by taking into account the softening of the one-pion exchange for this quantity as well as for all processes considered in our scenario.

As the nuclear matter equation of state (EoS), in [11] we used the Heiselberg-Hjorth-Jensen (HHJ) EoS [22] (with a fitting parameter $\delta = 0.2$) that fits the microscopic Akmal-Pandharipande-Ravenhall (APR) $A18 + \delta v + UIX^*$ EoS [23] for symmetric nuclear matter up to $4n_0$, where $n_0 = 0.16$ fm^{-3} is the nuclear saturation density. This yields an acceptable (although not perfect) fit of the APR EoS of neutron star matter for those densities. The maximum neutron star mass calculated with the HHJ($\delta = 0.2$) EoS, $M_{\rm max} = 1.94\ M_\odot$, proves to be smaller than the one calculated with the original APR EoS, $M_{\rm max} \simeq 2.2\ M_\odot$. However, the latter EoS becomes acausal for $n > 0.86$ fm^{-3}, whereas all HHJ($\delta \geq 0.13$) EoS respect causality at all densities. Recent measurements of two massive neutron stars, with $M_{1614} = 1.97 \pm 0.04\ M_\odot$ for PSR J1614-2230 [24] and $M_{0348} = 2.01 \pm 0.04\ M_\odot$ for PSR J0348-0432 [25], motivate us to use a stiffer EoS than that of HHJ($\delta = 0.2$) at large densities. In the present work we follow [12] where the EoS was modified in order to fulfill the new observational constraints on masses of neutron stars. To that end we incorporate excluded volume corrections in the HHJ($\delta = 0.2$) EoS such that it would remain unchanged for $n \lesssim 4n_0$ but would become stiffer for higher densities.

2. Nuclear medium cooling scenario

The nuclear medium cooling scenario worked out in Refs. [26, 27, 16, 28, 29] has been successfully applied to the description of the body of known surface temperature–age data of neutron stars [30, 31, 32, 33, 11].

We exploit the Fermi liquid approach, where the short-range interaction is treated with the help of phenomenological Landau-Migdal parameters, whereas long-range collective modes are explicitly presented. The most important effect comes from the mode with the pion quantum numbers treated explicitly, as it is a soft mode ($m_\pi \ll m_N$, with m_π (m_N) being the pion (nucleon) mass). The key effect is the softening of the pion mode with increasing density [15, 16]. Only with the inclusion of this softening effect the phase transition to a pion condensation state in dense nucleon matter may appear. Thus it is quite inconsistent to use FOPE model for description of NN-interaction and simultaneously include processes going on pion condensation.

This approach exploits a strong dependence of the main cooling mechanisms on the density and thus on the neutron star mass.

The insufficiency of the FOPE model for the description of the NN-interaction is a known issue. Actually, using the FOPE for the NN interaction amplitude, and simultaneously considering pion propagation as free, violates unitarity. Indeed, calculating the MU emissivity perturbatively one may use both the Born NN interaction amplitude given by the FOPE and the optical theorem, considering the imaginary part of the pion self-energy [26, 27, 28].

In the latter case, at low densities one needs to expand the exact pion Green's function to second order using for the polarization function $\Pi(\omega, k, n)$. The effective pion gap defined as $\omega^{*2} = -D_\pi^{-1}(\omega = 0, k = k_0)$. The polarization function $\Pi_0(\omega, k = k_0 \simeq p_{\mathrm{F},n}, n)$ yields a strong P-wave attraction. This attraction proves to be so strong that it would trigger a pion condensation instability already at low baryon densities of $n \sim 0.3\, n_0$, which is in disagreement with experimental data on atomic nuclei at the nuclear saturation density n_0. Note that the perturbative calculation contains no one free parameter. The paradox is resolved by observing that together with pion softening (i.e., a decrease of the effective pion gap $\omega^*(n)$ with increasing density for $n > n_{cr}^{(1)}$, $\omega^{*2}(n_{cr}^{(1)}) = m_\pi^2$) one needs to include a short-range repulsion arising from the dressed πNN vertices, $\Gamma(n) \simeq [1 + C(n/n_0)^{1/3}]^{-1}$ with $C \simeq 1.6$. This evaluation exploits an estimated value of the Fermi-liquid spin-spin Landau-Migdal parameter g and the Lindhard function taken in the limit of low transferred energy $\omega \ll p_{\mathrm{F},n}$. A consistent description of the NN interaction in matter should thus use a medium modified one-pion exchange (MOPE) interaction characterized by the fully dressed pion Green function, dressed vertices $\Gamma(n)$, and a residual NN interaction. We stress that dressing the pion mode is similar to the ordinary dressing of the photon mode in a plasma.

However, according to evaluations in [26, 16], the main contribution for $n > n_0$ is given by MOPE, whereas the relative contribution of the residual interaction diminishes with increasing density owing to polarization effects. Thus, in our simplified treatment the main dependence on the short-range interaction enters MOPE via the phenomenological vertex suppression factor $\Gamma(n)$.

The density dependence of the effective pion gap ω^* that we use for $n > n_{cr}^{(1)}$, taken to be 0.8 n_0, is demonstrated in Fig. 1 of [31]). Following the model used here and in [31, 32] within the HHJ($\delta = 0.2$) EoS, the pion condensation arises for neutron star masses $M \geq 1.32\ M_\odot$ (corresponding to the choice $n \geq n_{cr}^\pi = 3\, n_0$). To avoid misunderstanding we stress that, although to construct the curves $\omega^*(n)$ we used available experimental information and well established general principles [16], the quantitative density dependence of $\omega^*(n)$ remains essentially model dependent due to a lacking knowledge of the NN interaction in neutron star matter at large densities. Thus we hope that our successful description of the neutron star cooling may be helpful to correctly choose the parameterization of the interaction.

The direct Urca (DU) process, $n \to pe\bar{\nu}$, is too efficient for a description of the full set of

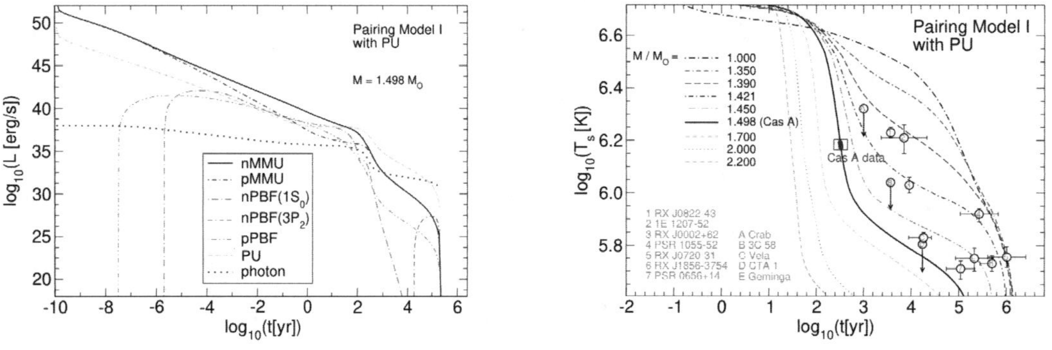

Figure 1. (Color online) Left panel: Individual contributions of the cooling processes, nMMU and pMMU, $1S_0$ pPBF and nPBF, $3P_2$ nPBF, PU, and surface photon emission, to the total stellar luminosity for the neutron star of mass 1.498 $M_\odot$ shown in the right panel of this figure. Modeling with PU exploits curves 1a+2 for the pion excitations, and 3 determining the pion condensate amplitude in Fig. 1 from [12], pairing gaps from model I. Right panel: Cooling of neutron stars in the nuclear medium cooling scenario, with different masses within model I for pairing and with PU, cf. also Fig. 17 of [31] and Fig. 1 of [11]. Data from Refs. [14, 9]. New data for Cas A from [13].

neutron star cooling data. Moreover, the DU process occurs only for high proton fractions, $x_p = n_p/n > 0.11\ldots0.14$. This process is practically excluded because of the choice of EoS. We adopted the HHJ($\delta = 0.2$) EoS for the description of the nucleon contribution. The energy density of nucleons is parameterized as follows

$$E_N = un_0 \left[m_N + e_{\rm B} u \frac{2+\delta-u}{1+\delta u} + a_{\rm sym} u^{0.6} (1-2x_p)^2 \right] , \tag{1}$$

where $u = n/n_0$, $e_{\rm B} \simeq -15.8$ MeV is the nuclear binding energy per nucleon, $a_{\rm sym} \simeq 32$ MeV is the symmetry energy coefficient and we chose $\delta = 0.2$. With these values of parameters one gets the best fit of APR (A18+δv+UIX*) EoS for symmetric nuclear matter up to $n \sim 4\ n_0$.

In the calculation of the neutrino emissivity of two-nucleon processes, e.g., $nn \to npe\bar{\nu}$, not only radiation from the nucleon legs but also from intermediate reaction states is allowed. For $n \gtrsim n_0$, the latter processes prove to be more efficient than the ordinary MU process from the legs. With such an interaction the ratio of the emissivity of the medium modified Urca (MMU) to the MU process, see [31, 32, 11],

$$\frac{\epsilon_\nu[\mathrm{MMU}]}{\epsilon_\nu[\mathrm{MU}]} \sim 3 \left(\frac{n}{n_0} \right)^{10/3} \frac{[\Gamma(n)/\Gamma(n_0)]^6}{[\omega^*(n)/m_\pi]^8} , \tag{2}$$

strongly increases with density for $n \gtrsim n_0$. For $n < n_{cr}^{(1)}$ we use ϵ_ν[MU] as in the minimal cooling scenario. Although an increase of the ratio of emissivities of the medium modified nucleon (neutron) bremsstrahlung process (MnB) to the unmodified bremsstrahlung (nB) is less pronounced, the MnB process, being not affected by the proton superconductivity, may yield a relatively large contribution in the region of a strong proton pairing. Note that being computed with values ω^* and Γ, which we use, the ratio of the MOPE NN cross section to that of the FOPE [28] proves to be σ[MOPE]/σ[FOPE] $\sim 1/3\ldots1/2$ for $n = n_0$ but it increases with increasing density. The subsequent increase of the cross section with density is due to the dominance of the softening of the pion mode owing to πNN and $\pi N\Delta^*(1236)$ P-wave attraction compared to the suppression of vertices owing to repulsive NN correlations [16, 28]. Thus the known suppression of the in-medium NN cross section at $n \lesssim n_0$ compared to that given by the FOPE [34, 35] does not conflict with a strong enhancement of the MMU emissivity with increasing density. Estimated strong density dependence of the in-medium neutrino-processes motivated authors of [26] to suggest that difference in surface temperatures of neutron stars is explained by different masses of the objects (that time only upper limits on surface temperatures were put). At the end, we should stress that in order to explain the cooling of both slowly and rapidly cooling stars one requires neutrino emissivities that differ by a factor $> 10^3$. Therefore, an uncertainty of the order of one in the emissivity of the processes does not affect the general cooling picture. In Fig.1 we show the time profiles of the individual contributions of the cooling processes. The simulation shows that the most active process in our scenario is the MMU.

3. Cooling of the young neutron star in CasA

The ingredients of the nuclear medium cooling scenario discussed above lead to neutron star cooling curves in Fig. 17 of Ref. [31], where model I for the proton gap has been adopted and the role of the thermal conductivity on the hot early stages of hadronic neutron star cooling was elucidated (see curves for $\kappa = 0.3$ in [31]). In Fig. 1 of [11] we redrew those cooling curves permitting readjustment of the thermal conductivity parameter. This allowed us to describe all cooling data including those for Cas A, known at that time.

4. Conclusions

We have shown in this paper that the nuclear medium cooling scenario allows one to nicely explain the observed rapid cooling of the neutron star in Cas A, as well as all other existing neutron star cooling data. As demonstrated already in [31] and then in [11], in our scenario the rapid cooling of very young objects like Cas A is mainly due to the efficient MMU processes, a very low (almost zero) value of the $3P_2$ neutron gap and a small thermal conductivity of neutron star matter. In the present work we do not use any artificial suppression parameter to demonstrate the effect of a small thermal conductivity caused by in-medium effects on Cas A cooling, but we *use the same values for the lepton thermal conductivity* as in Ref. [18]. The required smallness of thermal conductivity is provided by taking into account the collective effect of Landau damping.

We stress that, contrary to the minimal cooling models, within the nuclear medium cooling scenario we are trying to consistently include the most important collective effects in all relevant processes – the pion softening effect in the NN interaction amplitude [16], collective effects in the pair-breaking-formation processes [36], collective effects in the pairing gaps, the screening effect in the lepton contribution to the thermal conductivity [19] and a decrease of the nucleon contribution owing to the mentioned pion softening, etc. And we did not introduce any significant changes in our scenario developed in 2004 in [31], except for including the suppression of the thermal conductivity, now performed as in [18]. Thus in difference with other scenarios, explanation of the Cas A data straightly follows the predictions of our previous work [31].

Alternative explanations include the suggestion [37] that Cas A is a rapidly rotating star and during its spin-down the efficient DU process is switched on when the redistribution of matter leads to an increase of the central density beyond the DU threshold. Ref. [38] suggests that Cas A is a hybrid star. However, note that except our scenario other works aiming at a description of Cas A do not demonstrate their capability to obtain in the framework of the same assumptions an overall agreement with other available neutron star cooling data.

Further tests may be considered, such as a comparison of log N-log S distributions from population synthesis with the observed one for isolated neutron stars [39]. Also a continuation of the measurements of Cas A and new measurements of neutron star temperatures are welcome to discriminate between alternative cooling scenarios.

Acknowledgements
D.B. and H.G. acknowledge the stimulating and inspiring atmosphere at the workshop in Yerevan, as well as support for their participation from a grant by the Volkswagen Foundation. This work was supported by the NCN "Maestro" programme under contract number UMO-2011/02/A/ST2/00306. H.G. acknowledges support by the Bogoliubov-Infeld programme and by the Volkswagen Foundation under grant No. 85 182.

References
[1] H. Tananbaum, IAU Circ., **7246**, 1 (1999); J. P. Hughes, C. E. Rakowski, D. N. Burrows, and P. O. Slane, Astrophys. J. Lett. **528**, L109 (2000).
[2] W. B. Ashworth, Jr., J. Hist. Astron. **11**, 1 (1980).
[3] R. A. Fesen, M. C. Hammell, J. Morse, R. A. Chevalier, K. J. Borkowski, M. A. Dopita, C. L. Gerardy, S. S. Lawrence, J. C. Raymond, and S. van den Bergh, Astrophys. J. **645**, 283 (2006).
[4] W. C. G. Ho and C. O. Heinke, Nature **462**, 71 (2009).
[5] C. O. Heinke and W. C. G. Ho, Astrophys. J. Lett. **719**, L167 (2010).
[6] D. G. Yakovlev, A. D. Kaminker, O. Y. Gnedin and P. Haensel, Phys. Rept. **354**, 1 (2001).
[7] D. Page, U. Geppert and F. Weber, Nucl. Phys. A **777**, 497 (2006).
[8] D. Page, M. Prakash, J. M. Lattimer and A. W. Steiner, Phys. Rev. Lett. **106**, 081101 (2011).
[9] D. G. Yakovlev, W. C. G. Ho, P. S. Shternin, C. O. Heinke and A. Y. Potekhin, Mon. Not. Roy. Astron. Soc. **411**, 1977 (2011).
[10] P. S. Shternin, D. G. Yakovlev, C. O. Heinke, W. C. G. Ho and D. J. Patnaude, Mon. Not. Roy. Astron. Soc. **412**, L108 (2011).
[11] D. Blaschke, H. Grigorian, D. N. Voskresensky and F. Weber, Phys. Rev. C **85**, 022802 (2012).
[12] D. Blaschke, H. Grigorian and D. N. Voskresensky, Phys. Rev. C **88**, 065805 (2013).
[13] K. G. Elshamouty, C. O. Heinke, G. R. Sivakoff, W. C. G. Ho, P. S. Shternin, D. G. Yakovlev, D. J. Patnaude and L. David, Astrophys. J. **777**, 22 (2013).
[14] D. Page, J. M. Lattimer, M. Prakash and A. W. Steiner, Astrophys. J. Supp. **155**, 623 (2004); Astrophys. J. **707**, 1131 (2009).
[15] A. B. Migdal, Rev. Mod. Phys. **50**, 107 (1978).
[16] A. B. Migdal, E. E. Saperstein, M. A. Troitsky and D. N. Voskresensky, Phys. Rept. **192**, 179 (1990).
[17] A. Schwenk and B. Friman, Phys. Rev. Lett. **92**, 082501 (2004).
[18] P. S. Shternin and D. G. Yakovlev, Phys. Rev. D **75**, 103004 (2007).
[19] H. Heiselberg and C. J. Pethick, Phys. Rev. D **48**, 2916 (1993).
[20] P. Jaikumar, C. Gale and D. Page, Phys. Rev. D **72**, 123004 (2005).
[21] D. A. Baiko, P. Haensel and D. G. Yakovlev, Astron. Astrophys. **374**, 151 (2001); O. Y. Gnedin and D. G. Yakovlev, Nucl. Phys. **A582**, 697 (1995).
[22] H. Heiselberg and M. Hjorth-Jensen, Astrophys. J. **525**, L45 (1999).
[23] A. Akmal, V. R. Pandharipande and D. G. Ravenhall, Phys. Rev. C **58**, 1804 (1998).
[24] P. Demorest, T. Pennucci, S. Ransom, M. Roberts and J. Hessels, Nature **467**, 1081 (2010).
[25] J. Antoniadis, P. C. C. Freire, N. Wex, T. M. Tauris, R. S. Lynch, M. H. van Kerkwijk, M. Kramer and C. Bassa, Science **340**, 6131 (2013).
[26] D. N. Voskresensky and A. V. Senatorov, Sov. Phys. JETP **63**, 885 (1986); JETP Lett. **40**, 1212 (1984).
[27] D. N. Voskresensky and A. V. Senatorov, Sov. J. Nucl. Phys. **45**, 411 (1987); A. V. Senatorov and D. N. Voskresensky, Phys. Lett. B **184**, 119 (1987).
[28] D. N. Voskresensky, Lect. Notes Phys. **578**, 467 (2001).
[29] E. E. Kolomeitsev and D. N. Voskresensky, Phys. Atom. Nucl. **74**, 1316 (2011).
[30] C. Schaab *et al.*, Astron. Astrophys. **321**, 591 (1997).
[31] D. Blaschke, H. Grigorian and D. N. Voskresensky, Astron. Astrophys. **424**, 979 (2004).
[32] H. Grigorian and D. N. Voskresensky, Astron. Astrophys. **444**, 913 (2005).
[33] H. Grigorian, Phys. Rev. C **74**, 025801 (2006).
[34] D. Blaschke, G. Röpke, H. Schulz, A. D. Sedrakian and D. N. Voskresensky, Mon. Not. Roy. Astron. Soc. **273**, 596 (1995).
[35] C. Hanhart, D. R. Phillips and S. Reddy, Phys. Lett. B **499**, 9 (2001).
[36] E. E. Kolomeitsev and D. N. Voskresensky, Phys. Rev. C **77**, 065808 (2008); Phys. Rev. C **81**, 065801 (2010).
[37] R. Negreiros, S. Schramm and F. Weber, Phys. Lett. B **718**, 1176 (2013).
[38] A. Sedrakian, Astron. Astrophys. **555**, L 10 (2013).
[39] S. Popov, H. Grigorian, R. Turolla and D. Blaschke, Astron. Astrophys. **448**, 327 (2006).

2nd International Symposium on the Modern Physics of Compact Stars and Relativistic Gravity IOP Publishing
Journal of Physics: Conference Series **496** (2014) 012015 doi:10.1088/1742-6596/496/1/012015

Observational constraints of the compactness of isolated neutron stars

V Hambaryan, R Neuhäuser

Astrophysical Institute and University Observatory, Friedrich-Schiller University Jena, Schillergässchen 2-3, Jena Thüringen 07745, Germany

V Suleimanov, K Werner

Institute for Astronomy and Astrophysics, Kepler Center for Astro and Particle Physics, Eberhard Karls University, Sand 1, 72076 Tübingen, Germany

E-mail: valeri.hambaryan@uni-jena.de

Abstract. We report on our observational attempt to constrain the compactness of the isolated neutron stars via X-ray spin phase-resolved spectroscopy. There are seven thermally emitting neutron stars known from X-ray and optical observations, which are young (up to few Myrs), nearby (hundreds of pc), and radio-quiet with blackbody-like X-ray spectra. A model with a condensed iron surface and partially ionized hydrogen-thin atmosphere allows us to fit simultaneously the observed general spectral shape and the broad absorption feature (observed at 0.3 keV) in different spin phases. We constrain a number of physical properties of the X-ray emitting areas, including their temperatures, magnetic field strengths at the poles, and their distribution parameters. In addition, we place some constraints on the geometry of the emerging X-ray emission and the gravitational redshift of three isolated neutron stars.

1. Introduction: Thermally Emitting Isolated Neutron Stars (INSs)

The studies of INSs can provide an important input in our understanding of neutron stars and can put stringent constraints on the possible form of the equation of state of superdense matter. Observations and modeling of the thermal emission by INSs allow us to infer their surface temperatures and their total fluxes measured by a distant observer, as well as to estimate their real parameters, such as their apparent radii, provided the distances to the INSs are known accurately. On the other hand, the detection and identification of any absorption or emission feature in the spectrum or a rotational phase-resolved spectroscopy of INSs allow us to determine gravitational redshift, which in turn permits us to determine directly their mass-to-radius M/R ratio (see, e.g. [5, 6]).

The *ROSAT* X-ray observatory has discovered a small group (so far 7) nearby, thermally emitting and radio-quiet INSs which have the following common characteristics: (a) soft spectra, well described by blackbody radiation with the temperature in the range 60-120 eV, (b) absence of other spectral features, (c) no association with a known supernova remnant (SNR), (d) absence of radio emission or X-ray pulsations and, finally, (e) large ratio of the X-ray to optical emission f_x/f_{opt}. It has been suggested that these objects are either old INSs whose surface is reheated by accretion from the interstellar medium or young cooling stars which radiate away their thermal energy acquired at birth. Meanwhile, intensive X-ray observations using *XMM-Newton* and *Chandra* telescopes, as well as observations of these objects in optical and ultraviolet (UV), demand a revision of this origin of the thermal emission and provide further intriguing physical insight. Indeed, X-ray pulsations were found in six objects (with periods clustered in the range from 3 to 11 sec) and the period derivatives were determined for some of them. In the classical $P-\dot{P}$ diagram for the radio-pulsars these INSs occupy the region which is intermediate between radio pulsars and magnetars. Their inferred characteristic magnetic field strengths are above 10^{13} G.

Furthermore, imaging CCD-spectroscopy with XMM-Newton uncovered absorption features in at least three of INSs. It is likely that another three INSs feature similar absorption lines. At the current resolution ($E/dE \sim 60$) these lines are well described by Gaussian absorption lines. Their interpretation is not unique and they have been attributed to magnetically shifted atomic transitions, cyclotron resonances or an absorption feature from condensed surface (atmosphere). The inferred magnetic field strengths are again above 10^{13} G.

Another surprise came with the optical and ultraviolet observations of the fluxes from INSs, which were found to lie generally a factor of $\sim$ 5−10 above the extrapolation of the X-ray spectrum (the so-called "optical excess", see, e.g. [9]).

Here we present the results of a rotational phase-resolved X-ray spectroscopy of the three brightest INSs (RBS 1223, RX J0720.4−3125 and RX J1856.5−3754) based on the multi–epoch observations conducted by *XMM-Newton*.

2. Data analysis and results

High quality rotation phase resolved spectra are needed in order to fit using magnetized atmosphere models of neutron stars and to constrain their gravitational redshift [13, 6].

2.1. RBS 1223

RBS 1223 shows the highest pulsed fraction (13-42%, depending on energy band, see Fig. 1) and the strongest broad absorption feature [12] among thermally emitting INSs.

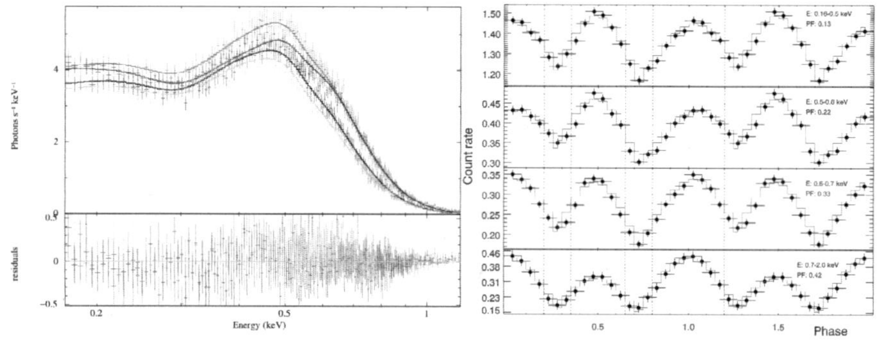

Figure 1. *XMM-Newton EPIC* pn co-added phase-averaged X-ray spectra (left panel) including primary and secondary peaks, first and second minima, and spin phase-folded light curves (right panel) in different energy bands of RBS 1223 combined from 12 pointed observations.

We used the data collected with *XMM-Newton* EPIC pn from the 12 publicly available (similar instrumental setup, i.e. Full Frame, Thin1 Filter) observations, in total comprising about 175 ks of effective exposure time. We extracted spin-phase resolved spectra with high S/N ratio (see Figs. 1 and 2) and fitted simultaneously with highly magnetized INS surface/atmosphere models [13]. These models are based on various local models and compute rotational phase dependent integral emergent spectra of INS using analytical approximations.

The basic model includes temperature/magnetic field distributions over isolated neutron star surface[1], viewing geometry and gravitational redshift. Three local radiating surface models are also considered, namely, a naked condensed iron surface [1] and partially ionized hydrogen model atmospheres, semi-infinite or finite atop of the iron condensed surface. The observed phase

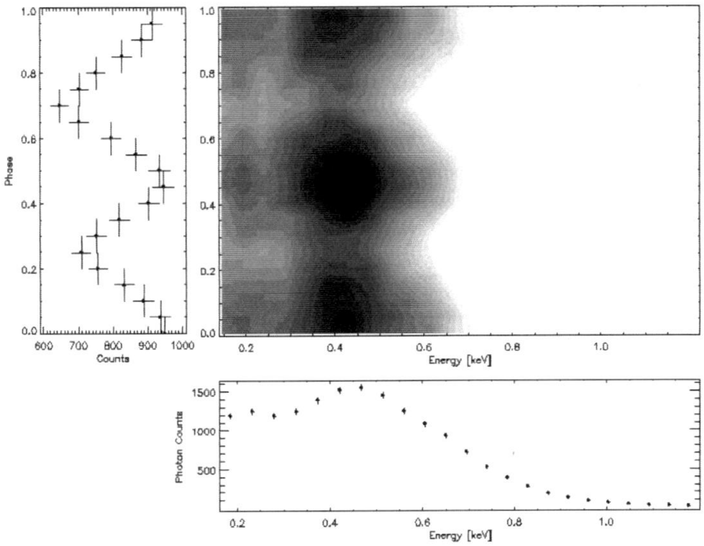

Figure 2. Energy-Phase image of RBS1223 combined from different observations. Rotational phase-folded light curve in the broad energy, 0.2-2.0 keV, band (left panel) and the phase averaged spectrum (bottom panel) are shown.

resolved spectra (i.e. energy-spin phase image, Fig. 2) of the INS RBS 1223 are satisfactorily fitted with two models, which have slightly different physical and geometrical characteristics of emitting areas. The fits show that the INS is an orthogonal rotator. The first model is parameterized by a Gaussian absorption line which is superimposed on a blackbody spectrum. The second model assumes a condensed iron surface above which there is a partially ionized, optically thin hydrogen atmosphere. Note, that the latter model is more physically motivated. We have additionally performed Markov Chain Monte Carlo (MCMC) fitting and estimated parameters and their uncertainties from posterior probability density functions.

2.2. RX J0720.4−3125

RX J0720.4−3125 is a special case among nearby thermally emitting INSs, because a clear variation in the spectrum has been detected by the *XMM-Newton* high-resolution spectral Reflection Grating Spectrometers (RGS) [15]. This detection was subsequently confirmed and analyzed by imaging-spectroscopic EPIC pn observations [3, 8] (see, e.g., Fig. 2).

Moreover, the *genuine* spin period of RX J0720.4−3125 turned out to be equal 16.78 sec (see Fig. 4) instead of 8.39 sec as reported in the literature (see, e.g., [2, 8]). Similar to the case of RBS 1223, the data from all observations of RX J0720.4−3125 conducted by *XMM-Newton*

[1] $T^4 = T^4_{\rm p1,2}\frac{\cos^2\theta}{\cos^2\theta + a_{1,2}\sin^2\theta} + T^4_{\rm min}$, $B = B_{\rm p1,2}\sqrt{\cos^2\theta + a_{1,2}\sin^2\theta}$, where the parameters $a_{1,2}$ are approximately equal to the squared ratio of the magnetic field strength at the equator to the field strength at the pole, $a_{1,2} \approx (B_{\rm eq}/B_{\rm p1,2})^2$. Using these parameters we can describe various temperature distributions, from strongly peaked ($a \gg 1$) to the classical dipolar ($a = 1/4$) and homogeneous ($a = 0$) ones.

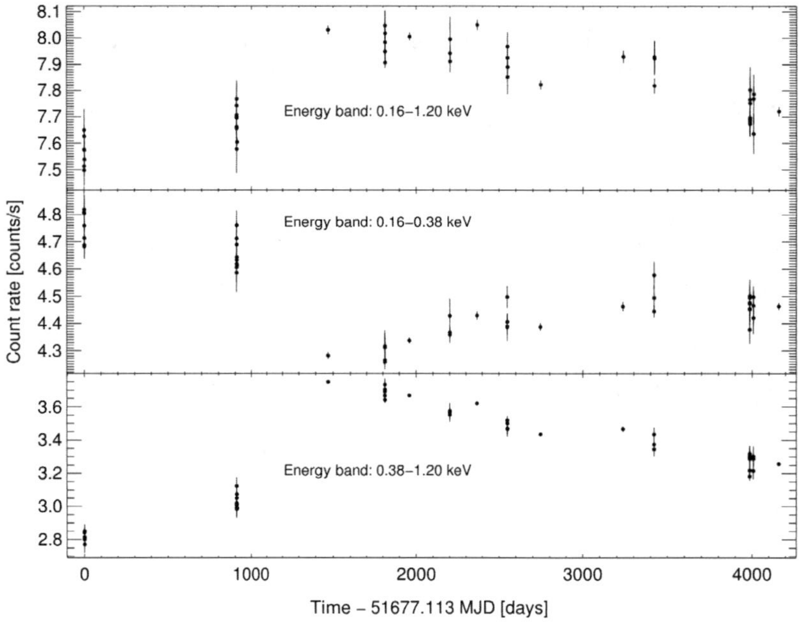

Figure 3. Observed count rates of RX J0720.4−3125 in the energy ranges 0.16−1.2 keV, 0.16−0.38 keV and 0.38−1.2 keV in different *XMM-Newton* EPIC pn observations.

EPIC pn with the same instrumental setup in 2000-2012 were reprocessed to form homogenous data set of solar barycenter corrected photon arrival times and phases were calculated according to the rotational period at that epoch. Taking into account aforementioned variations of RX J0720.4−3125 the whole data set, consisting of 16 different *XMM-Newton* pointings, was divided into five different subsets according to the brightness of the INS. Next, within each subset the spectra were co-added to form spin phase-resolved high S/N data groups and fitted simultaneously with the models described in the previous subsection. This procedure allowed us to estimate the gravitational redshift of RX J0720.4−3125, see Fig. 5.

2.3. RX J1856.5−3754

In contrary to the INSs RBS 1223 and RX J0720.4−3125, the X-ray spectrum of RX J1856.5−3754 does not show any significant absorption feature and the pulsed fraction is quite low ($\sim 1.5\%$). However, in this case there is a constraint on a viewing geometry from the observed bow-shock at this INS [10]. In addition, RX J1856.5−3754 is the nearest INS and the distance ($d = 123^{+11}_{-15}$ pc) is known with relatively good accuracy [16]. As in the previous cases of the INSs, the spin phase-resolved spectra were fitted in the same manner. The results of the estimation of gravitational redshift and therefore the compactness of this object, together with other studied INSs, are shown in Fig. 6. It is worthwhile to note that we obtained statistically acceptable fit also assuming a non-vanishing toroidal component of magnetic field of INS. This corresponds to the values of the parameter $a \gg 1$ which imply very small and hot emitting areas around the magnetic poles and yield small pulsed fraction, see footnote 1.

Nevertheless, our targeted parameter, the estimated gravitational redshift was approximately the same ($z = 0.22 \pm 0.07$, see also [7]) independent of the model. Note, also the relatively broad

Journal of Physics: Conference Series **496** (2014) 012015 doi:10.1088/1742-6596/496/1/012015

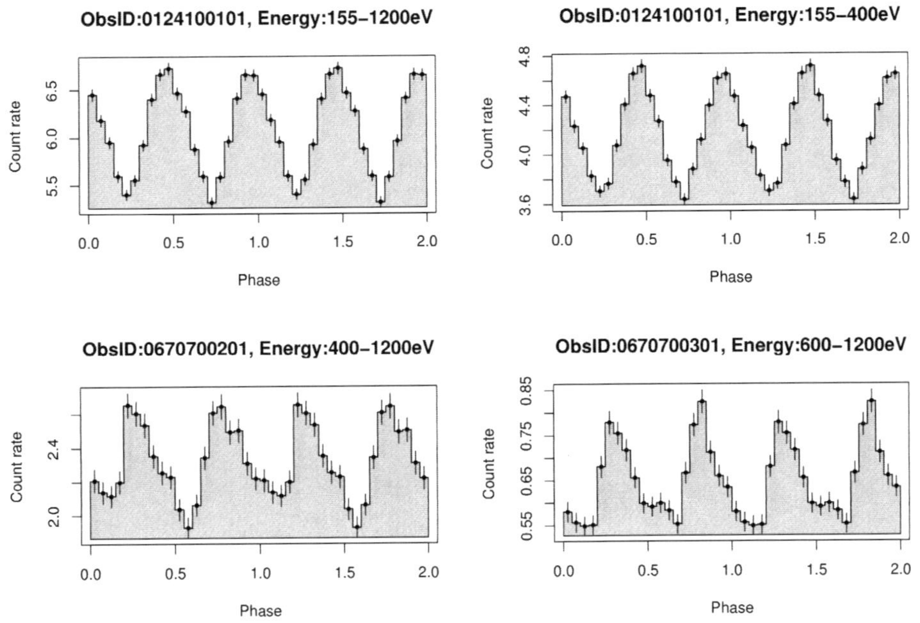

Figure 4. Phase-folded light curves with the *genuine* spin period (16.78 sec) of RX J0720.4−3125 in different *XMM-Newton* EPIC pn observations and energy bands.

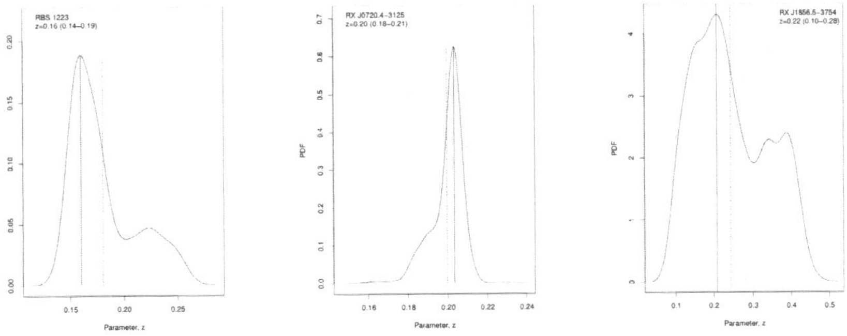

Figure 5. Probability density distribution of gravitational redshifts by Markov Chain Monte Carlo (MCMC) fitting with the model of strongly magnetized neutron star of condensed Iron surface and partially ionized Hydrogen thin atmosphere atop it for RBS 1223 (left panel), RX J0720.4−3125 (middle panel) and RX J1856.5−3754 (right panel). Dashed vertical line depicts the mean value, dotted vertical lines indicate the highest posterior probability density interval (95%).

posterior probability density function of the gravitational redshiht in comparison to the cases of RBS1223 and RX J0720.4−3125 (see Fig. 5).

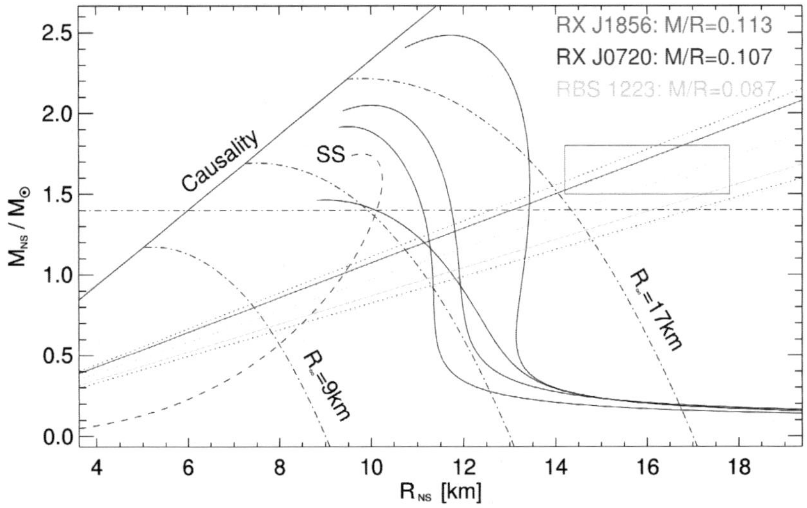

Figure 6. Mass-radius relations for several EOS ([4], thin solid curves), and a strange star (thin dash-dotted). Thick dashed: curve of constant $R^{\infty} = R/\sqrt{1 - 2GM/Rc^2} = 17$ km ([14].

3. Conclusions and outlook

To summarize, X-ray spin phase-resolved spectroscopic study of three thermally emitting INSs and the fit of highly magnetized atmospheric models allowed us to estimate their compactness, which are suggesting on a stiff equation of state (see, Fig. 6). More work for detailed spectral model computation will be certainly worth to do in the near future and application to the phase-resolved spectra of other INSs. In particular, analysis of the high resolution spectra observed by *XMM-Newton* and *Chandra* with possibly other absorption features [5, 11].

Acknowledgments

VH and VS acknowledge support by the German *Deutsche Forschungsgemeinschaft (DFG)* through project C7 of SFB/TR 7 "Gravitationswellenastronomie".

References

[1] van Adelsberg M, Lai D 2006 *MNRAS* **373** 1495
[2] Haberl F, Motch C, Buckley D A H, Zickgraf F J, Pietsch W. 1997 *A&A* **326** 662
[3] Haberl F., Turolla R., de Vries C. P., et al. 2006 *A&A* **451** L17
[4] Haensel P, Potekhin A Y, Yakovlev D G 2007 *ASSL* **326**
[5] Hambaryan V V, Neuhäuser R, Haberl F, Hohle M M, Schwope A 2009 *A&A* **497** L9
[6] Hambaryan V V, Suleimanov V, Schwope A P, Neuhäuser R et al. 2011 *A&A* **534**, 74
[7] Ho W C G, Kaplan D L, Chang P, van Adelsberg M, Potekhin A Y 2007 *MNRAS* **375** 821
[8] Hohle M M, Haberl F, Turolla R, Hambaryan V V, Zane Sm de Vries C P, Mendez M 2009 *A&A* **498** 811
[9] Kaplan D L , Kamble A , van Kerkwijk M H , Ho W C G 2011 *ApJ* **736** 117
[10] van Kerkwijk M H, Kulkarni S R 2001 *A&A* **380** 221
[11] Potekhin A Y 2010 *A&A* **518** 24
[12] Schwope A, Hambaryan V V, Haberl F, Motch C 2007 *ApSS* **308** 619
[13] Suleimanov V, Hambaryan V V, Potekhin A Y, et al. K 2010 *A&A* **522** 111
[14] Trümper J, Burwitz V, Haberl F, et al. 2003 *Nucl. Phys. B Proc. Suppl.* **132** 560
[15] de Vries C P, Vink J, Méndez M, & Verbunt F 2004 *A&A* **415** L31
[16] Walter F W, Eisenbeiss T, Lattimer J M, et al. 2010 *ApJ* **724** 669

2nd International Symposium on the Modern Physics of Compact Stars and Relativistic Gravity IOP Publishing
Journal of Physics: Conference Series **496** (2014) 012016 doi:10.1088/1742-6596/496/1/012016

Kerr-like behavior of orbits around rotating Newtonian stars

Włodek Kluźniak[1], **Dorota Rosińska**[2]
[1]Copernicus Astronomical Center, Bartycka 18, PL-00-716 Warszawa, Poland
[2]Institute of Astronomy, University of Zielona Góra, ul. Lubuska 2, PL-65-265 Zielona Góra, Poland

E-mail: wlodek@camk.edu.pl

Abstract. The innermost stable circular orbit and the splitting of epicyclic and orbital frequencies are among strong-field signatures of general relativity searched for in astronomical objects, particularly in the X-ray data accumulated in observations of neutron stars and black holes. It may come as a surprise that these effects are present in the Newtonian physics of rapidly rotating gravitating bodies, such as the classic Maclaurin spheroids.

1. Introduction

Investigations of orbital motion culminating after some millenia in Kepler's law, $P^2 \propto r^3$, predate the birth of physics. The discovery of extremely short periods in the X-ray emission of neutron stars has renewed interest in the laws governing orbital motion, as it was thought that the phenomenon of these fast quasi-periodic oscillations, as they are called, or kHz QPOs (see [1] for a review) arises in thin accretion disks, where the X-ray emitting fluid follows nearly circular orbits, with the consequence that on the one hand the fluid is nearly in hydrostatic equilibrium, being rotationally supported in the radial direction, while on the other it is in a state of differential rotation leading to vigorous dissipation of energy [2]. Similar effects have been discovered in close binaries harbouring black holes [3]. It was hoped that a close investigation of these quasi-periodic modulations of the X-ray flux may reveal characteristics of motion thought to be a signature of general relativity (GR), such as the innermost stable circular orbit (ISCO, more properly, the marginally stable orbit), or the splitting of orbital and epicyclic frequencies. The question whether such effects are a unique signature of GR is the topic of our contribution.

2. QPOs

Let us start with the case of the 0.11 Hz QPO in the X-ray flux of the cataclysmic variable SS Cygni in outburst [4]. No harmonics or subharmonics have been observed, unlike in X-ray pulsars, so this cannot be a rotating spot, whether on the surface of the white dwarf or in the accretion disk. Further, this 0.11 Hz solitary peak in the power-density spectrum (PDS) is not coherent (hence the name QPO), i.e., it cannot correspond to the (stable) rotation of the white dwarf. In a cataclysmic variable, relativistic corrections to orbital motion are expected to be negligible, with $GM/(rc^2) \sim 10^{-4}$, and this raises the very interesting question why a characteristic frequency should show up at all in the orbital motion of fluid in the accretion disk. After all, Keplerian frequency is scale-free (Fig. 1).

Journal of Physics: Conference Series **496** (2014) 012016 doi:10.1088/1742-6596/496/1/012016

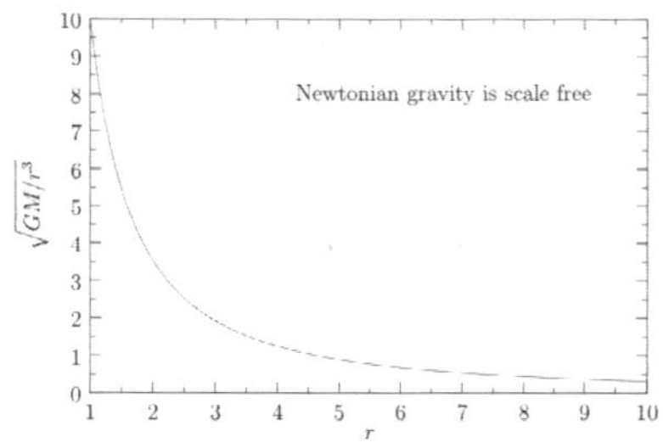

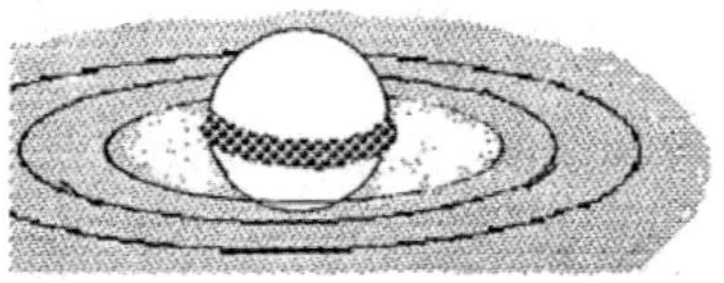

Figure 1. Keplerian frequency is scale-free, but a characteristic radius, e.g., the ISCO, implies a characteristic frequency.

The answer to this puzzle is that the radius of the white dwarf provides a characteristic scale, so the observed frequency should correspond to Keplerian motion close to the stellar surface, as indeed seems to be the case. In GR mass is equivalent to radius, therefore the mass of the gravitating object, be it a BH or neutron star provides another characteristic radius, $r_g = GM/c^2$, and in fact more than one, for there is also the ISCO radius (at $6r_g$ for the Schwarzschild metric). Thus, the maximum orbital frequency in a neutron star may occur at the marginally stable orbit, rather than at the stellar surface [5]. In the cartoon of Fig. 1, the accretion disk terminates close to the marginally stable orbit, well above the surface of a neutron star in a low-mass X-ray binary (LMXB). As remarked earlier, black holes (BH) exhibit similar QPOs, and since a BH has no surface the emission must come from the accretion disk. The BH QPOs occur at frequencies $\sim 100\,\mathrm{Hz}$, as would be expected if the frequency is suitably scaled up according to Kepler's law—the ISCO radius of a $6M_\odot$ BH being about 10^2 km, to a white dwarf's radius of 10^4 km.

3. Kepler's law and epicyclic frequencies

As is well known, Kepler's expression for orbital frequency, $\Omega_K(r) = 2\pi/P$, with the proportionality coefficient determined by Newton,

$$\Omega_K(r) = \sqrt{GM/r^3} \tag{1}$$

holds equally for a Newtonian $1/r$ potential and in the Schwarzschild metric, i.e., there is no difference between Newtonian mechanics and Einstein's general relativity in the orbital frequency of a test particle in stable circular motion around a spherically symmetric gravitating mass. However not all circular orbits are stable in GR, even in the spherically symmetric case. While angular momentum l of a test particle in circular orbit is a monotonically increasing function of the radius r in the Newtonian $1/r$ potential, so at all radii the condition for stability of orbits $dl/dr > 0$ is met, in the Schwarzschild metric l has a minimum at $r = 6GM/c^2$, and $dl/dr < 0$ for smaller values of r. Thus stable circular orbits in this metric exist only for $r > 6GM/c^2$.

GR effects become readily apparent for non-circular orbits, and this is the reason why one of the three classic tests of GR concerned the perihelion precession of the orbit of Mercury. The frequency of such a precession can be represented as the difference between the circular orbital frequency of Eq. (1) and radial epicyclic frequency ω_r (of "to and fro" motion), which is always lower (at least in the examples discussed here). Also of interest are slightly inclined, nearly circular, orbits, in which the motion can again be decomposed into circular motion and into vertical epicyclic ("up and down") motion, occuring at frequency, ω_z. For spherically symmetric bodies, all planes of motion (containing the center of the gravitating mass M) are equivalent, hence $\omega_z = \Omega_K$ for the Schwarzschild metric. However, rotation removes the degeneracy.

In the Kerr metric the expressions for the orbital and epicyclic frequencies [6] are all different:

$$\omega_r^2/\Omega_o^2 = 1 - 6/r + 8j/r^{3/2} - 3j^2/r^2,$$

$$\omega_z^2/\Omega_o^2 = 1 - 4j/r^{3/2} + 3j^2/r^2, \tag{2}$$

where $-1 < j < 1$ is the dimensionless BH spin, i.e., the Kerr parameter, the orbital frequency being $\Omega_o = c^3[GM(r^{3/2} + j)]^{-1}$, with r expressed in units of GM/c^2. Note that $\omega_r^2 < \omega_z^2 < \Omega_o^2$ for prograde motion ($j > 0$). The location of the marginally stable orbit [7] corresponds to the zero of radial epicyclic frequency. Newton's $1/r$ potential is fully degenerate and therefore bound orbits are closed ($\omega_z = \omega_r = \Omega_K$, as can be checked by taking the limit of $j = 0$ and $r \to \infty$).

4. Diskoseismology

The presence of the ISCO, occuring as it does at the zero of ω_r, has an interesting consequence: the radial epicyclic frequency necessarily has a maximum in the region of stable circular orbits (at $r = 8GM/c^2$ in the Schwarzschild metric). This fact may have profound astronomical consequences.

Being bodies of fluid close to hydrostatic equilibrium, accretion disks are capable of many modes of oscillation, whose study in GR was pioneered in [8]. As it turns out, the frequency of g-modes must necessarily be less than ω_r. As $\omega_r \to 0$ on both sides of its maximum (i.e., at the ISCO and at $r \to \infty$) g-modes of any admissible frequency are trapped between two radii on two sides of the maximum of ω_r. In effect, the radial epicyclic frequency forms a cavity for g-modes, consequently these have a discrete (and rather stable) frequency spectrum [8]. Another interesting mode, the c-mode [6] with frequency $\approx \Omega_o - \omega_z$, corresponds to the Lense-Thirring precession. For a more comprehensive discussion see the reviews [9, 10].

If these theoretical results do indeed correspond to the observed QPOs, the latter are probably a manifestation of GR effects [11]. However, it would be premature to claim that, for instance, the presence of a maximum in the radial epicyclic frequency is unique to GR. Seemingly same phenomena may have very different origins and, as we show below, at least some properties of GR orbits have direct analogues in purely Newtonian gravity.

5. Orbital properties of rapidly rotating stars

While the external metric of non-rotating stars is given by Schwarzschild, the orbits around rotating stars cannot be described by those of the Kerr metric. For slowly rotating stars, the metric is known analytically [12], and through first order in the stellar angular momentum ("spin") the Hartle-Thorne metric coincides with the Kerr one. However, higher order terms carry information about the multipoles of mass distribution, and for high spins these significantly affect the properties of stellar orbits, in particular the epicyclic frequencies. The shape of the $\omega_r(r)$ curve for rapidly rotating massive quark stars implies that rotation-induced oblateness pushes the ISCO out to larger radii [13]. Similar, although less pronounced, effects have been noted in rotating neutron star models [14]. To demonstrate that rotation-induced flattening is indeed responsible for these effects we have investigated models of quark stars, which allow computations both at usual stellar (about Solar) masses and at very low masses (quark matter, if it exists, is self-bound).

Fig. 2 shows the epicyclic and orbital frequencies around two $1.4M_\odot$ quark stars modeled with a simple MIT bag equation of state

$$P = (\rho - \rho_0)c^2/3,$$

with $\rho_0 = 4.2785 \times 10^{14}\,\mathrm{g/cm^3}$. The GR calculations were carried out with the RNS code [15]. Note that the frequencies in the plots are reduced by a factor of 2π with respect to the angular frequencies discussed previously ($2\pi\nu = \omega$). The vertical scale is given in kHz, the horizontal in units of gravitational radius. The darker (leftmost) curves correspond to a stellar rotation rate of 600 Hz. This is a modest rotation rate, and qualitatively the behavior is quite similar to the Kerr case for $j << 1$, eq. (2): there is only a very slight splitting of the vertical epicyclic frequency (dashed curve) from the orbital one (solid curve), and the radial epicyclic frequency (dotted curve) goes to zero close to the Schwarzschild ISCO radius of $6M$ (c.f., Kerr metric figures in [6]).

Based on the Kerr behavior, one could expect that doubling the stellar spin would result in an increased splitting of vertical epicyclic and orbital frequencies preserving the $\omega_z^2 < \Omega_{\rm o}^2$ relation, an increase of the maximum value or ω_r, and a decrease of the ISCO radius (location of zero of ω_r). Instead, when the stellar rotation rate is increased to 1165 Hz, the maximum value of ν_r decreases, the ISCO is pushed out to larger radii (the sparser, rightmost, dotted curve), while the vertical epicyclic frequency (uppermost dashed curve) is now larger than the orbital frequency (the upper of the solid curves) in the radius range depicted in the figure. In fact, this is the behaviour expected for flattened fluid configurations.

We have investigated [16] the frequencies of Maclaurin spheroids, the classic figures of equilibrium of rotating uniform density fluid configurations, and found that $\omega_r^2 + \omega_z^2 = 2\Omega_{\rm o}^2$, implying that $\omega_z^2 > \Omega_{\rm o}^2$. Further, at high ellipticities of the spheroids, a marginally stable orbit appears outside the equator, and is pushed out to larger and larger radii as the ellipticity (or equivalently the rotation rate) of the spheroid increases. Having computed quark star models at extremely low masses of $0.001M_\odot$, we are in a position to compare the numerical models with the analytic results for Maclaurin spheroids. The results are shown in Fig. 3, the dots corresponding to the numerical results [17], and the curves to the analytic formulae [16]. Angular frequencies squared have been plotted against the radius in units of the equatorial radius of the fluid configuration/star. Clearly there is excellent agreement between numerical and analytic models in the Newtonian limit, that is to say, between the GR results of the RNS code in the low-mass limit and the fully Newtonian analytic calculations for the Maclaurin spheroids.

Note that although the results presented in Fig. 3 are Newtonian (unlike those of Fig. 2), they include the presence of a marginally stable orbit [18, 19, 20] (at about 1.15 stellar radii, for the selected ellipticity of 0.954), a maximum in the radial epicyclic frequency, and the splitting of orbital and epicyclic frequencies.

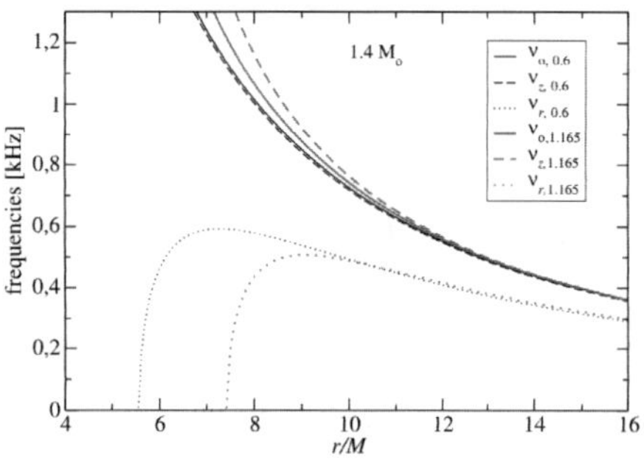

Figure 2. The orbital and epicyclic frequencies for two models of a $1.4M_\odot$ quark star in GR, rotating at 600 Hz (in black), and at 1165 Hz (in red) [17].

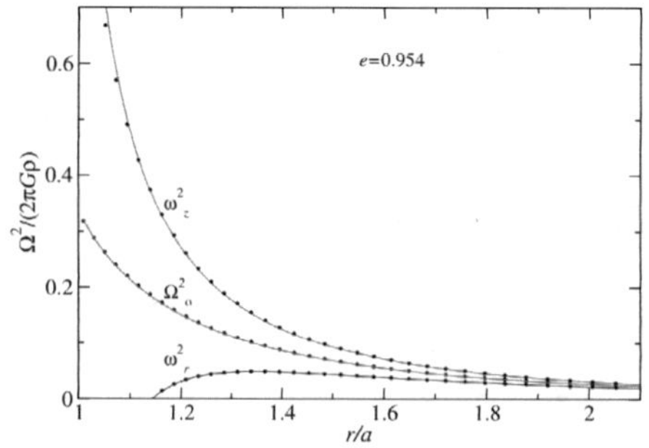

Figure 3. Comparison of analytic results for the Maclaurin spheroid (curves) [16] with numerical results of the RNS code for a $0.001M_\odot$ quark star (dots) [17].

6. Discussion

Astrophysical phenomena, such as X-ray emissions of LMXBs and their modulations (kHz QPOs) can, and should, be understood in terms of our best theory of gravity, Einstein's GR. However, some authors have tried to infer the correct theory of gravity from X-ray observations, and that may be premature, at the current state of observational techniques and theory of accretion.

A case in point is reverberation mapping. It has been suggested that once we have direct evidence of the size of the BH hole accretion disk, comparison with the angular frequencies inferred from observations will provide a quantitative test of GR. In fact, what would be tested in such a situation, is Kepler's law. We note that from the point of view of physics there is no difference between testing the law in the Solar system and in motion around a Schwarzschild black hole, as the problem is scale-free (Fig. 1, left panel). In fact, Kepler's law, with Newton's constant of proportionality, has been tested over some seven orders of magnitude in radius and mass, if one takes into account the motion of the Apollo spacecraft around the moon and that of the Earth around the Sun, and more if one remembers the outer planets. LMXBs are fairly tight binary systems of about Solar mass, with periods of days, hours or even eleven minutes. The "dynamical" masses of BHs in such systems have been derived by extrapolating Kepler's law from the Solar system to smaller distances, by about four orders in magnitude. Describing the fluid orbits in the accretion disk of a BH in a LMXB is a further extrapolation by another four orders in distance. We find it surprising that this last extrapolation should be subjected to attempted experimental verification (although there is no harm in that), while the first extrapolation by as many orders of magnitude is accepted without as much as a second thought.

It has been hoped that some qualitative features of GR, such as the presence of the marginally stable orbit may be tested in LMXBs [21], but we have shown that the same effect is also present in the Newtonian Maclaurin spheroids. The tasks ahead of us are more difficult than expected.

Acknowledgments

This work was supported in part by grants 2013/08/A/ST9/00795, N N203 511238 of the Polish National Center for Science, and POMOST/2012-6/11 Foundation of Polish Science.

References

[1] van der Klis M 2000 *Ann. Rev. Astron. & Astroph.* **38** 717
[2] Shakura NI, Sunyaev RA 1973 *Astron. & Astroph.* **24** 337
[3] McClintock JE and Remillard RA 2006 Compact stellar X-ray sources ed W Lewin and M van der Klis Cambridge Astrophysics Series No. 39 (Cambridge UK: Cambridge University Press) p 157–213
[4] Mauche C *Astrophys. J.* **580** 423
[5] Kluźniak W and Wagoner RV 1985 *Astrophys. J.* **297** 548
[6] Perez C, Silbergleit AS, Wagoner RV and Lehr DE 1997 *Astrophys. J.* **476** 589
[7] Bardeen JM, Press WH and Teukolsky SA 1972 *Astrophys. J.* **178** 347
[8] Okazaki, AT, Kato S and Fukue J *Pub. Astron. Soc. Jap.* **39** 457
[9] Kato S 2001 *Pub. Astron. Soc. Jap.* **53** 1
[10] Wagoner RV 1999 *Phys. Rep.* **311** 259
[11] Silbergleit AS, Wagoner RV and Ortega-Rodríguez M 2001 *Astrophys. J.* **548** 335
[12] Hartle JB and Thorne KS 1968 *Astrophys. J.* **153** 807
[13] Stergioulas N, Kluźniak W and Bulik T. 1999 *Astron. & Astroph.* **352** L116
[14] Kluźniak W, Abramowicz MA, Kato S, Lee WH and Stergioulas N 2004 *Astrophys. J.* **603** L89
[15] Stergioulas N and Friedman JL 1995 *Astrophys. J.* **444** 306
[16] Kluźniak W and Rosińska D 2013 *Month. Not. R. Astron. Soc.* **434** 2825
[17] Gondek-Rosińska D, Kluźniak W, Stergioulas N, Wiśniewicz M 2014 *submitted*
[18] Amsterdamski P, Bulik T, Gondek-Rosińska D and Kluźniak W 2002 *Astron. & Astroph.* **381** L21
[19] Kluźniak W, Bulik T, Gondek-Rosińska D 2001 *Exploring the gamma-ray universe* ed B Battrick, A Gimenez, V Reglero and C Winkler, ESA SP-459 (Noordwijk: ESA Publications Division) p 301–304
[20] Zdunik JL and Gourgoulhon E 2001 *Phys. Rev. D.* **63** 087501
[21] Kluźniak W, Michelson, P, Wagoner, RV 1990 *Astrophys. J.* **358** 538

Journal of Physics: Conference Series **496** (2014) 012017 doi:10.1088/1742-6596/496/1/012017

Magnetosonic waves in the crust of a neutron star

D M Sedrakian, A S Harutunyan and M V Hayrapetyan

Department of Physics, Yerevan State University, Alex Manoogian 1, Yerevan, 370025, Armenia

E-mail: dsedrak@ysu.am

Abstract. The equations of magnetohydrodynamics are used to show that the energy released at the inner surface of the crust of a neutron star generates magnetosonic wave beams that propagate to the star's surface. These equations can be linearized under the conditions appropriate for the matter in the crust of neutron stars and for the frequency range $10^7 - 10^{11}$ Hz. The solutions describe a beam of standing wave with approximately constant transverse cross-section (radius). The outer base of this beam on the star's surface is a source of radio emission. Electrical currents are excited in this source and it becomes an antenna that emits radio waves into the circumstellar space. The intensity of the radio emission decreases at higher frequencies, so that the spectrum of emitted radiation by pulsars is limited from above ($\omega \leq 10^{11}$ Hz).

1. Introduction

Observations of more than 1700 pulsars have raised many questions about the structure of neutron stars and the processes taking place in them. One major unsolved problem is the mechanism of the radio emission from pulsars, whose radio luminosity is of the order of $10^{26} - 10^{30}$ erg/s which is a fraction $10^{-4} - 10^{-6}$ of the total loss of rotational energy of a pulsar. If one assumes that the source of the radioemission is the rotational energy of a pulsar, then one needs to identify the physical mechanism by which the rotational kinetic energy of the pulsar is converted into the energy of electromagnetic radiation in order to explain the features of this radioemission. In some earlier papers [1, 2, 3, 4] a possible source for the radioemission was taken to be the energy of the magnetic field drawn from the rotational energy of the star. The neutron star has a magnetic field on the order of $10^{12} - 10^{13}$ G. This field has a complicated structure and is produced by the superposition of a residual magnetic field and the magnetic field generated during the transition of the rotating nuclear component of a neutron star (neutrons and protons) into a superfluid state [5, 6, 7]. It has been shown in Refs. [3, 4] that a "magnetic spot" with a radius on the order of 300 m develops at the boundary of the superfluid core and crust of a neutron star. The magnetic spot forms due to the formation of a vortex free zone with a thickness on the order of 5 m at the boundary of the crust of the neutron star [8]. The energy released in this region excites a magnetosonic wave which propagates toward the star's surface in the form of a cylindrical beam. In general, the radius of the beam can vary as it approaches the star's surface. On reaching the surface, the wave beam is reflected, so that a transversely limited standing magnetosonic wave is formed over the width of the star's crust. Magnetosonic waves transfer the released energy to the surface of the neutron star where it is subsequently emitted in the form of radiowaves which are generated on the star's surface.

2. MHD equations and evolution equation

The behavior of the magnetoelastic plasma in the crust of a neutron star can be described by a system of MHD equations, which takes into account the viscous forces and the finite electrical and thermal conductivities of matter:

$$\frac{\partial\rho}{\partial t}+\vec{\nabla}\left(\rho\vec{V}\right)=0, \tag{1}$$

$$\rho\frac{d\vec{V}}{dt}-\frac{1}{4\pi}\left(\vec{H}\vec{\nabla}\right)\vec{H}=-\vec{\nabla}\left(P+\frac{H^2}{8\pi}\right)+\eta\Delta\vec{V}+\left(\xi+\frac{\eta}{3}\right)\vec{\nabla}\left(\vec{\nabla}\vec{V}\right), \tag{2}$$

$$\frac{d\vec{H}}{dt}-\left(\vec{H}\vec{\nabla}\right)\vec{V}+\vec{H}\left(\vec{\nabla}\vec{V}\right)=\nu_m\,\triangle\,\vec{H}, \tag{3}$$

$$\rho T\frac{ds}{dT}=-\left(\vec{\nabla}\vec{q}\right)-\pi_{\alpha\beta}\frac{\partial V_\alpha}{\partial x_\beta}+\frac{j^2}{\sigma}, \tag{4}$$

$$\rho T\frac{ds}{dt}=\frac{1}{\gamma-1}\left(\frac{dP}{dt}-c_s^2\frac{d\rho}{dt}\right). \tag{5}$$

Here ρ, P and T are the density, pressure and temperature of matter, s is the entropy per unit mass, $\vec{V}$ is the average velocity of matter, $\vec{j}$ is the current density, $\vec{q}$ is the heat flux, $\vec{H}$ is the magnetic field, η and ξ are the coefficients of the shear and bulk viscosities, $\pi_{\alpha\beta}$ is the viscous stress tensor, σ is the electrical conductivity, $\nu_m=c^2/4\pi\sigma$ is the magnetic viscosity, c_s is the speed of sound, $\gamma=4/3$ is the adiabtic index for an ideal ultrarelativistic gas.

Let us consider the propagation of magnetosonic waves in the plasma of the crust of a neutron star of finite width l. We assume that the magnetic field in the crust is directed along the y axis, and that the x axis is directed along the radius from the surface to the star's center. The time variation in the magnetic field, density of matter and pressure are given in the form

$$\vec{H}'=\vec{H}_0+\vec{h},\qquad\rho=\rho_0+\rho',\qquad P=P_0+P', \tag{6}$$

where $\vec{H}_0$, ρ_0 and P_0 are the equilibrium values of these quantities, while $\vec{h}$, ρ' and P' are their perturbations. It is assumed that the magnetic field perturbation $\vec{h}$ is also directed along the y axis. The solution of the MHD equations for the perturbations are quasimonochromatic waves [9, 10, 11] and each perturbation can be written as the sum of two oppositely propagating wave beams. We choose as the unknown function the component of the velocity of matter, $V_x\equiv u$,. With this choice we obtain

$$u\left(x,y,z,t\right)=u_1\left(\tau_1,\tau_1',y,z\right)+u_2\left(\tau_2,\tau_2',y,z\right), \tag{7}$$

where

$$\tau_{1,2}'=\frac{l\mp x}{c_n},\qquad\tau_{1,2}=\tau_{1,2}'-t \tag{8}$$

are the eikonals of waves moving to the right and left, and c_n is the velocity of the wave, which is given by

$$c_n^2=c_s^2+c_A^2,\qquad c_A^2=\frac{H_0^2}{4\pi\rho}. \tag{9}$$

Here c_A is the Alfven speed. The perturbations in the unknown physical quantities are related to V_x by the following equations

$$V_y=V_z=0,\quad h_x=h_z=0,\quad h_y=\mp\frac{H_0}{c_n}V_x,\quad\rho'=\mp\frac{\rho_0}{c_n}V_x,\quad P'=c_s^2\rho'. \tag{10}$$

The equations for the $u_{1,2}$ separate [9, 10] and they can be found solving the following equation

$$\frac{\partial^2 u}{\partial \tau' \partial \tau} + \frac{c_n^2}{2}\left(\frac{\partial^2 u}{\partial r^2} + \frac{1}{r}\frac{\partial u}{\partial r}\right) - \frac{\partial u}{\partial \tau}\frac{d \ln \Phi}{d\tau'} = -\frac{1}{c_n}\frac{\partial}{\partial \tau}\left(\Gamma u \frac{\partial u}{\partial \tau} + D\frac{\partial^2 u}{\partial \tau^2}\right), \tag{11}$$

which is called the evolution equation. In Eq. (11) the subscripts 1 and 2 on the quantities u, τ, τ' and Φ are omitted and $r = \sqrt{y^2 + z^2}$ is the cylindrical coordinate. The coefficient Γ arises from nonlinear effects and D is the coefficient of dissipation. These coefficients are defined as

$$\Gamma = \frac{\gamma + 1}{2}\frac{c_s^2}{c_n^2} + \frac{3}{2}\frac{c_A^2}{c_n^2}, \tag{12}$$

$$D = -\frac{1}{2c_n}\left\{\frac{1}{\rho}\left(\xi + \frac{4}{3}\right) + \frac{c_A^2}{c_n^2}\nu_m + \frac{(\gamma - 1)^2 \kappa T}{\rho c_n^2}\right\}, \tag{13}$$

where κ is the thermal conductivity. The function Φ in the evolution equation is related to the inhomogeneity of the medium and represent the one-dimensional ray solution in the direction perpendicular to the wave front [11]. The nonliear term in Eq. (11) causes expansion or contraction of the transverse dimensions of the wave beams that are the solutions of this equation. In addition, because of this term and the finite size of the crust of the neutron star, a monochromatic wave excited in this medium can excite waves at higer harmonics. But as a wave propagates in the crust of a neutron star, each successive harmonic is weaker in intensity than its predecessor. Therefore we can write the solution of the Eq. (11) in the form of two terms with first and second harmonics:

$$u = \frac{1}{2}\left(v_1 e^{i\omega t} + v_2 e^{2i\omega t}\right) + c.c. \tag{14}$$

Substituting this solution in Eq. (11) and equating the coefficients of the first and second harmonics, we obtain two equations for the unknowns v_1 and v_2:

$$i\omega\frac{\partial v_1}{\partial \tau'} + \frac{c_n^2}{2}\left(\frac{\partial^2 v_1}{\partial r^2} + \frac{1}{r}\frac{\partial v_1}{\partial r}\right) - i\omega v_1 \frac{d \ln \Phi}{d\tau'} = \frac{\Gamma\omega^2}{2c_n} v_1^* v_2 + i\omega\frac{\omega^2 D}{c_n}, \tag{15}$$

$$2i\omega\frac{\partial v_2}{\partial \tau'} + \frac{c_n^2}{2}\left(\frac{\partial^2 v_2}{\partial r^2} + \frac{1}{r}\frac{\partial v_2}{\partial r}\right) - 2i\omega v_2 \frac{d \ln \Phi}{d\tau'} = \frac{\Gamma\omega^2}{2c_n} v_1^2 + v_2\frac{8i\omega^3 D}{c_n}. \tag{16}$$

Let us assume that the average value of τ' satisfies the condition

$$|v_2| \ll \frac{c_n}{\tau' \Gamma \omega}. \tag{17}$$

In this case we can neglect the nonlinear term in Eq. (15). On the other hand, since v_2 is a second order perturbation, it is natural to assume that the first term on the right hand side of Eq. (16) is of the order as the first term on the left, i.e.

$$\frac{\omega\,|v_2|}{\tau'} \sim \frac{\Gamma\omega^2\,|v_1^2|}{c_n}. \tag{18}$$

On replacing τ' by the characteristic value l/c_n, instead of Eqs. (17) and (18) we obtain

$$\left|\frac{v_2}{v_1}\right| \sim \frac{\Gamma\omega l}{c_n^2}|v_1| \ll 1. \tag{19}$$

If this condition is met, then Eq. (15) can linearized and the second harmonic can neglected in Eq. (14). As we show below, condition (19) is well satisfied for the crust of a neutron star, so the contribution of the second and higher harmonics to the solution of Eq. (11) will also be small. This means that only the first term [which is determined using the linearized Eq. (15)] is retained in the desired solution.

The solution of Eq. (11) derived in Refs. [11, 12] and has the form

$$\begin{aligned} u_{1,2} &= \frac{\Phi\left(\tau'_{1,2}\right) b_{1,2}}{2f\left(\tau'_{1,2}\right)} \\ &\times \exp\left\{-\frac{r^2}{2r_0^2 f^2\left(\tau'_{1,2}\right)} + i\left[\varsigma\left(\tau'_{1,2}\right) + \frac{r^2}{2Q\left(\tau'_{1,2}\right)}\right] + i\omega\tau'_{1,2} - \omega^2 \int_0^{\tau'_{1,2}} \frac{D}{c_n} d\tau\right\} + c.c. \end{aligned} \quad (20)$$

where the additional factor $\exp\left[i\varsigma\left(\tau'_{1,2}\right) + ir^2/2Q\left(\tau'_{1,2}\right)\right]$ takes into account the changes in the transverse dimensions of the wave beam, r_0 is the radius which defines the initial transverse cross-section of the beam, and $f\left(\tau'\right) \approx 1$ is the dimensionless radius of the beam [11, 12]. Here we have restored the subscripts on $u_{1,2}$ and the other functions. We have also written two subscripts on the function b; the relationship between b_1 and b_2 can be found from the boundary condition at the star's surface, i.e., for $x = 0$. Because the outer boundary of the crust is a free sufrace, the appropriate boundary condition is $p' = 0$, which together with Eq. (10) yields

$$\rho'|_{x=0} = \left(-\frac{\rho_0}{c_n} u_1 + \frac{\rho_0}{c_n} u_2\right)|_{x=0} = 0, \quad (21)$$

i.e., $u_1(0) = u_2(0)$. If we also note that $\tau'_1|_{x=0} = \tau'_2|_{x=0} = l/c_n$, then we obtain $b_1 = b_2 \equiv b$, and for the modulus of u we have

$$|u| = \frac{2\Phi\left(l/c_n\right) b}{f\left(l/c_n\right)} \exp\left\{-\frac{r^2}{2r_0^2 f^2\left(l/c_n\right)} - \omega^2 \int_0^l \frac{D}{c_n^2} dx\right\}. \quad (22)$$

As magnetosonic wave propagates the magnetic field perturbation excites electric currents in the crust. Since $\vec{j} = (c/4\pi)\,\mathrm{curl}\vec{h}$, while $\vec{h}$ is directed along the y axes and depends only on x, only the component j_z of the vector $\vec{j}$ is nonzero. For the current density at the star's surface we have

$$\begin{aligned} j_z(r,t) &= \frac{cH_0}{4\pi} \frac{i\omega b\Phi\left(l/c_n\right)}{c_n^2 f\left(l/c_n\right)} \times \\ &\exp\left\{-\frac{r^2}{2r_0^2 f^2\left(\tau'_{1,2}\right)} + i\omega\left(\frac{l}{c_n} - t\right) - \omega^2 \int_0^l \frac{D}{c_n^2} dx + i\left[\varsigma\left(l/c_n\right) + \frac{r^2}{2Q\left(l/c_n\right)}\right]\right\} + c.c., \end{aligned} \quad (23)$$

and, finally, for the modulus of the current density we have

$$|j_z| = \frac{c}{4\pi} \frac{\omega H_0}{c_n^2} |u|, \quad (24)$$

where $|u|$ is given by Eq. (22).

The solutions for $|u|$ and $|j_z|$ show that thes quantities depend on the amplitude of the perturbation in the velocity of the matter b and, therefore, on the perturbation in the magnetic field at the boundary of the core and crust of a neutron star.

3. Estimating b from observations of the pulsar radio emission

The electric current density excited on the star's surface is proportional to b, which is the amplitude of the perturbation in the velocity of the plasma at the inner boundary of the crust. This quantity can be estimated using the observed total radio emission intensity of pulsars. Since the transverse dimensions of the wave beam in the crust do not vary essentially $(f \approx 1)$, the transverse dimensions of the region where the currents are excited is also of the order of r_0. This region becomes the source of the radio emission from the pulsar. We assume that this source is a disk of radius r_0 and thickness λ, where λ is the wavelength of the emitted wave. A volume on the order of λ^3 will radiate coherently. The intensity of the radio emission from this volume is given by the dipole radiation formula

$$I_1 = \frac{2}{3c^3}\left|\ddot{\vec{d}}\right|^2, \tag{25}$$

where $\vec{d} = \sum e_k \vec{r}_k$ is the dipole moment of this volume. Here e_k and $\vec{r}_k$ are the charge and radius-vector of the k-th particle and the sum is taken over the particles contained in this volume. Thus,

$$\dot{\vec{d}} = \sum e_k \vec{V}_k \approx \lambda^3 \sum e_\alpha n_\alpha \vec{V}_\alpha = \vec{j}\lambda^3, \tag{26}$$

where the subscript α refers to the type of particles (electrons and ions) and $\vec{j}$ is the average current density in the volume λ^3. Differentiating Eq. (26) with respect to time, we finally obtain $\ddot{\vec{d}} = \dot{\vec{j}}\lambda^3$. Substituting the expression for $\left|\ddot{\vec{d}}\right|$ in Eq. (25), we obtain

$$I_1 = \frac{2\lambda^6}{3c^3}\left|\dot{\vec{j}}\right|^2. \tag{27}$$

The number of coherently radiating volumes in the source (the disk) will be $N \sim (r_0/\lambda)^2$. The total intensity of the radio emission is the sum from the coherently radiating volumes, so that

$$I = I_1 N = \frac{2}{3c^3}\left|\dot{\vec{j}}\right|^2 \lambda^4 r_0^2. \tag{28}$$

Here $\left|\dot{\vec{j}}\right|$ is determined from Eq. (24), i.e.,

$$\left|\dot{\vec{j}}\right| = \frac{c}{4\pi}\frac{\omega^2 H_0}{c_n^2}\left|u\left(l/c_n\right)\right|, \tag{29}$$

so that

$$I = \frac{2\pi^2 c}{3}\left(\frac{cH_0 r_0}{c_n^2}\right)^2 \left|u\left(l/c_n\right)\right|^2, \tag{30}$$

where $\lambda = 2\pi c/\omega$. With Eq. (30) we can estimate $|u|$ at the star's surface for an observed total intensity of the radio emission from a pulsar of $I \sim 10^{30}$ erg/s:

$$\left|u\left(l/c_n\right)\right| \sim b\Phi\left(l/c_n\right) \sim 10^{-8} c_n. \tag{31}$$

From this, we determine the range of frequencies for which condition Eq. (19) is satisfied:

$$\omega \ll \frac{c_n^2}{\Gamma l\,|u|}. \tag{32}$$

Since $\Gamma \sim 1$, Eq. (31) implies that the condition (32) takes the form $\omega \ll 10^{12}$ Hz. We see that this condition is well satisfied within the range of frequencies of the radio emission from pulsars, $10^7 \lesssim \omega \lesssim 10^{11}$ Hz. Since $|u| = b\Phi$ has a maximum at the star's surface $(x = 0)$, the condition (32) will also be satisfied in the crust of the neutron star, where the wave beam propagates.

4. Conclusion

The MHD equations were used to show that the energy release on the interface between the core and the crust of a neutron star leads to generation of magnetosonic wave beams that propagate toward the star's surface. It was shown that for radio frequencies $10^7 - 10^{11}$ Hz and for the conditions corresponding to the matter in the crust these wave beams are linear. A wave beam which is reflected from the surface of the star forms a standing wave in the crust and a disk shaped region develops on the surface of the crust within which electric currents are excited by the magnetosonic wave. These currents convert a limited region of the star's surface into an antenna that emits radio waves into the circumstellar space. The observed total intensity of the radio emission from a star has been used to estimate the initial amplitude of the magnetosonic wave. The obtained values of the amplitude show that our analysis is consistent with the linear solution of the evolution equation. This solution shows that the wave beams undergo dissipation in the star's crust. The dissipation is greater at higher frequencies. As a result, the waves with frequencies $\omega \geq 10^{11}$ Hz do not reach the star's surface, so they cannot be emitted in the form of radio waves.

Acknowledgments

One of the authors (M. V. H.) was supported by the grant No. Az:85182 of the Volkswagen Stiftung (Hannover, Germany).

References

[1] Sedrakian D M 1989 *Astrophysics* **30** 337
[2] Sedrakian D M 1989 *Astrophysics* **31** 476
[3] Sedrakian D M, Sedrakian A D 1991 *Sov. Phys. - JETP* **73** 193
[4] Sedrakian D M, Hayrapetyan M V 2012 *Astrophysics* **55** 377
[5] Pines D 1980 *Usp Fiz Nauk* **131** 479
[6] Beskin V S, Gurevich A V and Istomin Y N 1986 *Soviet Physics Uspekhi* **29** 946
[7] Sedrakian D M, Shakhabasyan K M 1991 *Soviet Physics Uspekhi* **34** 555
[8] Sedrakian D M 2000 *Astrophysics* **43** 275
[9] Bagdoev A G, Shekoyan A V 1999 *Acoustical Physics* **45** 149 (in Russian)
[10] Bagdoev A G, Sedrakian D M 2001 *Astrophysics* **44** 113
[11] Bagdoev A G 1981 Wave propagation in continuous media, Yerevan, Academy of Science, Armenian SSR (in Russian)
[12] Sedrakian D M, Harutunyan A S and Hayrapetyan M V 2013 *Astrophysics* **56** 229

2nd International Symposium on the Modern Physics of Compact Stars and Relativistic Gravity IOP Publishing
Journal of Physics: Conference Series **496** (2014) 012018 doi:10.1088/1742-6596/496/1/012018

Time-dependent Ginzburg-Landau equations for rotating two-flavor color superconductors

K M Shahabasyan, M K Shahabasyan
Physics Department, Yerevan State University, Yerevan, Armenia

A D Sedrakian
Institute for Theoretical Physics, J. W. Goethe University, D-60438 Frankfurt am Main, Germany

Abstract. We discuss a time-dependent generalization of the stationary Ginzburg-Landau theory for two-flavor color superconducting quark matter and its modification in the presence of rotation. General expressions are obtained for the relaxation time-scales of the order parameter and color-magnetic fields and for the dissipative function, which obtains contributions from the relaxation of the order parameter and Ohmic dissipation. We also obtain a stationary equation that governs the penetration of the color-electric field in the color superconductor.

1. Introduction

Dense cores of massive compact stars may contain quark matter, which at low temperatures can form a color superconductor [1, 2, 3, 4, 5]. Recent confirmation of the existence of two-solar mass compact stars indicates that the densities in the cores of some compact stars could be sufficiently large to cross the critical density of deconfinement from the nuclear to quark matter. Observationally, color superconducting quark matter can manifest itself in many aspects of compact star phenomenology, such as the magnetic field evolution, their cooling and rotational dynamics. In this paper we develop a formalism suitable for semi-phenomenological treatments of time-dependent problems in the color superconducting phases of compact stars.

The Ginzburg-Landau (hereafter GL) theory of superconductivity has been very successful in describing semi-phenomenologically the properties of a variety of condensed matter systems, such as the electronic superconductors, liquid ^{3}He phases, etc. It has been applied to color superconducting matter both in the early stages of development [1] and more recently in Refs. [6, 7, 8, 9, 10]. The latter references discuss in detail the equilibrium GL theories for the key, robust color superconducting phases, such as the two-flavor color superconducting phase (2SC) and the three flavor color-flavor-locked (CFL) phase.

In this contribution we review our recent work on non-equilibrium Ginzburg-Landau approach to the 2SC color superconducting matter. The method is based on the idea of introducing a time-dependent term in the GL equations, which takes into account the relaxation of the order parameter of the condensate. This term has a universal form, but it contains a phenomenological coefficient, which encodes the microphysics of transport and dissipation in the condensate under consideration. We will discuss and estimate the value of this phenomenological term in the case of the 2SC condensate. The 2SC phase features a condensate of quarks paired in color and flavor anti-triplet state, *i.e.,* the red up quarks pair with green down quarks, etc, whereas the blue color remains unpaired. The pairing energy scale given by the gap parameter is of the order of tens of MeV, *i.e.,* is well above the temperatures in mature compact stars. The situation is complicated by the mixing of the electromagnetic and strong gauge fields. Below, we will assume that the blue quarks are unpaired and can scatter the Cooper pairs inelastically.

2. Time-independent GL equations for the 2SC phase

We start by reviewing the stationary limit of the time-dependent GL (hereafter TDGL) equations. We work in the laboratory frame and assume that the color superconductor is rotating with respect to this frame with some angular velocity $\mathbf{\Omega}$. The GL functional $\mathcal{F}$ is defined as

$$\mathcal{F} = \mathcal{F}_n + \int \left[2\alpha dd^* + \beta\left(dd^*\right)^2 + 2\gamma\left|\left(\hbar\boldsymbol{\nabla} + i\frac{q}{c}\boldsymbol{A}_x\right)d\right|^2 + \frac{1}{8\pi}\left(\mathrm{rot}\ \boldsymbol{A}_x\right)^2 + \frac{1}{8\pi}\left(\mathrm{rot}\ \boldsymbol{A}_y\right)^2\right]dV, \tag{1}$$

where $\mathcal{F}_n$ is the free energy of the normal state, d is the (complex) condensate order parameter for pairing with zero total angular momentum, the phenomenological coefficients α, β and γ and the charge of a quark Cooper pair q are defined below. Here $\boldsymbol{A}_x$ and $\boldsymbol{A}_y$ are the "rotated" fields, as introduced in Ref. [11]. Writing the gauge sector of the theory in terms of "rotated" fields, instead of original physical electromagnetic $\boldsymbol{A}$ and color $\boldsymbol{A}_8$ fields, has the advantage that the massive and massless bosonic modes of the model are separated from the outset. The mixing relations between the fields are given by [11]

$$\boldsymbol{A}_x = -\sin\theta_M\boldsymbol{A} + \cos\theta_M\boldsymbol{A}_8, \tag{2}$$

$$\boldsymbol{A}_y = \cos\theta_M\boldsymbol{A} + \sin\theta_M\boldsymbol{A}_8, \tag{3}$$

where the mixing angle is given in terms of electrical charge e and strong coupling constant g as

$$\cos\theta_M = \frac{\sqrt{3}g}{\sqrt{e^2+3g^2}}, \tag{4}$$

and $q = \sqrt{e^2+3g^2}/3$ is the charge of Cooper pair, e is the ordinary electrical charge. The GL equations are the stationary points of Eq. (1) with respect to the variations of the condensate conjugate order parameter and the gauge fields. These variations define the equation for the condensate order parameter [6]

$$\alpha d + \beta\left|d\right|^2 d - \gamma\left(\hbar\boldsymbol{\nabla} + i\frac{q}{c}\boldsymbol{A}_x\right)^2 d = 0, \tag{5}$$

and the equations which defines the color-magnetic currents [6]

$$\boldsymbol{j}_x = 2i\hbar\gamma q\left(d^*\boldsymbol{\nabla}d - d\boldsymbol{\nabla}d^*\right) - 4\gamma\frac{q^2}{c}\left|d\right|^2\boldsymbol{A}_x, \qquad \boldsymbol{j}_y = 0. \tag{6}$$

The coefficients appearing in the GL functional (1) have been derived from the microscopic theory and are given as [1]

$$\alpha = \frac{(T-T_c)}{T_c}\nu, \quad \beta = \frac{7\zeta(3)}{8\left(\pi k_B T_c\right)^2}\nu, \quad \gamma = \frac{p_F^2}{6\mu^2}\beta = \frac{7\zeta(3)}{16(\pi k_B T_c)^2}\frac{n_b}{\mu}, \tag{7}$$

where T_c is the critical temperature, μ is the chemical potential, p_F is the quark Fermi momentum, $\nu = \mu p_F/\pi^2\hbar^3 c^2$ is the density of states at the Fermi surface, n_b is the baryon number density. Here and below we assume that the quarks are ultra-relativistic. The relation between the phenomenological time-independent GL theory for the 2SC phase and the underlying microscopic theory was also established in Ref. [7] from the weak-coupling perturbative one-gluon exchange interaction of QCD and in Ref. [8] from the Nambu-Jona-Lasinio model. Both approaches lead to the same expressions for the coefficients α, β and γ given in Eq. (7). Furthermore, several authors discussed the possibility of formation of color-magnetic flux tubes and type-II superconductivity of 2SC color superconductor [6, 7, 8, 9, 10]. It was shown in Ref. [10] that for reasonably small values of the gap parameter $\Delta \geq 25$ MeV the color-magnetic flux tubes are energetically favorable.

The solution of Eq. (5) for a homogeneous superconducting state is

$$d_0 = \left(-\frac{\alpha}{\beta}\right)^{1/2} = \left[\frac{8}{7\zeta(3)}\left(\frac{T_c - T}{T_c}\right)\right]^{1/2} \pi k_B T_c \tag{8}$$

which is precisely the same expression as in the GL theory for non-relativistic electronic superconductors [12]. The phenomenological coefficients of these theories differ, however, by the fact that the density of states of non-relativistic and ultra-relativistic system differ. To obtain the non-relativistic counterparts of the coefficients α, β, and γ in Eq. (7) one has to replace the chemical potential of ultra-relativistic quarks μ by the electron mass m_e and the baryon number density n_b by the electron density n_e. This duality is natural because pairing in both cases occurs in a state with zero total angular momentum.

Next we consider the effect of rotation on the 2SC condensate. For that purpose we will write below the momentum density in the frame rotating with an angular velocity $\mathbf{\Omega}$. We assume that the rotation is slow enough and super-currents are such that the Galilean transformation rules are sufficient to establish the connection between the velocities in the laboratory and rotating frames. Within the context of compact stars the rotating frame is conveniently defined in terms of the crustal component of the star and all the components coupled to it on short dynamical time-scales. All these components thus execute rigid-body rotation. The 2SC phase is embedded in the fluid of ultra-relativistic electrons, which form a normal Fermi-liquid that ensures the charge neutrality of the systems. The electron fluid is coupled to the crustal components of the star on short time-scales via the magnetic field. Since the blue quark component is non-superfluid, it contributes to the normal component because it is coupled to the electron liquid on short time-scales electromagnetically.

Denoting the velocity of the normal component of the 2SC phase as $\boldsymbol{v}_n = [\mathbf{\Omega} \times \mathbf{r}]$ we obtain, for example, the momentum density in the rotating frame $\boldsymbol{p}' = \rho_s (\boldsymbol{v}_s - \boldsymbol{v}_n)$, or explicitly

$$\boldsymbol{p}' = \frac{8}{3}\gamma\mu\hbar|d|^2\left[\boldsymbol{\nabla}\chi - \frac{q}{\hbar c}\left(\boldsymbol{A}_x + \frac{2\mu}{3qc}[\mathbf{\Omega} \times \mathbf{r}]\right)\right]. \tag{9}$$

The order parameter is taken here in the following form $d = |d|e^{-i\chi}$, the superfluid mass density of the 2SC condensate is $\rho_s = 16\gamma\mu^2 |d|^2 /9$. Therefore the rotation renormalizes the vector potential according to

$$\boldsymbol{A}_x \to \boldsymbol{A}'_x \equiv \boldsymbol{A}_x + \frac{2\mu}{3qc}[\mathbf{\Omega} \times \mathbf{r}]. \tag{10}$$

The GL equations (5) and (6) in the rotating frame are obtained in a similar fashion via the replacement $\boldsymbol{A}_x \to \boldsymbol{A}'_x$. The energy of the rotating system described by the corresponding GL equations will be minimized at the cost of generation of the London field $\boldsymbol{B}_L$, defined through the condition $\boldsymbol{A}'_x = 0$, *i.e.*,

$$\boldsymbol{B}_L = -\frac{4\mu}{3qc}\mathbf{\Omega}. \tag{11}$$

Numerically $B_L \ll H_{c1}$, where H_{c1} is the critical field for formation of color-magnetic flux tubes, *i.e.*, the rotation can not generate color magnetic flux tubes (see aslo Ref. [7]).

Once more we can establish a duality between the 2SC superconductor and nonrelativistic electrons. Indeed, the London field (11) has already been obtained in the context of nonrelativistic electronic superconductors [13] and the proton superconducting fluid in rotating neutron stars [14]. The London field (11) reduces to the conventional London field of a nonrelativistic electron superconductor on replacing $q/2 \to -|e|$ and $\mu/3 \to m_e$. Furthermore, the same replacement allows one to recover the stationary GL equations for rotating nonrelativistic electronic superconductors, derived in Refs. [15, 16], from Eqs. (5) and (6), after the replacement $\boldsymbol{A}_x \to \boldsymbol{A}'_x$.

3. Time-dependent GL equations for the 2SC phase

The purpose of this section is to generalize the stationary GL equations of the previous section to the time-dependent situation. In doing so, we shall continue to use the analogy (duality) between the ordinary electronic superconductors and the 2SC quark condensate. As well known, the TDGL equations are valid in the vicinity of the critical temperature T_c. An additional constraint, in the non-stationary case, is that the deviations from the equilibrium are small, so that the heat bath remains essentially in equilibrium. For ordinary superconductors this condition is fulfilled only for gapless superconductors [17, 18]. In gapless superconductors one or several mechanisms destroy the coherence of Cooper pairs. As a result, the energy gap in the spectrum disappears but the order parameter and super-current remain intact. An example of such a mechanism is the inelastic interaction of electrons with phonons. In the case of ordinary superconductors, the condition for applicability of the TDGL theory is $\tau k_B T_c \gg \hbar$, where τ is the characteristic pair-breaking time. This time can be identified with the inelastic electron-phonon collision time. For these equations to be valid the condition $\tau\Delta \ll \hbar$ must also be fulfilled, where Δ is the pairing gap [19]. While the stationary GL equations can be derived from the underlying microscopic BCS theory, the derivation and justification of their time-dependent analogues is possible only within a certain class of theories, one example being the dirty superconductors.

We now establish the TDGL equations for the 2SC color superconductor, by using the analogy to the ordinary dirty superconductors. Because a 2SC superconductor contains an admixture of blue quarks we anticipate that there will be some degree of inelastic scattering among the Cooper pairs and the unpaired blue quarks. Less important scattering is expected due to the electromagnetic interactions of the Cooper pairs with the electrons. Combined, these interaction will be the source of destruction of the coherence among the quark Cooper pairs [10]. The physical picture is analogous to the case where impurities act to destroy the coherence among the electronic Cooper pairs in dirty superconductors. The equation for the condensate wave-function of a uniformly rotating 2SC condensate is [20] is

$$-\Gamma\left(\frac{\partial}{\partial t} - \frac{iq}{\hbar}\phi'_8\right) d = \alpha d + \beta\,|d|^2\, d - \gamma_d\left(\hbar\boldsymbol{\nabla} + \frac{iq}{c}\boldsymbol{A}'_x\right)^2 d, \tag{12}$$

where the field potentials ϕ'_8 and A'_x *are*

$$\phi'_8 = \phi_8 - \frac{\mu}{3qc^2}\Omega^2 r^2, \quad \boldsymbol{A}'_x = \boldsymbol{A}_x + \frac{2}{3}\frac{\mu}{qc}\left[\boldsymbol{\Omega}\times\mathbf{r}\right], \tag{13}$$

where ϕ_8 is the scalar potential of the color-electric field of the eight gluon. Here the coefficients α and β have their stationary values given in Eq. (7), while the non-stationary processes parameter [18] Γ and coefficient [21] γ_d for systems with strong inelastic scattering are given by

$$\Gamma = \frac{\pi\hbar\nu}{8k_BT_c}, \qquad \gamma_d = \frac{\pi\nu D}{8\hbar k_BT_c}. \tag{14}$$

Of course, the density of state and the diffusion coefficient in the latter equation should be taken those for the quark 2SC superconductor, *i.e.*, $\nu = \mu p_F/\pi^2\hbar^3 c^2$ and $D = v_F^2\tau_q/3$, where τ_q is the time-scale for the inelastic scattering of Cooper pairs with blue quarks and electrons. We expect that this time-scales should be roughly of the order of the momentum relaxation time-scale τ_s of two counter-streaming quark matter beams [22]. The second TDGL equation for the net current is given by [20]

$$\boldsymbol{j}_d = 2i\gamma_d\hbar q\left[d^*\boldsymbol{\nabla}d - d\boldsymbol{\nabla}d^*\right] - \frac{4q^2\gamma_d}{c}\left|d\right|^2\boldsymbol{A}'_x - \sigma_q\boldsymbol{E}', \tag{15}$$

here we have defined the short-hand notation

$$\boldsymbol{E}' \equiv \frac{1}{c}\frac{\partial\boldsymbol{A}_x}{\partial t} + \boldsymbol{\nabla}\phi_8 + \frac{2\mu}{3qc^2}\Omega^2\mathbf{r}, \tag{16}$$

where σ_q is the electrical conductivity of normal quark matter [22]. As has been pointed out in Ref. [23] the electric mixing angle θ_D in the 2SC phase vanishes. Consequently, while the vector potential experiences a rotation due to the non-vanishing magnetic mixing angle θ_M, see Eq. (4), the rotated electric scalar potential remains equal to ϕ_8, *i.e.*, the scalar potential of the eighth gluon. The relaxation times for the order parameter and current are

$$\tau_d = \frac{\Gamma}{|\alpha|}, \qquad \tau_j = \frac{\sigma_q}{4\gamma_d q^2|d|^2}. \tag{17}$$

The dissipative function, which corresponds to Eq. (12), is

$$W = 2\Gamma\left|\frac{\partial d}{\partial t} - \frac{iq}{\hbar}\phi'_8 d\right|^2 + \sigma_q\boldsymbol{E}'^2. \tag{18}$$

The first term in Eq. (18) describes the energy loss due to the relaxation of order parameter, while the second one is associated with Joule heating by normal currents. Upon writing the condensate order parameter in terms of its modulus and phase we rewrite the dissipative function in the form

$$W = 2\Gamma\left[\left(\frac{\partial|d|}{\partial t}\right)^2 + |d|^2\left(\frac{\partial\chi}{\partial t} + \frac{q}{\hbar}\phi'_8\right)^2\right] + \sigma_q\boldsymbol{E}'^2, \tag{19}$$

which separates the contributions due to the relaxation of the magnitude and phase of the order parameter. Another interesting expression is obtained if we use the relations

$$\frac{2iq}{\hbar}\left(d\frac{\partial\mathcal{F}}{\partial d} - d^*\frac{\partial\mathcal{F}}{\partial d^*}\right) = \mathrm{div}\boldsymbol{j}_s, \quad -\mathrm{div}\boldsymbol{j}_s = \mathrm{div}\boldsymbol{j}_n = \sigma_q\mathrm{div}\boldsymbol{E}', \tag{20}$$

where in the second relation we used the conservation of the net current. Writing again the condensate order parameter in terms of its modulus and phase and substituting $\boldsymbol{E}'$ from Eq. (16) we obtain

$$\frac{4\Gamma q}{\hbar}\left(\frac{\partial\chi}{\partial t} + \frac{q}{\hbar}\phi'_8\right)|d|^2 = \sigma_q\mathrm{div}\left(\frac{1}{c}\frac{\partial\boldsymbol{A}_x}{\partial t} + \boldsymbol{\nabla}\phi'_8\right). \tag{21}$$

In the stationary limit $\partial\chi/\partial t = 0 = \partial\boldsymbol{A}_x/\partial t$ and we find a London-type equation for the electric potential

$$\lambda_E^2\boldsymbol{\nabla}^2\phi'_8 - \frac{|d|^2}{d_0^2}\phi'_8 = 0, \tag{22}$$

where we defined a new length

$$\lambda_E = \left(\frac{\hbar^2\sigma_q}{4\Gamma q^2|d|^2}\right)^{1/2}, \tag{23}$$

which can be interpreted as the penetration depth of the static color-electric field into the 2SC phase.

We now estimate time-scales for which our TDGL equations are applicable. For the zero temperature energy gap of the order of 25Mev critical temperature $T_c = 0.57\Delta_0 = 14.2$ MeV. The condition of weak pair breaking [18] $\tau k_B T_c \gg \hbar$ is fulfilled if the inelastic quark-quark collision time $\tau > 4.3 \times 10^{-19}$ s.

To summarize, the stationary Ginzburg-Landau equations for the 2SC quark superconductors can be extended to the non-stationary regime. The resulting time-dependent GL equations are suitable for a range of problems related to the relaxations of the order parameter and color-magnetic fields for small perturbations from equilibrium. Some general expression were derived without specifying details of the problem, such as the geometry or the initial conditions. These expression include the time-scales of relaxation of the order parameter and color-magnetic field, the dissipative function of the system, as well as the penetration depth of the color-electric field in the quark superconductor. The obtained equations can be used in the modeling of the 2SC phase in massive compact stars [24], where this phase occupies a substantial volume of the star.

Acknowledgments

Two of us (K. M. S. and A. D. S.) acknowledge the support through a collaborative research grant No. 85182 of the Volkswagen Foundation (Hannover, Germany). Two of us (K. M. S. and M. K. S.) acknowledge the support from State Committee on Science (Armenia) via Grant No. 11-1c107.

References

[1] Bailin D and Love A 1984 Phys. Rept. **107** 325
[2] Alford M G, Rajagopal K, and Wilczek F 1998 Phys. Lett. **B422** 247
[3] Rapp R, Schafer T, Shuryak E V and Velkovsky 1998 Phys. Rev. Lett. **81** 53
[4] Carter G W and Diakonov D 1999 Phys. Rev. **D60** 016004
[5] Blaschke D and Roberts C D 1998 Nucl. Phys. **A642** 197
[6] Blaschke D and Sedrakian D M, nucl-th/0006038.
[7] Iida K and Baym G 2002 Phys. Rev. **D65** 014022
[8] Giannakis I and Ren H-c 2002 Phys. Rev. **D65** 054017
[9] Giannakis I and Ren H-c 2003 Nucl. Phys. **B669** 462
[10] Alford M and Sedrakian A 2010 J. Phys. **G37** 075022
[11] Alford M, Berges J and Rajagopal K 2000 Nucl. Phys. **B571** 269
[12] Gorkov L P 1959 Zh. Eksp. Teor. Fiz. **36** 1918
[13] London F 1950 Superfluids, v. I, Wiley New York
[14] Sedrakian D M and Shahabasyan K M 1980 Dokl. Akad. Nauk Arm. SSR **70** 28
[15] Verkin B I and Kulik I O 1971 Zh. Eksp. Teor. Fiz. **61** 2067
[16] Sedrakian D M, Mkrtchyan G S and Shahabasyan K M 1976 Izvestia Akad. Nauk Arm. SSR Fiz. **11** 385
[17] Gorkov L P and Eliashberg G M 1968 Zh. Eksp. Teor. Fiz. **54** 612
[18] Gorkov L P and Kopnin N B 1975 Usp. Fiz. Nauk **116** 413
[19] Kopnin N B 1995 Vortices in Type II Superconductors p. I (Landau Inst. Theor. Phys., Moscow)
[20] Shahabasyan K M, Shahabasyan M K and Sedrakian A D 2013 Astrophysics **56** 255
[21] Saint-James D, Sarma G and Thomas E J 1969 Type II Superconductivity, Pergamon, London
[22] Heiselberg H and Pethick C J 1993 Phys. Rev. **D48** 2916
[23] Schmitt A, Wang Q and Rischke D H 2004 Phys. Rev. **D69** 094017
[24] Bonanno L and Sedrakian A 2012 Astron. Astrophys. **539** A16

Journal of Physics: Conference Series **496** (2014) 012019 doi:10.1088/1742-6596/496/1/012019

Coupled post-glitch response of the crust and interior of neutron stars

C. A. van Eysden
NORDITA, Roslagstullsbacken 23, SE-10691 Stockholm, Sweden

Abstract.
The spin-up of magnetized plasma and its role in the post-glitch response of neutron stars was first studied over thirty years ago, where it was demonstrated that co-rotation between crust and plasma is established rapidly (within a few seconds) by a process analogous to Ekman pumping in a viscous fluid. However, if the magnetized plasma is considered to be ideal, conservation of energy implies that the final state cannot be co-rotation. Using an exact analytical solution for the coupled motion of the crust and plasma, we demonstrate that the system oscillates persistently and explore the consequences for neutron star observations.

Introduction

Pulsars are rapidly rotating neutron stars that emit radio waves along their magnetic dipole axis, which is misaligned with the rotation axis, creating a lighthouse effect that is observed at earth as pulsations. The pulse arrival time of these objects can be measured with extremely high precision, however, about 5% of the known pulsar population is occasionally observed to 'glitch', where the pulse frequency undergoes a sudden increase, followed by a recovery. The glitch event itself occurs over a timescale of less than a minute [2], while the recovery typically takes days to weeks [3]. The increase in pulse frequency during a glitch ranges from one part in 10^{-4} to 10^{-9}, and is typically 10^{-6} and 10^{-8} in the Crab and Vela pulsars respectively [4, 3].

The mechanism for driving a pulsar glitch is still not completely understood, but it is generally attributed to an avalanche unpinning event of pinned superfluidity in the neutron star crust [5, 6, 9]. The core of the star, which consists of a neutron superfluid and a plasma of electrons and superconducting protons, responds to the glitch via magnetic, viscous and mutual friction forces. Ref. [10] showed that, for typical pulsars, the magnetic tension couples the plasma and the crust on a timescale of $\sim 2\,\mathrm{s}$, bringing the two components into co-rotation. As such, the crust and the plasma are usually assumed to be rigidly locked together in models of pulsar glitch recovery.

In this article, we re-visit the problem of the response of a magnetised plasma to an impulsive acceleration of the container and the analysis of [10]. We are interested in particular in two aspects: (i) what is the coupled response of container, and (ii) why is the final state of the plasma and container that of co-rotation? The first is motivated by recent studies of glitch recovery that calculate the motion of the neutron star crust in response to the interior fluid for comparison to observational data, which can potentially extract information about the transport coefficients of the interior fluid of the neutron star [4]. The second is motivated by the fact that the magnetised plasma is assumed to be ideal, making this problem qualitatively different from

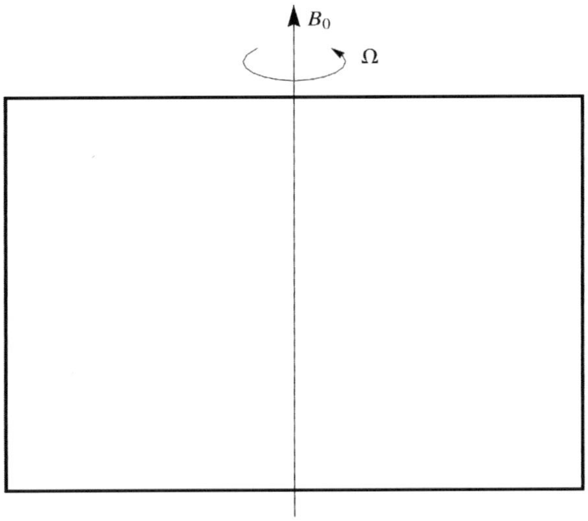

Figure 1. To investigate the coupled response of a magnetised plasma and its container, we consider cylidrical geometry, with a uniform magnetic field aligned with the rotation axis.

classical spin up where there is dissipation. Conservation of energy and momentum arguments suggest that this system should be executing torsional oscillations, similar to those identified in magnetar quasi-periodic oscillations Ref. [11].

The model

To study the coupled response of a magnetised plasma and its container, we consider the cylindrical geometry illustrated in figure 1. For $t < 0$, the container and fluid are rotating uniformly with angular velocity Ω about the vertical axis, denoted $\hat{\mathbf{k}}$. The fluid and container also possess a uniform magnetic field of strength B_0, aligned with the rotation axis. At $t = 0$, the angular velocity of the container is impulsively increased to $\Omega + \delta\Omega(0)$ and the container and fluid are left to respond self-consistently. In contrast to previous studies, we also consider the motion of the container, so that $\delta\Omega$ is a function of t.

The equations describing the system are those of magnetohydrodynamics. In general, $\delta\Omega(0)/\Omega$ is much smaller than unity, so the equations can be linearised. Also, since the magnetic diffusivity is extremely small in neutron stars it is neglected. The scaled, linearised equations are

$$\frac{\partial \delta\mathbf{v}}{\partial t} + 2\hat{\mathbf{k}} \times \delta\mathbf{v} = -\nabla\delta P - \xi^2 \hat{\mathbf{k}} \times (\nabla \times \delta\mathbf{B}) + E\nabla^2 \delta\mathbf{v}\,, \tag{1}$$

$$\nabla \cdot \delta\mathbf{v} = 0\,, \tag{2}$$

$$\frac{\partial \delta\mathbf{B}}{\partial t} = \nabla \times \left(\delta\mathbf{v} \times \hat{\mathbf{k}}\right)\,. \tag{3}$$

where we define the dimensionless coefficients

$$E = \frac{\nu}{L^2\Omega}\,, \quad \xi = \frac{1}{L\Omega}\sqrt{\frac{B_0^2}{\mu_0 \rho}}\,. \tag{4}$$

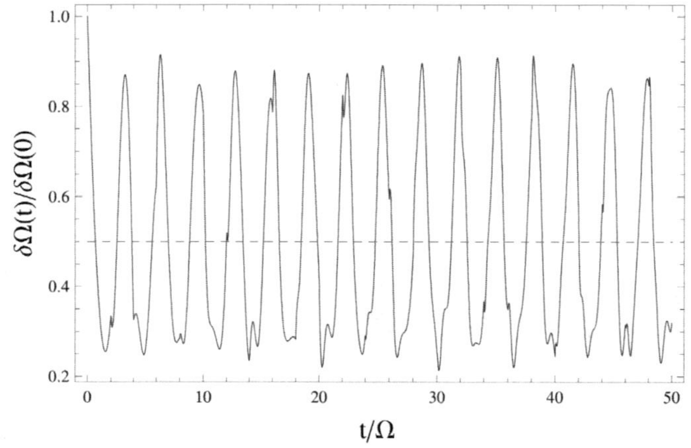

Figure 2. Motion of the container for slow rotation, i.e., $\xi = 10$, $E = 0$ and $K = 1$.

The first is the usual Ekman number [12] and denotes the relative strength of viscous forces to inertial forces. The second, defined by Ref. [10] is the ratio of magnetic forces to inertial forces. We consider the infinite parallel plate geometry (containers of infinite aspect ratio) as in the classic studies of spin up [12, 10]. The usual no-slip boundary conditions apply to the fluid at the upper and lower plates

$$\delta \mathbf{v} = \mathbf{x} \times \delta\Omega(t)\hat{\mathbf{k}}\,. \tag{5}$$

As there is no magnetic diffusivity the magnetic field is flux-frozen, and Eq. (5) also satisfies the continuity requirements for the magnetic field at the boundary. The evolution of $\delta\Omega(t)$ is determined by considering a torque balance on the container, and evaluating the hydrodynamic torque using the stress tensor of the fluid.

The solution to this system of equations is obtained using Laplace transform techniques, the details of which will be presented elsewhere. However, in contrast to previous studies, we make no approximations regarding the magnitude of E and ξ, and study the behaviour of the system for a range of these parameters.

Results

First, let us consider the limit $\xi \gg 1$, $E = 0$, i.e., when the rotational period is much longer than the Alfvén crossing time and the fluid is inviscid. We also take $K = 1$, where K is the ratio of the moments of inertia of the fluid (rotating as a rigid body) to that of the container. The motion of the container is given in Fig. 2. We find that the solution oscillates, as expected, where the restoring force for the oscillations is provided by the magnetic field line tension. Unexpectedly, however, we find that these oscillations are not smooth, but incoherent. There are two reasons for this. First, the container is coupled to a spectrum of Alfvén waves in the fluid, and as a result behaves like a mass coupled to many smaller masses via springs. Importantly, the system is not chaotic (the equations are linear) or random, and the complicated motion arises from the many degrees of freedom in the system. Second, the eigenvalues of the system are solutions to a complicated transcendental equation and are unevenly spaced. This means that the nodes of different modes do not necessarily coincide, therefore the modes do not add to form a coherent, periodic result. This effect becomes more pronounced at larger times, when small misalignments of nodes have accumulated into larger phase differences between the modes.

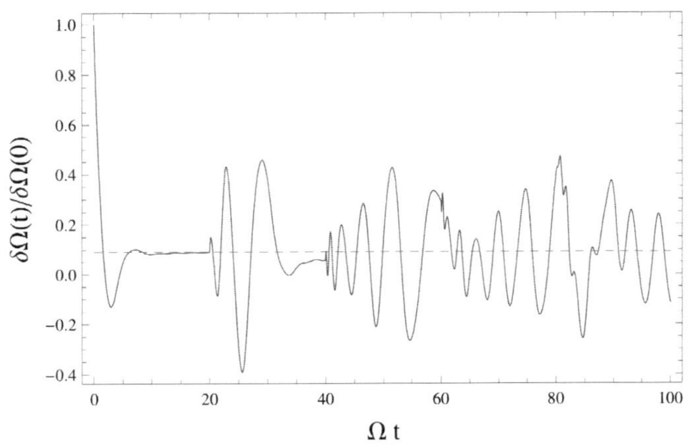

Figure 3. Motion of the container for fast rotation, i.e., $\xi = 0.1$, $E = 0$, $K = 10$.

In Fig. 3, we consider a faster rotating container, with numbers closer to those corresponding to neutron stars, $\xi = 0.1$, $E = 0$ and $K = 10$. The rotation period is now much faster than the Alfvén crossing time. We find that the behavior identified by Ref. [10] is observed initially, namely, the container overshoots, oscillates briefly before settling to a state of co-rotation with the interior fluid. This is achieved by a mechanism closely analogous to Ekman pumping in a viscous fluid, in which a secondary flow circulates fluid from boundary layers at the container walls into the interior, spinning up the fluid [12]. However, co-rotation only persists until a dimensionless time of 20, at which point oscillations of the nature of those observed in Fig. 2 are present. The onset of these oscillations correspond exactly with the Alvfén crossing time of the container, when the magneto-inertial wave excited by the impulsive acceleration of the container reaches the opposing boundary and is reflected. At short times, when this wave has only propagated a short distance from the boundary, it facilitates the Ekman-like circulation identified by Ref. [10] in an analogous manner to the viscous boundary layer in classic Ekman pumping. After the Alfvén crossing time, however, it induces oscillations like in a slowly rotating container. These oscillations are not identified by Ref. [10], who studies the limit $\xi \to \infty$, because the magneto-inertial oscillations take an infinitely long time to traverse the container.

In Fig. 4, we present the solution for numbers expected in a neutron star, namely, $\xi = 0.001$, and $K = 50$. We also consider the inviscid limit, $E = 0$. The top-left panel shows the response for five Alfvén crossing times, while the remaining panels show short sections of the top-left panel in greater detail. We find that the qualitative features in Fig. 4 are present and exaggerated further. Oscillations onset after the Alfvén crossing time, and become messier at later times. The amplitude of oscillation is approximately one tenth of the glitch amplitude, and will onset approximately 20 s after the glitch for a pulsar with $\Omega = 100\,\mathrm{rad\,s^{-1}}$. Note, however, that in spherical geometry the Alfvén crossing time varies with latitude and is much shorter near the equator, resulting in quantitatively different behavior. Also note that because we have neglected other components of the star, i.e., the neutron superfluid, that glitch recovery is almost immediate, namely, 2 s for a pulsar with $\Omega = 100\,\mathrm{rad\,s^{-1}}$.

In Fig. 5, we consider the same parameters as those in Fig. 4, but include viscosity, taking $E = 10^{-7}$. We find similar qualitative behavior to that seen in Fig. 4, however the oscillations are significantly damped and have a maximum amplitude of approximately 1/200 of the glitch size. The time taken for the oscillations to become messy is also much longer. In Fig. 6, we plot the response to the event shown in Fig. 5 at four later times. Each panel has the same

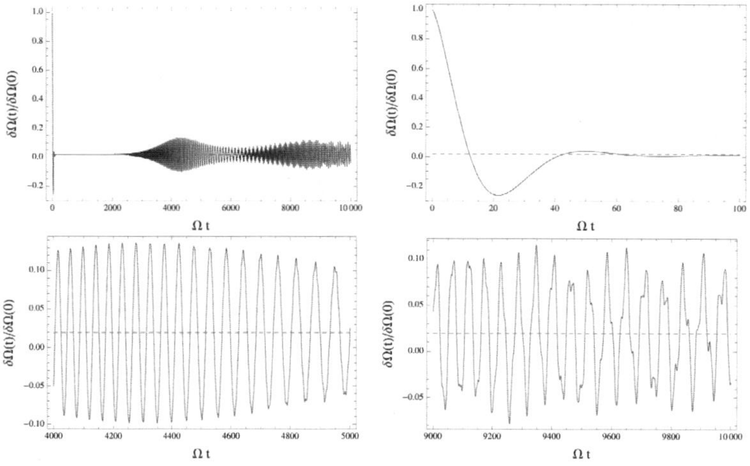

Figure 4. Motion of the container for typical neutron star numbers, i.e., $\xi = 0.001$, $E = 0$, $K = 50$.

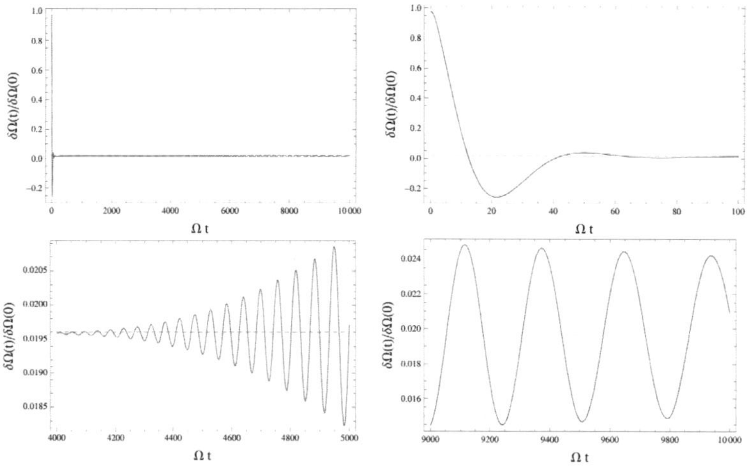

Figure 5. Same as in Fig. 4, but with $E = 10^{-7}$.

scale on the time and amplitude axes for comparison. Clearly, as time progresses, the oscillation amplitude decreases while the period becomes longer. This is a result of viscosity damping the higher wavelength modes faster, so that at later times only the long wavelength (corresponding to long period) persist. Examining the top-right panel, the oscillation has an amplitude roughly a factor of 1/2000 of the glitch amplitude and a frequency of 20 s at a time of 2.78 hrs after the glitch, assuming $\Omega = 100\,\mathrm{rad\,s^{-1}}$.

Conclusions

We have shown that in response to a glitch, the crust and interior plasma are expected to execute torsional oscillations. The state of co-rotation between the crust and plasma only persists until the Alfvén crossing time, when magneto-inertial waves excited by the glitch have traversed the

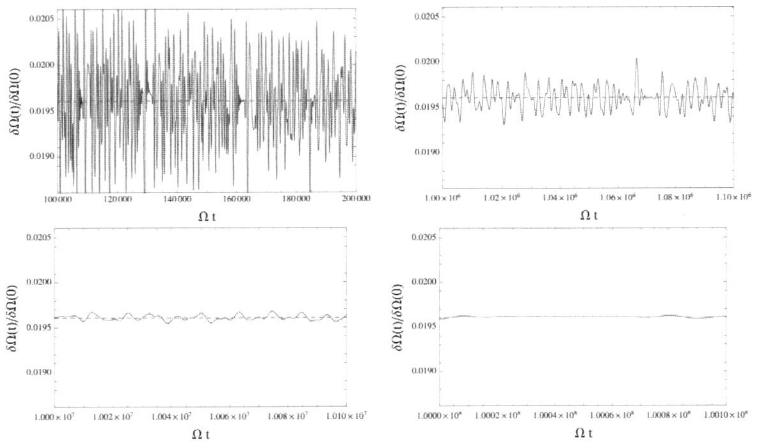

Figure 6. Same result as in Fig. 5 at later times.

star and are reflected from the opposing boundary. The period of oscillation for a typical neutron star is predicted to be of the order of seconds. The oscillations are damped by viscosity, which establishes co-rotation on the viscous diffusion time. The oscillations are incoherent, and may be a contributing factor in timing noise. The model considered here neglects the important dynamics of the neutron superfluid in both the crust and the core, which is expected to result in the longer time-scales observed in glitch recovery, and which may affect the predictions presented here. This will be considered in future work.

References

[1] Dodson R G, McCulloch P M and Lewis D R 2002 *The Astrophysical Journal* **564** L85–L88 arXiv:astro-ph/0201005
[2] Dodson R G, McCulloch P M and Lewis D R 2002 *The Astrophysical Journal* **564** L85–L88 arXiv:astro-ph/0201005
[3] Espinoza C M, Lyne A G, Stappers B W and Kramer M 2011 arXiv:1102.1743
[4] van Eysden C A and Melatos A 2010 *Monthly Notices of the Royal Astronomical Society* **409** 1253–1268 arXiv:1007.4360
[5] Anderson P W and Itoh N 1975 *Nature* **256** 25–27
[6] Melatos A and Warszawski L 2009 *The Astrophysical Journal* **700** 1524–1540 arXiv:0904.2998
[7] Warszawski L and Melatos A 2011 *Monthly Notices of the Royal Astronomical Society* **415** 1611–1630 arXiv:1103.6090
[8] Easson I 1979 *The Astrophysical Journal* **228** 257–267
[9] Warszawski L and Melatos A 2011 *Monthly Notices of the Royal Astronomical Society* **415** 1611–1630 arXiv:1103.6090
[10] Easson I 1979 *The Astrophysical Journal* **228** 257–267
[11] van Hoven M and Levin Y 2011 *Monthly Notices of the Royal Astronomical Society* **410** 1036–1051 arXiv:1006.0348
[12] Greenspan H P and Howard L N 1963 *Journal of Fluid Mechanics* **17** 385–404

Journal of Physics: Conference Series **496** (2014) 012020 doi:10.1088/1742-6596/496/1/012020

Cosmological models in conformal representations of Jordan theory

R. M. Avakyan, E. V. Chubaryan, G.H. Harutyunyan, A. V. Hovsepyan, A. S. Kotanjyan

Abstract. This paper is devoted to the investigation of conformally-related variants of the modified tensor-scalar Jordan theory on the example of determination of the comparative characteristics of the model Universe in "Einstein" and "proper" frames. Within the framework of this model, we consider the possibility for the accelerated expansion of the Universe at the recent epoch.

Yerevan State University, 1 Alex Manoogian Street, 0025 Yerevan, Armenia

E-mail: rolavag@ysu.am, echub@ysu.am, hagohar@ysu.am, ahovs@mail.ru, anna.kotanjyan@ysu.am

1. Introduction

The role of conformal transformations in gravitational theories has been discussed by H. Weyl [1], W. Pauli [2], A.Z. Petrov [3], R. Dicke [4], etc (for a recent review see [5]). In the Jordan-Brans-Dicke (JBD) theory the specificity of the scalar field is particularly evident in the conformal representations, where the scalar field is separated from the metric part of the gravitational field, and as it turns out, it may play a definite role in the phenomenon of acceleration of the expanding Universe. In the absence of matter fields, JBD theory is conformally-invariant [6]. Under certain conformal transformations it takes a standard "Einstein" form. In this case, the energy-momentum tensor for minimally or conformally coupled scalar field is added to the energy-momentum tensor of matter. There are clear advantages to have this conformal relation: one can use solutions already obtained for one representation to generate solutions of another, equivalent, representation.

The present paper is devoted to the theoretical study of the evolution of the Universe within the framework of various variants of JBD theory. We consider the possibility for the accelerated expansion of the Universe at the recent epoch. By taking into account that in General Relativity the arguments in favor of the cosmological constant are convincing enough, we introduce a similar quantity in JBD theory, assuming that the corresponding field is a scalar one. However, the latter is non-dynamical and is governed by the gravitational scalar.

2. The cosmological problem with the cosmological scalar $\varphi = \Lambda y/y_0$

The modified version of the Jordan theory, in addition to the metric tensor, contains a scalar field $y(x)$ in its gravitational sector. The corresponding action functional has the form [7]

$$W = \frac{1}{c}\int \left\{ -\frac{y}{2k}\left[R + 2\Lambda\frac{y}{y_0} - \zeta g^{\mu\nu}\frac{y_\nu y_\mu}{y^2} \right] \right\} \sqrt{-g}d^4x \tag{1}$$

where $k = 8\pi/c^2$, y_0 is a constant, ζ is the dimensionless parameter of the theory and we use the notation $y_\mu = \partial_\mu y$. After the conformal transformation $\widetilde{g}_{\mu\nu} = g_{\mu\nu} y/y_0$, the action (1) takes the form [8]

$$\widetilde{W} = \frac{1}{c}\int \left[-\frac{y_0}{16\pi}\left(\widetilde{R} + 2\Lambda\right) + \frac{1}{2}\widetilde{g}^{\mu\nu}\Phi_\mu\Phi_\nu \right]\sqrt{-\widetilde{g}}d^4x, \tag{2}$$

were the new scalar field Φ is defined by the relation

$$\Phi_\mu = \frac{y_\mu}{y}\sqrt{\frac{(3+2\zeta)y_0}{16\pi}}. \tag{3}$$

The action (2) coincides with the Einstein action in the presence of a minimally-coupled scalar field ("Einstein" frame of the Jordan theory) [8]. The corresponding cosmological problem, described by the Friedmann-Roberston-Walker (FRW) line element

$$dS^2 = dt^2 - a^2(t)[dr^2 + r^2(d\theta^2 + \sin^2\theta d\phi^2)], \tag{4}$$

has been discussed in [11]. In the present paper we compare the results obtained for the model with a minimally coupled scalar field with the model in the presence of a self-consistent scalar field in the case of action (1) and for the metric

$$dS^2 = d\tau^2 - R^2(\tau)[dr^2 + r^2(d\theta^2 + \sin^2\theta d\phi^2)]. \tag{5}$$

The time coordinates and the scale factors in two different representations are related by

$$R(\tau) = (y/y_0)^{-1}a(t), \quad d\tau = (y/y_0)^{-1/2}dt. \tag{6}$$

As in the paper [9], we solve the standard cosmological problem with the notations $H = \dot{R}/R$, $\psi = \dot{y}/y$, $q = \ddot{R}R/R^2$, where the dot stands for the derivative with respect to τ. The set of field equations reads

$$\frac{1}{yR^3}\frac{d(\dot{y}R^3)}{d\tau} = \frac{2}{3+2\zeta}\left(\varphi - y\frac{d\varphi}{dy}\right), \tag{7}$$

$$2\dot{H} + 3H^2 - \psi H + \zeta\psi^2/2 - \varphi = 0, \tag{8}$$

$$3H^2 - \zeta\psi^2/2 + 3\psi H - \varphi = 0. \tag{9}$$

By taking into account that in the problem under consideration $\varphi = \Lambda y/y_0$, the equation (7) is reduced to $\ddot{y}/y = -3\psi H$, or equivalently

$$\dot{\psi} + \psi^2 + 3H\psi = 0. \tag{10}$$

From here, for the evolution of the expanding Universe ($H > 0$) we get

$$\ddot{y}/y < 0 \Rightarrow \dot{\psi}/\psi^2 < -1. \tag{11}$$

Further, subtracting (9) from (8), we get

$$2\dot{H} + \zeta\psi^2 - 4\psi H = 0. \tag{12}$$

Defining a new function $\dot{y} = z(y)$, the latter is reduced to the Euler equation [10]:

$$y^2 z'' - 2z'y - 3\zeta z/2 = 0, \tag{13}$$

with the solution [11]

$$\psi = \dot{y}/y = c_1 (y/y_0)^{\frac{1+\sigma}{2}} + c_2 (y/y_0)^{\frac{1-\sigma}{2}}, \quad \sigma = \sqrt{3(3+2\zeta)}. \tag{14}$$

After the substitution of (14) into (7) and (9), we obtain the relation between the constants of integration c_1 and c_2:

$$c_1 c_2 = -\Lambda/(3+2\zeta), \tag{15}$$

from which it follows that c_1 and c_2 have opposite signs, when $\Lambda > 0$, and one of these constants is equal to zero, when $\Lambda = 0$.

By taking into account that

$$\ddot{y}/\dot{y} = c_1 \left(\frac{3+\sigma}{2}\right) y^{\frac{1+\sigma}{2}} + c_2 \left(\frac{3-\sigma}{2}\right) y^{\frac{1-\sigma}{2}} < 0, \tag{16}$$

we conclude that $c_1 < 0$, $c_2 > 0$. These constants are determined from the condition $y = y_0$ at $t = 0$:

$$\psi_0 = -|c_1| + c_2, \ |c_1| c_2 = \Lambda/(3+2\zeta). \tag{17}$$

As a result one finds the following relations

$$|c_1| = -\psi_0/2 + \sqrt{\psi_0^2/4 + \Lambda/(3+2\zeta)}, \ c_2 = \psi_0/2 + \sqrt{\psi_0^2/4 + \Lambda/(3+2\zeta)}. \tag{18}$$

By taking into account that our ultimate aim is to compare the obtained results with the corresponding problem for the action (2), let us present (14) in the form

$$\psi = c_2 (y/y_0)^{\frac{1-\sigma}{2}} \left[1 - (|c_1|/c_2)(y/y_0)^{\sigma}\right]. \tag{19}$$

After the integration of this we find the following expression

$$\left(\frac{y}{y_0}\right)^{\sigma/2} = \frac{1}{\alpha} \tanh\left(\frac{\sigma}{2}\sqrt{\frac{\Lambda}{2\zeta+3}}t + \delta\right) = \frac{1}{\alpha} \tanh\left(\frac{3}{2} H_0 \sqrt{\overset{0}{\Omega}_\Lambda} t + \delta\right). \tag{20}$$

Here we have introduced the notations

$$\alpha = \frac{|c_1|}{c_2} < 1, \quad e^{2\delta} = \frac{1+\alpha}{1-\alpha}, \quad \overset{0}{\Omega}_\Lambda = \frac{\Lambda}{3H_0^2}. \tag{21}$$

From (10) one can see that

$$\dot{\psi}/\psi + \dot{y}/y + 3\dot{R}/R = 0, \tag{22}$$

from which for the function R/R_0 one obtains

$$\left(\frac{R}{R_0}\right)^3 = \frac{\psi_0}{\psi}\frac{y_0}{y}. \tag{23}$$

For the numerical calculations of the parameters H, Ω_Λ, Ω_y, q it is convenient to use the following combination of equations (10) and (12):

$$\frac{d}{dt}\left(\frac{\dot{\psi}}{\psi}\right) - \dot{\psi} - \frac{4+3\zeta}{2}\psi^2 = 0. \tag{24}$$

With the definition $z(\psi) = \dot{\psi}/\psi$, the equation (24) is reduced to a known class of differential equations (see for example, [10]): $zz'(\psi) - z - (4+3\zeta)\psi/2 = 0$. Taking $z = u\psi$, the variables are separated and we find

$$\psi^2 = 4\left(\sigma - 1 + 2u\right)^{\frac{1-\sigma}{\sigma}}\left(\sigma + 1 - 2u\right)^{-\frac{1+\sigma}{\sigma}}. \tag{25}$$

In order to find the Einsteinian limit ($\zeta \to \infty$), we present (25) in the form

$$\left(\frac{\psi}{2}\right)^{2\sigma} = \frac{\sigma-1}{\sigma+1}\frac{1}{(\sigma^2-1)^{\sigma}}\left(1+\frac{|u|}{1-\sigma}\right)^{1-\sigma}\left(1+\frac{|u|}{1+\sigma}\right)^{-1-\sigma}, \tag{26}$$

from which it follows that $\lim\limits_{\sigma\to\infty}\psi = 1/\sigma \to 0$.As a result, the solution of the equation (26) can be written as

$$u = \frac{\dot{\psi}}{\psi^2} = \frac{1}{2} - \frac{\sigma}{2}\frac{1+\alpha\left(y/y_0\right)^{\sigma}}{1-\alpha\left(y/y_0\right)^{\sigma}}. \tag{27}$$

3. Determination of the integral parameters

It is easy to see that all relevant functions are expressed in terms of the ratio ψ/H, which can be directly determined from (10):

$$H/\psi = -\left(1+u\right)/3. \tag{28}$$

Then, from (12), for the deceleration parameter one finds

$$q = 1 + \frac{\dot{H}^2}{H^2} = 1 + \frac{2\psi}{H} - \frac{\zeta}{2}\frac{\psi^2}{H^2}. \tag{29}$$

From (9) we have

$$1 - \frac{\zeta}{6}\frac{\psi^2}{H^2} + \frac{\psi}{H} = \frac{\varphi}{3H^2} \equiv \Omega_\Lambda, \tag{30}$$

where Ω_Λ is the effective energy contribution of the cosmological scalar. It should be noted that (30) allows us to obtain ψ_0/H_0 for $y = y_0$:

$$\overset{0}{\Omega}_\Lambda = \frac{\Lambda}{3H_0^2} = 1 - \frac{\zeta}{6}\frac{\psi_0^2}{H_0^2} + \frac{\psi_0}{H_0}. \tag{31}$$

which is needed for the calculation of $\alpha = |c_1|/c_2$ from (18). For the evaluation of H/H_0 we use (28):

$$\frac{H}{H_0} = -\frac{\psi}{H_0}\frac{1+u}{3},$$

and the behavior of R/R_0 is determined from (23).

By taking into account that we consider the late stages of the Universe expansion, the expression (20) for large positive values of the argument can be presented in the form

$$\left(y/y_0\right)^{\sigma/2} \approx \left(1 - 2e^{-2x}\right)/\alpha, \tag{32}$$

where $x = (3/2)\,H_0\sqrt{\overset{0}{\Omega}_\Lambda}t + \delta$ and we determine the relation between conformally related time coordinates by (6). As a result one gets $\tau = \alpha^{1/\sigma}t$.

In figures 1-4 we have plotted the time evolution of the functions H/H_0, R/R_0, q, Ω_y, Ω_Λ for the values of the parameter $\zeta = 50$ and $\zeta = 1000$.

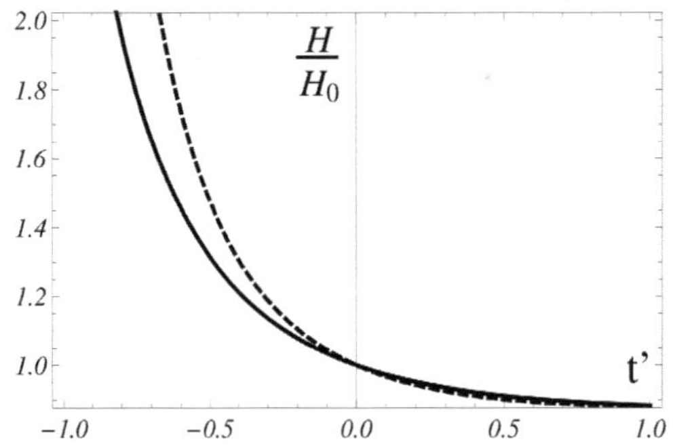

Figure 1. Time dependence of the Hubble parameter H/H_0. The time is measured in units $5 \cdot 10^9$ years. Dashed and full curves correspond to values $\zeta = 50$ and $\zeta = 1000$, respectively.

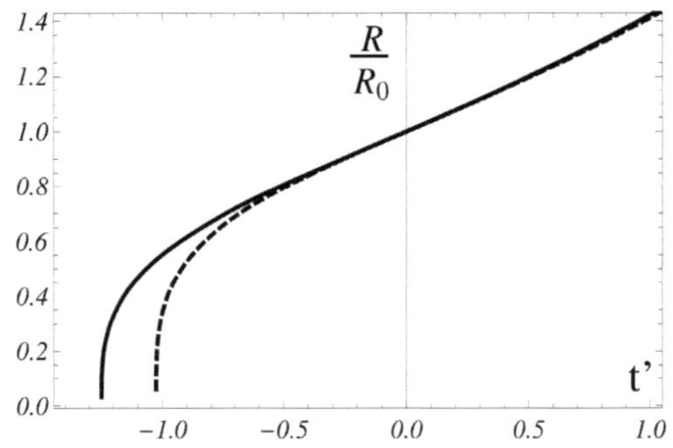

Figure 2. Time dependence of the scale factor R/R_0 for $\zeta = 50$ (dashed curve) and $\zeta = 1000$ (full curve).

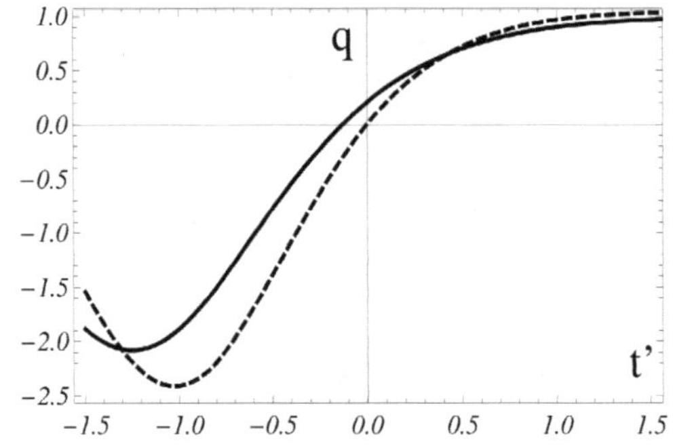

Figure 3. Time dependence of the "deceleration" parameter q for $\zeta = 50$ (dashed curve) and $\zeta = 1000$ (full curve).

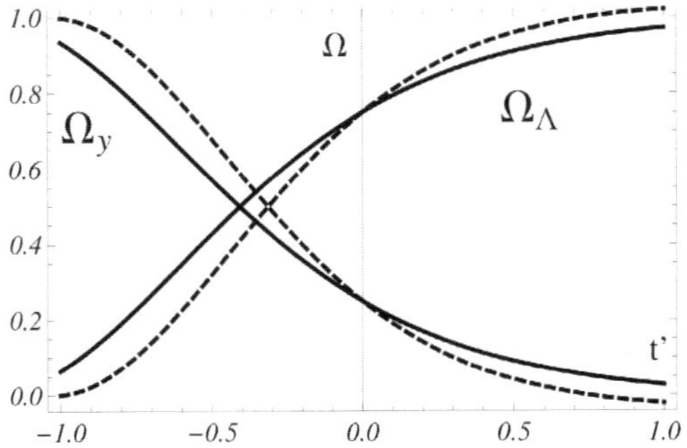

Figure 4. Time dependence of the energy contribution of the scalar field, Ω_y, and of the field, corresponding to cosmological scalar, Ω_Λ. Dashed and full curves correspond to $\zeta = 50$ and $\zeta = 1000$, respectively.

4. Nonminimally coupled scalar field in the presence of $\varphi = \Lambda$

In this section we consider the problem described by the action functional [7]- [13]

$$W = \int \left\{ -y \left[(R + 2\Lambda) + \zeta g^{\mu\nu} \frac{\nabla_\mu y \nabla_\nu y}{y} \right] \right\} \sqrt{-g} d^4 x. \tag{33}$$

Taking the line element in the FRW form (4), we introduce the cosmological functions

$$H = \frac{\dot{a}}{a},\ \psi = \frac{\dot{y}}{y},\ q = \frac{a\ddot{a}}{\ddot{a}},\ \Omega_\Lambda = \frac{\Lambda}{3H^2}. \tag{34}$$

The system of equations in convenient combinations is the following:

$$3H^2 - \psi^2\zeta/2 + 3\psi H = \Lambda \tag{35}$$

$$3H\psi - \zeta\psi^2 - 3qH^2 = -\frac{2\Lambda\zeta}{3 + 2\zeta}, \tag{36}$$

$$\frac{d(ya^3)}{dt} = \frac{2\Lambda}{3 + 2\zeta} ya^3. \tag{37}$$

In order to solve the equation (37) it is resonable to calculate the derivative of the expression

$$\frac{d(ya^3)}{dt} = ya^3 (\psi + 3H). \tag{38}$$

By taking into account (37), we can see that

$$\dot{\psi} + \psi^2 + 3H\psi = \frac{2\Lambda}{3 + 2\zeta}. \tag{39}$$

Now, combining the equations (35) and (36), we obtain

$$3\dot{H} + 9H^2 + 3H\psi = 2\Lambda\frac{3 + 4\zeta}{3 + 2\zeta}, \tag{40}$$

and, as a result,

$$\frac{d^2(ya^3)}{dt^2} - 2\Lambda y a^3 \frac{4+3\zeta}{3+2\zeta} = 0. \tag{41}$$

The solution of the equation (41) can be written in the form

$$ya^3 = c_1 \cosh At + c_2 \sinh At = c \cosh\left(At + \alpha\right), \text{ for } c_1 > c_2, \tag{42}$$

$$ya^3 = c_1 \cosh At + c_2 \sinh At = c \sinh\left(At + \alpha\right), \text{ for } c_1 < c_2, \tag{43}$$

where $A^2 = 2\Lambda\left(3+4\zeta\right)/\left(3+2\zeta\right)$. In the case (43), $ya^3 = c\sinh\left(At+\alpha\right)$, the integration constants are determined from the following relations

$$\psi_0 + 3H_0 = A \coth\alpha \tag{44}$$

$$y_0 a_0^3 = c \sinh\alpha. \tag{45}$$

From the equation (37) we have

$$\dot{y}a^3 = \dot{y}_0 a_0^3 + \frac{2\Lambda c}{A\left(3+2\zeta\right)}\left[\cosh\left(At+\alpha\right) - \cosh\alpha\right]. \tag{46}$$

From here it follows that

$$\psi = \frac{\psi_0 + H_0\sqrt{\frac{6\overset{0}{\Omega}_\Lambda}{(4+3\zeta)(3+2\zeta)}}\left[\sinh At + \coth\alpha\left(\cosh At - 1\right)\right]}{\cosh At + \coth\alpha \sinh At}. \tag{47}$$

Now, by using equations (38), (44) and (47), we get

$$\frac{H}{H_0} = \frac{\sqrt{\frac{6\overset{0}{\Omega}_\Lambda}{(4+3\zeta)(3+2\zeta)}}\left[\left(\zeta+1\right)\left(\sinh At + \coth\alpha\left(\cosh At - 1\right)\right)\right] + 1}{\cosh At + \coth\alpha \sinh At}. \tag{48}$$

At the initial time $t = 0$ from (35) one can find the ratio $\psi_0/\ H_0$:

$$\frac{\psi_0}{H_0} = \frac{3}{\zeta}\left(1 \pm \sqrt{1 + \frac{2}{3}\zeta\left(1 - \overset{0}{\Omega}_\Lambda\right)}\right). \tag{49}$$

From (44) and (49) it is easy to find the parameter α:

$$\coth\alpha = \frac{1}{\zeta\sqrt{2\overset{0}{\Omega}_\Lambda}}\sqrt{3\frac{3+2\zeta}{4+3\zeta}}\left(1 \pm \sqrt{1 + \frac{2}{3}\zeta\left(1 - \overset{0}{\Omega}_\Lambda\right)} + \zeta\right). \tag{50}$$

The expression for the scale factor $a\left(t\right)$ is obtained by integrating (48):

$$\frac{a}{a_0} = \left[\cosh At + \coth\alpha \sinh At\right]^{\frac{1+\zeta}{4+3\zeta}} \left[\frac{\sqrt{\frac{cth\alpha+1}{cth\alpha-1}}e^{At} - 1}{\sqrt{\frac{cth\alpha+1}{cth\alpha-1}} - 1}\,\frac{\sqrt{\frac{cth\alpha+1}{cth\alpha-1}} + 1}{\sqrt{\frac{cth\alpha+1}{cth\alpha-1}}e^{At} + 1}\right]^{\frac{1-(\zeta+1)\sqrt{\frac{6\overset{0}{\Omega}_\Lambda}{(4+3\zeta)(3+2\zeta)}}\coth\alpha}{\sqrt{cth^2\alpha-1}}}. \tag{51}$$

The deceleration parameter q can be found combining the equations (36) and (35):

$$q = 6\overset{0}{\Omega}_\Lambda \frac{1+\zeta}{3+2\zeta}\left(\frac{H_0}{H}\right)^2 - \frac{\psi}{H} - 2. \quad (52)$$

The energy contribution of the scalar field at an arbitrary time is determined from the relation

$$\Omega_y = 1 - \Omega_\Lambda = 1 - \overset{0}{\Omega}_\Lambda \left(\frac{H_0}{H}\right)^2 = \frac{\zeta}{6}\left(\frac{\psi}{H}\right)^2 - \frac{\psi}{H}. \quad (53)$$

where

$$\Omega_\Lambda = 1 - \overset{0}{\Omega}_\Lambda \left(\frac{H_0}{H}\right)^2 = \frac{\zeta}{6}\left(\frac{\psi}{H}\right)^2 - \frac{\psi}{H}. \quad (54)$$

We will not consider the case of (42), as in the problem under consideration $\tanh\alpha$ and $\coth\alpha$ take the same numerical value (50) which is greater than one.

In figures 5-8 we have displayed the time dependence of the functions $H(t)/H_0$, $q(t)$, $\Omega_y(t)$ and $a(t)/a_0$. As before, the time is measured in units of t' defined as:

$$At = H_0\sqrt{6\overset{0}{\Omega}_\Lambda \frac{4+3\zeta}{3+2\zeta}} = t'\sqrt{\frac{2(4+3\zeta)}{3(3+2\zeta)}}, \quad (55)$$

(in the Einsteinian limit one has $At = t'$).

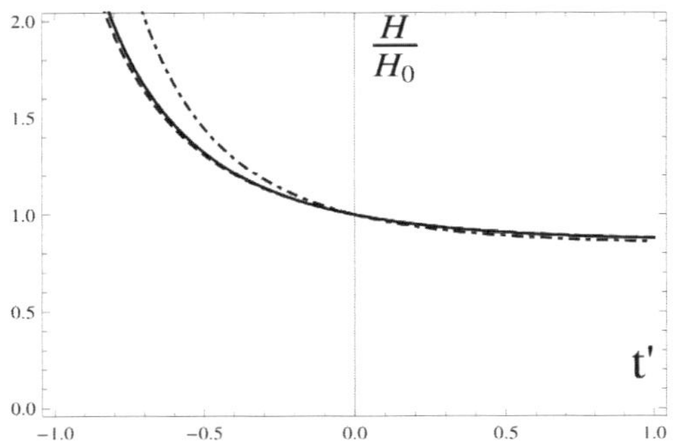

Figure 5. Time dependence of $H(t')/H_0$ for $\zeta = 50$ (upper curve in the left sector), $\zeta = 500$ and $\zeta = 1000$ (upper and lower curves in the converging pair).

5. Conclusion

As a result, for the time dependence of the parameters in the model of the Universe similar qualitative features are obtained as in [8]. The quantitative differences are related to the form of the potential energy for the scalar field with general conclusion confirming the essential dependence of the influence of scalar field from the interaction parameter ζ [12]. On the example of the Universe acceleration (see figure 3) it is clearly seen that for large positive values of ζ (of the order 10^3) the acceleration practically coincides with that for General Relativity. For smaller values of ζ, the influence of the scalar field on the expansion rate is stronger and this leads to the corresponding decrease of q.

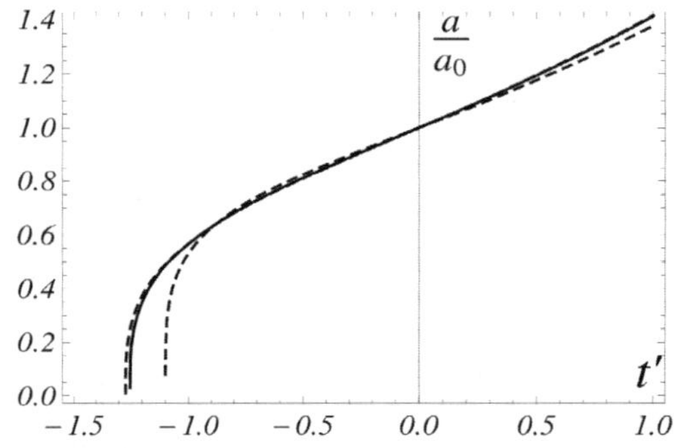

Figure 6. Time dependence of the scale factor $a(t')/a_0$ for the values $\zeta = 50$ (dashed curve), $\zeta = 500$ (full curve in the converging pair), $\zeta = 1000$ (dashed curve in the converging pair).

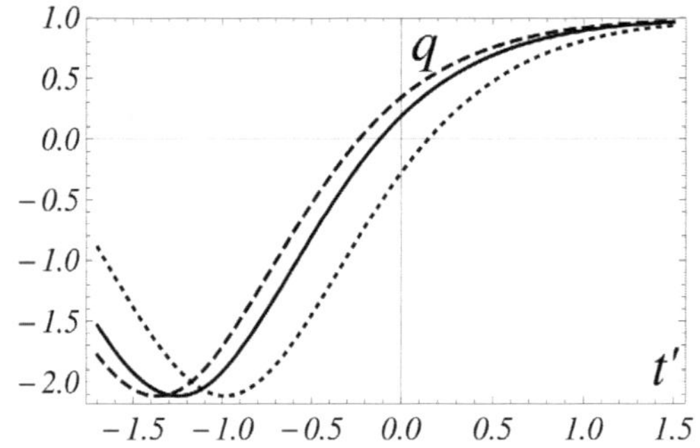

Figure 7. Time dependence of $q(t')$ (in units $5 \cdot 10^9$ years) for $\zeta = 500$ and for $\overset{0}{\Omega}_\Lambda = 0.8$ (dashed curve), $\overset{0}{\Omega}_\Lambda = 0.75$ (full curve), $\overset{0}{\Omega}_\Lambda = 0.6$ (dotted curve).

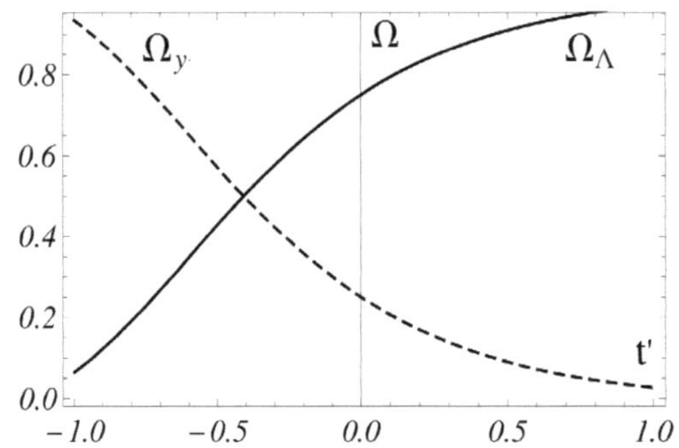

Figure 8. Time dependence of the energy contributions $\Omega_y(t)$ (dashed curve) and $\Omega_\Lambda(t)$ (full curve) for $\zeta = 500$.

For nonminimally coupled scalar field in the presence of $\varphi = \Lambda$ we consider the proper representation of the Jordan modified theory, taking into account the presence of a cosmological scalar $\varphi = \Lambda$. The results obtained for different values of the dimensionless parameter ζ (50, 500, 1000) are presented in figures 5- 8. These results show that the behavior of the curves for $\zeta = 500$ and $\zeta = 1000$ is practically indistinguishable, in contrast to the case $\zeta = 50$. This confirms the proximity of the Jordan theory and General Relativity for large values of ζ [14]. The results described above show that the role of scalar fields in variants with small ζ is different from that in General Relativity, providing a possibility to adjust the deceleration parameter q in comparing with the observational data. Figure 7 shows the time dependence of the deceleration parameter for different values of the energy contribution of the cosmological scalar. Figure 8 displays the energy contributions of the scalar field and the cosmological scalar typical for all versions of the theory. In comparing the obtained results with other models for a flat universe within tensor-scalar theory [15] a well-defined tendency of development is observed: the energy contributions of ordinary matter and the scalar field steadily converge to zero, whereas the contribution due to the presence of the cosmological parameter grows.

The work was supported by State Committee Science MES RA, within the frame of the research project No. SCS 13-1C040.

References

[1] Weyl H 1923 *Raum, Zeit, Materie* (Berlin: Springer-Verlag)
[2] Pauly W 1981 *Theory of Relativity* (New-York: Dover Publications)
[3] Petrov A Z 1966 *New Methods in General Relativity Theory* (Moscow: Nauka).
[4] Dicke V 1962 Phys. Rev. **125** 2163.
[5] Clifton T Ferreira P G Padilla A and Skordis C 2012 Phys. Rept. **513** 1
[6] Jordan P 1955 *Scwerkraft und Weltall* (Braunschweig).
[7] Avakyan R M Harutyunyan G H and Papoyan V V 2005 Astrophysics **48** 455.
[8] Abrahamyan L Avakyan R M Harutyunyan G H Hovsepyan A V and Chubaryan E V 2013 Astrophysics **56** 471.
[9] Avakyan R M Harutyunyan G H Hovsepyan A V and Kotanjyan A S 2013 Astrophysics **56** 617.
[10] Kamke E 1976 *Handbook on Ordinary Differential Equations* (Moscow: Nauka).
[11] Avakyan R M Harutyunyan G H 2005 Astrophysics **48** 633.
[12] Harutyunyan G H and Papoyan V V 2002 Physics of Particle and Nuclei **33** (7) 114.
[13] Avakyan R M Harutyunyan G H and Papoyan V V 2008 Astrophysics **51** 151.
[14] Will C M 1993 *Theory and Experiment in Gravitational Physics*, (Cambridge: Cambridge University Press)
[15] Avakyan R M Harutyunyan G H and Ovsepyan A V 2010 Astrophysics **53** 317.

Journal of Physics: Conference Series **496** (2014) 012021 doi:10.1088/1742-6596/496/1/012021

Gravitational collapse and the vacuum energy

M. Campos

Physics Department, Roraima Federal University, RR Brasil

E-mail: `miguel.campos@pq.cnpq.br`

Abstract. To explain the accelerated expansion of the universe, models with interacting dark components (dark energy and dark matter) have been considered recently in the literature. Generally, the dark energy component is physically interpreted as the vacuum energy of the all fields that fill the universe. As the other side of the same coin, the influence of the vacuum energy on the gravitational collapse is of great interest. We study such collapse adopting different parameterizations for the evolution of the vacuum energy. We discuss the homogeneous collapsing star fluid, that interacts with a vacuum energy component, using the stiff matter case as example. We conclude this work with a discussion of the Cahill-McVittie mass for the collapsed object.

1. Introduction

In recent years there have been a number of important discoveries relevant to the fields of cosmology and relativistic astrophysics. A couple months ago the Nobel prize for 2013 was awarded for the prediction and discovery of the Higgs boson. The Higgs boson has a special place in the inflationary model of the early universe. The research of Vera Rubin, which predicted the presence of dark matter in the galaxies through the study of the orbital velocities of the interstellar matter in the galaxies, has been substantiated through the discovery of the accelerated expansion of the universe. Indeed, the data from distant supernovae that took place at two different times provide the evidence for the accelerated expansion of the universe.

Within the framework of general relativity, we must provide, quite generally, a negative pressure to the cosmic fluid in order to explain the increase of the expansion velocity of the universe. Among the several models that can explain the accelerated expansion of the universe the most popular one is the Λ-CDM (cosmological constant plus cold dark matter) model. Alternative models consider a time-dependent cosmological term, in order to avoid the discrepancy between the values of acceleration measured by the cosmological observations and those estimated from the quantum field theory. Independent of the details of these models, the cosmological term in the framework of the general relativity theory is generally interpreted as the vacuum energy of the all fields that fill the universe.

We view the cosmic expansion and the gravitational collapse as different sides of the same coin. From this standpoint, a study of the influence of the vacuum energy on the gravitational collapse of a fluid star emerges naturally. Thus, in this contribution, we study the final fate of a collapsing star in a model which contains vacuum energy and a matter fluid. We will not specify the matter components individually. Apart from this, we will find the collapsed mass using the Cahill-McVittie definition.

2. Basic equations of the model

In this study we assume that the space-time is homogeneous and isotropic and is governed by the metric

$$ds^2 = dt^2 - R^2\left(dr^2 + r^2 d\Omega^2\right), \tag{1}$$

where $d\Omega^2 \equiv d\theta^2 + \sin^2\theta d\varphi^2$, and the function R depends only of the time. In this aspect we follow the pioneering collapse model described by Oppenheimer [1].

The exterior star space-time does not have a direct influence on the collapse process. Hence, the interior space-time of the star is the important geometric object of our study, which is governed by the metric given by Eq. (1).

The appearance of the apparent horizon is the main characteristics that decides on the final outcome of the collapse process which could be either a black hole or a naked singularity. If the apparent horizon is formed before the singularity, the final fate of the collapse process is a black hole. In the opposite case we have the formation of a naked singularity.

Let us begin this work by considering the Einstein's field equations. For a fluid star with pressure and density and for the space-time described by the metric above, they can be written as:

$$G^{\mu\nu} = \frac{8\pi G}{c^4}\left[T^{\mu\nu} + \frac{\Lambda}{8\pi G}g^{\mu\nu}\right], \tag{2}$$

where $G^{\mu\nu}$ is the Einstein tensor, $T^{\mu\nu}$ is the energy-momentum tensor, G is the newtonian gravitational constant, and c is the light velocity. On the other hand, according to the Bianchi identities the covariant derivative of the Einstein tensor is identically null ($G^{\alpha\beta}_{;\beta} = 0$). Consequently, the Einstein field equations can be written in the form of a conservation equation

$$u_\mu T^{\mu\nu}_{;\nu} = -u_\mu\left(\frac{\Lambda g^{\mu\nu}}{8\pi G}\right)_{;\nu}, \tag{3}$$

where the coupling of the stellar fluid to the vacuum component assumes the form

$$\dot{\rho}_f + 3\frac{\dot{R}}{R}\left(\rho_f + P_f\right) = -\dot{\rho}_v, \tag{4}$$

where u_μ, ρ_f, ρ_v and P_f are the four-velocity of an element of the star fluid, the energy density of the material component, the vacuum energy density (ρ_Λ), and the fluid's material pressure, respectively. For the energy momentum tensor of the fluid we adopt the popular perfect fluid from

$$T^{\mu\nu} = (\rho + P)u^\mu u^\nu - Pg^{\mu\nu}, \tag{5}$$

and we assume the usual form of the equation of state for the vacuum component $P_\Lambda = -\rho_\Lambda$. This form was introduced by Y. Zeldovich [2] in the discussion of the properties of a Higgs field. An interesting discussion of the inclusion of the Higgs field into the theory of elementary particles and the vacuum energy can be found in Ref. [3].

At this point we can write Einstein's field equations in the following form

$$\frac{8\pi G}{c^4}\rho_f + \Lambda(t) = 3H^2, \tag{6}$$

$$\frac{8\pi G}{c^4}p_f - \Lambda(t) = -2\dot{H} - 3H^2, \tag{7}$$

where H is the Hubble function ($H = \dot{R}/R$).

To integrate Einstein's field equations we need to establish an expression for the vacuum component. This is in full analogy to the $\Lambda(t)$-CDM cosmological models, which require a law for the decay of the vacuum.

3. Gravitational collapse and the vacuum energy

There are a large number of models in the literature, which propose different $\Lambda(t)$ decay laws, see for example Refs. [4], [5], and citations therein. In this work, we begin our study by using the law proposed by Wang and Meng [6]. The decay law for the vacuum energy that interacts with the matter fluid, given by Eq. (4), is model independent. The authors assume that the energy density of the matter component will dilute at a rate that is different from the usual one $\rho_f = \rho_{f0}R^{-3}$, namely $\rho_f = \rho_{f0}R^{-3+\epsilon}$. The parameter ϵ represents the deviation from the standard case, where the interaction between the matter fluid and the vacuum energy does not exist. In some sense, this assumption is independent of the adopted model, but in several cases it can mimic a specific model. For example, substituting for the parameter $\epsilon = 3\beta$, we recover the homogeneous and isotropic collapse of dust, where the decaying vacuum energy is proportional to the square of Hubble function. This model was studied in Ref. [7]. The relation between the approach adopted by Wang and Meng and the model with Λ-term which depends on the square of the Hubble function can be see by integrating the energy conservation equation, given by Eq. (4), assuming that the matter fluid and the vacuum energy components have identical functional dependences.

In this brief study, we will exemplify the collapse process, using the equation of state $P_f = \rho_f$, which corresponds to the "stiff" matter. Naturally, this equation of state is more realistic than that of the dust fluid form the point of view of the collapse process. Hence, redefining the parameter $\epsilon = 6\beta - 3$, we obtain the collapse process for the stiff matter case, with the decay of the vacuum energy proportional to the square of the Hubble function, namely $\Lambda = 3\beta H^2$.

With this dependence for the vacuum energy density, and using the Einstein field equations, a direct integration gives

$$R(t) = R_i[3(1-\beta)H_i(t_c - t)]^{\frac{1}{3(1-\beta)}} , \tag{8}$$

where t_c is the collapse time and we used the initial conditions $R(t=0) = R_i$, $H(t=0) = -H_i$. In the Fig. 1 we display the influence of the vacuum component on the scale factor. We can

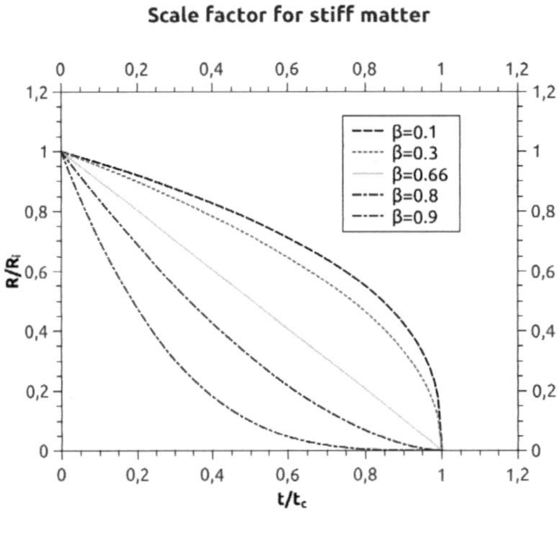

Figure 1. The evolution of the scale factor for stiff matter. We consider different values for the β-parameter. Note the interesting change of concavity that occurs at $\beta = 2/3$. The concavity is linked to the second derivative of the scale factor and it correspond to the acceleration of the fall of comoving fluid element.

write an expression for the collapse time, using the scale factor above and recalling that we are dealing with stiff matter case, in the following form

$$t_c = \frac{H_i^{-1}}{3(1-\beta)} \,. \tag{9}$$

Next we ask what is in the final stage of the collapse process - a black hole or a naked singularity? To answer this question we look for the apparent horizon formation moment, that are space-like surfaces with future pointing towards the converging null geodesics on both sides of the surface [8]. If the apparent horizon dresses the singularity we have the formation of a black hole. If the apparent horizon is formed after the singularity is reached, we have the appearance of a naked singularity (see Fig. 2).

The condition for the appearance of the apparent horizon is given by [8]

$$K_{,\alpha}K_{,\beta}g^{\alpha\beta} = \left(r\dot{R}\right)^2 - 1 = 0 \,, \tag{10}$$

where $()_{,x} = \frac{\partial}{\partial x}$ and $K(t,r) = rR(t)$.

Since initially the fluid star is not trapped the condition

$$K_{,\alpha}K_{,\beta}g^{\alpha\beta} = \left[r\dot{R}\left(t_i\right)\right]^2 - 1 < 0 \,, \tag{11}$$

must be obeyed, which implies that $0 < K_i H_i < 1$.

Finally, to calculate the collapsed mass we use the Cahill-McVittie definition [9], which is given by

$$m(t,r) = \frac{1}{2}K\left(1 + K_{,\alpha}K_{,\beta}g^{\alpha\beta}\right) = \frac{1}{2}K\dot{K}^2 \,. \tag{12}$$

Summarizing, the apparent horizon condition, which is given by Eq. (10), and the mass function

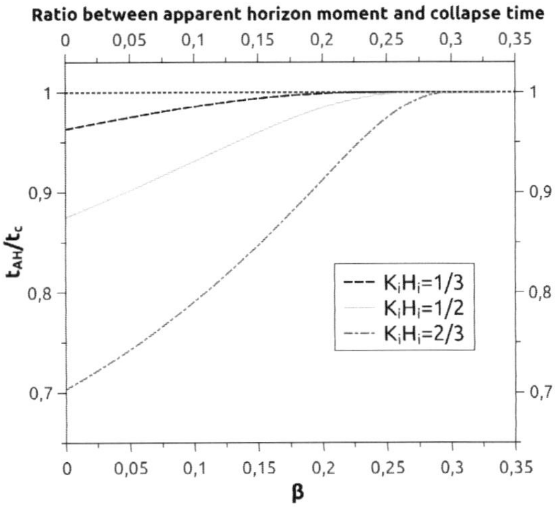

Figure 2. Ratio between the apparent horizon appearance moment and the collapse time for stiff matter. In the considered interval of the β-parameter the cosmic censorship conjecture is valid and the singularity is dressed. Hence, the final fate of the collapse process is a black hole.

Journal of Physics: Conference Series **496** (2014) 012021 doi:10.1088/1742-6596/496/1/012021

for the stiff matter, for a homogeneous and isotropic collapse, are given by:

$$\dot{K} = K_i H_i \left[1 - 3(1-\beta)H_i t_{AH}\right]^{\frac{3\beta-2}{3(1-\beta)}} = 1, \tag{13}$$

$$m(r,t) = \frac{1}{2}K_i^3 H_i^2 \left\{1 - 3(1-\beta)H_i t\right\}^{\frac{2\beta-1}{1-\beta}}, \tag{14}$$

where t_{AH} is the apparent horizon formation moment:

$$t_{AH} = \frac{H_i^{-1}}{3(1-\beta)}\left[1 - (K_i H_i)^{\frac{3(1-\beta)}{2-3\beta}}\right], \tag{15}$$

or equivalently [see Eq. (9)]

$$\frac{t_{AH}}{t_c} = \left[1 - (K_i H_i)^{\frac{3(1-\beta)}{2-3\beta}}\right]. \tag{16}$$

Now we turn to the determination of the collapsed mass. The total collapsed mass can be defined through the value of the mass at the moment of formation of the apparent horizon (t_{AH}). In other words, the collapsed mass is truly the upper limit of the matter that can be accumulated inside the surface determined by the apparent horizon. Therefore, we write the expression for the collapsed mass as

$$M(\tau_{AH}) = \frac{1}{2}K_i^{3/2} H_i^{1/2} (K_i H_i)^{\frac{3\beta/2}{2-3\beta}}, \tag{17}$$

where $M_s(\tau_{AH}) = \frac{1}{2}K_i^{3/2}H_i^{1/2}$ is the mass of a pure stiff fluid (fluid without interaction with the vacuum energy component, that is $\beta = 0$). The expression above can be obtained by substituting Eq. (15) into Eq. (14).

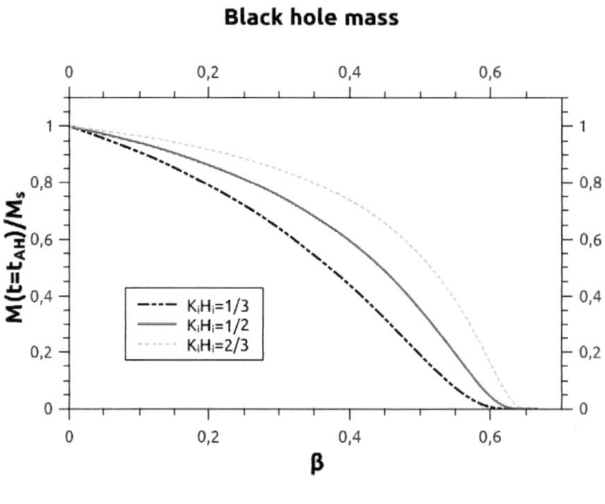

Figure 3. Collapsed mass of the black hole formed. We display the ratio between the mass of the black hole formed and the mass of the pure stiff fluid (without influence of the vacuum component). In the interval considered for the β-parameter the cosmic censorship conjecture remains valid. This is also seen in the graph for the apparent horizon moment (Fig. 2). We adopt the mass definition by Cahill-McVittie. The mass of the black hole formed diminishes with increase of the vacuum energy.

4. Final remarks

We discussed the influence of the vacuum energy on the gravitational collapse of a stiff fluid. We begun the discussion by adopting an approach which is analogous to the one adopted in the work of Wang and Meng [6]. However, because the parameter that represents the deviation from the standard case (which is the case without the influence of the vacuum energy and with null pressure) can mimic several laws for the cosmological term, we were able to find the relation between the Wang-Meng model and the model which has the Λ-term proportional to the square of the Hubble function.

For the stiff matter case, we establish the value for the β-parameter for which the cosmic censorship conjecture is valid, namely $\beta = 2/3$. This value appears in the graph for the scale factor (Fig. 1), where we observed that the change of the concavity of the graph is related to the acceleration of the fall of the comoving fluid element. In the graph for the apparent horizon formation condition (Fig. 2) we see that for $\beta > 2/3$ the singularity is not dressed.

Finally, in the graph for the black hole mass (Fig. 3), we note that for $\beta = 2/3$ we can obtain a black hole with null mass, that is a consequence of the inclusion in the energy momentum tensor of the vacuum component, and more specifically, of the vacuum pressure. Note that, null concavity is attained at the same value of the β-parameter as in the graph for the scale factor. Naturally, the relation between both is an old acquaintance of ours: the Newton's second law that relates the mass and acceleration. The statements above can be made more clear by rewriting the Einstein field equations as

$$\frac{\ddot{R}}{R} = -\frac{4\pi G}{c^4}(\rho + 3P) \; .$$

This expression is null for stiff matter and $\beta = \frac{2}{3}$.

On the another hand, it is also interesting to compare the collapse time for a dust fluid and the stiff matter, which are given by [7] [see Eq. (9)]

$$t_c(\text{dust}) = \frac{2}{3H_1(1-\beta)}, \qquad t_c(\text{stiff}) = \frac{1}{3H_1(1-\beta)}. \tag{18}$$

The difference between the both collapse times, furnish, theoretically, the necessary time for a dust fluid to reach the stiff matter state. Note that the time for dust cloud to become a ball of stiff matter is equal to the time needed for the stiff ball collapse.

References

[1] Oppenheimer J R and Synyder H 1939 *Phys. Rev.* **55** 455
[2] Zeldovich Y 1968 *Soviet Physics* **11** 381
[3] Longair M S 2008 *Galaxy Formation* Springer-Verlag Berlin Heidelberg
[4] Overduin J M and Cooperstock F I 1998 *Phys. Rev. D* **58**, 043506
[5] Overduin J M 1999 *Astrophys. J.* **517**, L1
[6] Wang P and Meng X 2005 *Class. Quant. Grav.* **22**, 283
[7] Campos M and Lima J A S 2012 *Phy. Rev. D* **86** 043012
[8] Hawking S and Ellis G F R 1973 *The large scale structure of space-time* Cambridge University Press Cambridge
[9] Cahill M E and McVittie G C 1970 *J. Math. Phys.* **11** 1382

Journal of Physics: Conference Series **496** (2014) 012022 doi:10.1088/1742-6596/496/1/012022

Microscopic approach to string gas cosmology

Oleg Evnin

Department of Physics, Faculty of Science, Chulalongkorn University, Thanon Phayathai, Pathumwan, Bangkok 10330, Thailand

E-mail: oleg.evnin@gmail.com

Abstract. In this contribution to the proceedings of the Conference on Modern Physics of Compact Stars and Relativistic Gravity in Yerevan, Armenia (September 18-21, 2013), I review recent work attempting to give a fundamental definition to string evolution in a dynamical, fully compact universe, and present a sketch of how the resulting formalism can be used for addressing questions of phenomenological significance in the field of string gas cosmology.

1. Introduction

Considerations of the dynamics of string gases in fully compact cosmological spaces have produced a number of attractive phenomenological conjectures, starting with the pioneering work [1] (for reviews, see [2, 3]). Besides providing a possible alternative to the inflationary pardigm in relation to the horizon problem, etc, the set-up is quite unique in its potential ability to explain the number of macroscopic spatial dimensions of the universe, which equals three observationally. As emphasized in [1], the special kinematics of stretched string scattering is three dimensions is likely to enforce expansion of exactly three macroscopic dimensions, no matter what additional compact spatial dimensions are present in the universe. (Note that 'strings' in our present context mean the fundamental quantum relativistic strings of string theory, rather than effective solitonic strings in field theories, for example. Though similar effects may possibly be induced by other stringy states, the work reviewed here focuses on string gas cosmology scenarios based on fundamental strings.)

Since the introduction of the key ideas in [1], much work has been done in an attempt to implement this type of string cosmology scenarios at a more quantitative level, with mixed results. For a sampling of publications, see [4, 5, 6, 7], with more references to earlier work available in the reviews [2, 3]. The most recent publications claim a successful implementation of the scenario, with only three macroscopic spatial dimensions expanding.

A common feature of most of the above considerations is that they rely on a 'macroscopic' thermodynamic description of the string gas (followed by numerical simulations of the cosmological expansion it induces). As one is talking about universes whose size is of order of the string scale and whose energy density is of order of the string scale (corresponding to the Hagedorn temperature of string theory), one is typically dealing with just a few string quanta moving in a string-sized universe. One is then, at best, in a 'mesoscopic' regime, where the validity of thermodynamical treatment would require extra justification. Coupled with the inherent complexity of the string dynamics and the difficulty of giving its precise thermodynamical description, this may give some attraction to a fully 'microscopic' alternative: working with the string gas in terms of pure quantum states.

Journal of Physics: Conference Series **496** (2014) 012022 doi:10.1088/1742-6596/496/1/012022

Even apart from the practical advantages or disadvantages of such a fully microscopic approach to string gases in small compact universes, it is important to ascertain that there is a way to implement this type of fundamental description in lines with the general principles of string theory. As it turns out, naïve applications of the usual perturbative string theory in fully compact spaces fail for a number of reasons. The most prominent of these failures are the infrared divergences arising when all spatial dimensions have been compactified (and related to the special kinematics in 0+1 dimensions). These divergences are intimately related [8] to those emerging in the soliton recoil problem in field theory [9, 10] and the D0-brane recoil problem in string theory [11, 12, 13]. The infrared divergences are a reflection of the gravitational backreaction induced by even one string moving in a compact cosmological space (in a non-compact space, this backreaction does not affect the asymptotic regions, hence the background remains close to the unperturbed one almost everywhere). Due to the large ratio of the Planck and string scales, the divergences only affect the string perturbative expansion starting from the first subleading order (again, this is similar to the recoil problems). One needs to obtain a reformulation of the usual perturbative string theory free from the infrared problems, if one is to have a fundamental microscopic description of strings in a fully compact universe.

The problem of reorganizing the usual string perturbation theory in compact spaces has been considered in [14]. It has been suggested that a functional integral over the modes of the background receiving large backreaction should be introduced in addition to the usual functional integral over string worldsheets. Such a treatment deviates from the usual paradigm of formulating string theory in a fixed classical background, with massless string quanta representing small deformations of this background. One reason for this unusual treatment is that, in the presence of strong backreaction, deviations from the unperturbed background are not small. This type of modified string perturbation theory is modelled on the solution to the D0-brane recoil problem in [13], which is in turn modelled on the Christ-Lee perturbation theory for the field-theoretical soliton recoil problem [10]. Note that, technically, these discussions are phrased for bosonic strings moving in a toroidal universe, though the core issues are completely general and in no way dependent on the presence or absence of supersymmetry, or the detailed topology, but only on compactness of the spatial manifold.

In this contribution, I shall review the formalism of [14], in a way somewhat different from how it was presented in the original publication. I will then give a brief sketch of some practical computations this formalism can be used for.

2. The formalism

I shall start by simply presenting the proposed expression for the evolution amplitude, subsequently providing justifications and clarifications. For a more systematic and detailed approach, one can consult [14].

What amplitudes could one define in our cosmological setting? Since the space is fully compact, the usual notion of S-matrix clearly does not apply. One should therefore resort to some sort of finite-time transition amplitudes. The question of finite-time transition amplitudes in string theory has not been settled even in non-compact Minkowski space, but a formal definition has been given, for example, in [15]. One can try to extend this definition to the case when the background is dynamical rather than fixed. Due to the time-reparametrization invariance of the background space-time, it will only make sense to talk about the amplitudes for the universe to have two cross-sections with specified spatial properties and string contents (without any specification of the 'moments of time' when such configurations occur, since that notion is not reparametrization-invariant). This is very similar to the kind of amplitudes considered in minisuperspace models of quantum cosmology [16, 17, 18, 19].

We shall integrate over the following class of backgrounds for the metric, the 2-form and the dilaton (the justification comes from analyzing which massless field modes contribute to the

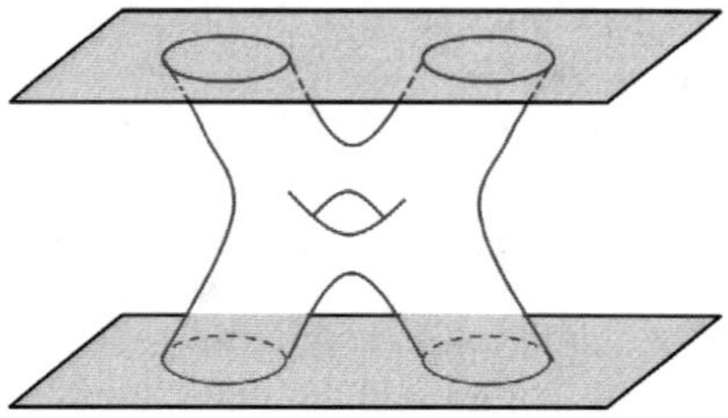

Figure 1. A string worldsheet cut by the two observation moments in a dynamical universe.

infrared divergences in a naïve formulation of perturbative string theory in a fixed space-time background):

$$\begin{aligned} ds^2 &= -G_{00}(t)dt^2 + 2G_{0i}(t)dtdx^i + G_{ij}(t)dx^i dx^j, \qquad x^i \in [0, 2\pi), \\ B_{\mu\nu} &= B_{\mu\nu}(t), \qquad \Phi = \Phi(t). \end{aligned} \tag{1}$$

In the language of the minisuperspace literature, this is a compactified Bianchi I universe coupled to the dilaton and the 2-form. G_{00} and G_{0i} can be gauged away completely, except for a single gauge-invariant number, $\int dt\sqrt{G_{00}}$. We shall suppress $B_{\mu\nu}$ and Φ in the formulas and discussions below for compactness, but it is understood that they are also included, whenever the metric G_{ij} appears. One finally specifies the values of the background fields (1), which we denote $G^{(1)}$ and $G^{(2)}$, on the two observation hypersurfaces, together with the (loop-like) configurations of closed strings, which we denote $\ell^{(1)}$ and $\ell^{(2)}$, on the same observation hypersurfaces. (A pictorial representation of the slice of the worldsheet between the two observation hypersurfaces is given in Fig. 1.) One then defines the following transition amplitude.

$$\langle G^{(2)}, \ell^{(2)} | G^{(1)}, \ell^{(1)} \rangle = \int_{-\infty}^{\infty} dT \int_{\substack{G_{ij}(0)=G^{(1)}_{ij} \\ G_{ij}(T)=G^{(2)}_{ij}}} \mathcal{D}G_{ij}(t)\, e^{iS_{\rm ms}} \left(\prod_{\ell^{(1)}_m \ell^{(2)}_n} \int_{Diff(S^1)} \right) \int_{\ell^{(1)}_m \ell^{(2)}_n} \mathcal{D}X e^{-S_X}, \tag{2}$$

Here, $S_{\rm ms}$ is the low-energy string theory action with the background (1) substituted in it, and S_X is the (non-linear sigma-model) action for a string worldsheet propagating in (1). Both actions can be found in chapter 3 of [20], for instance.

Note that (2) has a rather peculiar structure, compared to ordinary quantum-mechanical transition amplitudes ("integral of the exponential of the action constrained by the boundary conditions"). Namely, one has to integrate over T, the end-time of the experiment, and over all reparametrizations of the boundary loops of the string worldsheet ($\ell^{(1)}$ and $\ell^{(2)}$), which is represented by $\int_{Diff(S^1)}$ in (2). The first of the two integrals is mandated by the time-reparametrization invariance of the background, and indeed it eliminates the unphysical dependence on the observation time. Such amplitude structures have been thoroughly considered in the minisuperspace literature, see [17, 19]. The integral over boundary reparametrizations is mandated by the diff-invariance on the string worldsheet, and eliminates the unphysical dependence on how the boundary loops of the string worldsheet are parametrized. Such integrals have appeared in considerations of off-shell string amplitudes in Minkowski space, see [15].

Having formally defined the amplitude in a closed form we shall now review the problems in conventional perturbative string theory applied to compact spaces that lead one to consider such amplitudes, as well as various consistency checks that the amplitude (2) satisfies.

2.1. Motivation

The functional integration over a class of backgrounds in the basic amplitude (2) may appear as a rather odd departure from the usual perturbative string theory (which only includes a functional integral over worldsheets), and it is important to explain why the usual formulation of string theory fails for compact spaces.

The most conservative approach to string theory in compact spaces would be to consider a toroidal space with a non-compact time and a flat Minkowski metric, and study how strings propagate in this background (this is simply a full spatial compactification of the usual string theory in Minkowski space). Naturally, amplitudes will be given by a sum over worldsheet topologies, with higher topologies (worldsheets with more handles) weighted by higher powers of the string coupling. The first sub-leading correction will be given by worldsheets of toroidal topology, and by the general rules of string theory, these amplitudes must be integrated over the modular parameter of the torus. There is a limit in this modular parameter integration that corresponds to the torus becoming thin and long. In this limit, an infrared divergence will arise in a fully compact target space, invalidating string perturbation theory.

We shall not derive the divergence accurately here, and only give a qualitative hint on how it arises. (The reader may consult [14] for further details.) Basically, the limit of the long thin torus worldsheet can be expressed through sphere diagrams augmented by an additional loop with different string states running in the loop. In particular, there will be contributions from the zeroth Kaluza-Klein (uniform) modes of massless string fields on the spatial torus running in the loop. The contribution of the infrared region of the loop momentum to these amplitudes will simply be proportional to

$$\int\limits_{k\approx 0} \frac{dk}{k^2}, \tag{3}$$

where $1/k^2$ is nothing but the massless particle propagator. This expression will be divergent on account of the propagator singularity at $k = 0$. (If more than one spatial dimension were decompactified, the above integral would be convergent around $k = 0$, because of the measure dk. The case of exactly one spatial dimension is subtle, see [21]. There are no divergences from large values of k in any situation in string theory, because of the known UV-finiteness.)

The divergence we have displayed is very robust, and essentially depends on kinematics in $0 + 1$ dimensions. Identical divergences arise in the D0-brane recoil problem [11, 13].

2.2. Justification

Having observed the divergence trouble in the usual string perturbation theory put on a compact space, one might look for possible cures. An important source of inspiration comes from the Christ-Lee perturbation theory for recoiling field-theoretical solitons [10], which cures divergences of similar algebraic structure and physical origin by reorganizing the perturbative expansion of field theory in such a way that a functional integral over the center-of-mass position of the soliton appears explicitly. Since in field theory the perturbative expansion is derived from the underlying Lagrangian formulation, the justification of the Christ-Lee perturbation theory is exact.

In string theory, one does not have the advantage of being able to derive the perturbative expansion from a more fundamental underlying framework. Nonetheless, one might try to imitate the Christ-Lee perturbation theory in the language of perturbative string theory and see whether it works. For the problem of D0-brane recoil, this was done in [13], and it was shown that the divergences cancel as a result. The amplitude (2) is an attempt to extend the same kind of treatment to the case of strings in a compact target space. (The similarities of the infrared divergences one has to deal with in all these cases have already been pointed out.)

One might find it conterintuitive that an integral over (a class of) background deformations is included in the amplitude (2), whereas it is usually thought that such modes are already described by excitations of strings. The conventional intuition is, however, unlikely to apply to the particular modes we have included. For example, the usual string-theoretic description of

Journal of Physics: Conference Series **496** (2014) 012022 doi:10.1088/1742-6596/496/1/012022

the initial and final states of the zeroth Kaluza-Klein modes of the massless fields would be in terms of massless string quanta in different on-shell states. However, if all spatial dimensions are compactified, zeroth Kaluza-Klein modes have zero momenta in all spatial directions, and hence on-shell their momentum is exactly zero. It is difficult to imagine how one would describe initial and final states of the minisuperspace cosmology (1) in terms of different numbers of massless string quanta at exactly zero momentum. Clearly, something is wrong with the usual string theory conventions when the number of non-compact dimensions becomes small!

2.3. Finiteness properties

The ultimate check for whether the proposed amplitudes (2) make sense is in the way they resolve the infrared issues of the usual perturbative string theory on a compact space.

For the case of D0-brane recoil, which has been used here as a prototype for our treatment of the compact space case, the resolution is quite straightforward [13]. The modification of the formalism we introduce (functional integration over a set of background modes) produces a new perturbatively small divergence on the lowest genus worldsheet that cancels the divergence in the modular integration on the next-to-lowest genus worldsheet inherited from the standard string theory. Such a cancellation pattern is known as the Fischler-Susskind mechanism.

The situation for the present amplitudes (2) is somewhat more subtle, in the sense that they are essentially finite-time transition amplitudes (there is no obvious way to talk about infinite time scattering in a compact space). For finite time transition amplitudes, one cannot expect infrared divergences in the first place (the space is compact and the time is cut off to a finite duration). Hence, amplitudes (2) are automatically finite in this straightforward naïve sense.

If one looks a little deeper, one discovers that there are extremely abrupt dependences on the worldsheet and modular integration cut-offs in different contributions to (2), inherited from the corresponding infrared divergences on an infinite time interval. These abrupt dependences would require cut-offs to be sent to numbers of the sort of 10^{-100} for perfectly reasonable amplitudes before a cut-off-independent limit is reached (in each of the terms of the amplitude separately). However, the dangerous cut-off-dependent terms cancel between the different contributions to the amplitude exactly in the manner of the Fischler-Susskind mechanism described above.

Overall, one discovers that (2) addresses the infrared problems of the usual formulation of string theory on a compact space. Technical details can be recovered from [14].

3. The observables

Armed with the basic definition of transition amplitudes (2), it is natural to ask how they can be used to address questions of phenomenological significance in string gas cosmology, either analytically or numerically.

Even though much of the work in [14] analyzing the validity of the newly constructed formalism was at first sub-leading order in the string coupling (torus topology worldsheets), and hence technically rather involved, one should expect that the most important practical qualitative predictions will emerge at leading order (sphere topology worldsheets) and will be computationally much more straightforward in a number of ways. The general idea is to specify a certain set of strings on the two observation hypersurfaces (Fig. 1) and see how their presence affects the cosmological expansion, i.e., how the amplitudes for cosmological evolution without any strings get modified when strings are present.

Further simplifications arise from the fact that, for transitions between universes whose size is comparable to the string scale, which is the regime relevant in string cosmology, the functional integral over the background modes in (2) is in a semi-classical regime. (This is a reflection of the large ratio between the Planck and string scales in perturbative string theory.) The leading contribution to the amplitude (2) will then arise from strings propagating in a fixed classical background connecting the initial and final background configurations. Furthermore, it is likely that an adequate estimate for the worldsheet functional integral will be given simply be the exponential of the minimal possible area of the worldsheet connecting the initial and final configurations of the strings. The background solutions for Bianchi I gravity-dilaton system are known as Kasner-dilaton solutions, and with the above picture in mind, one should expect the leading string backreaction effects to be expressible through fairly straightforward geometrical quantities in the Kasner-dilaton spacetimes.

It remains to be seen up to which point the computations sketched here can be carried out in practice, and what light they shed, for example, on the question of the number of macroscopic dimensions expanding. (I.e., one may start with a string-sized universe configuration, in which the expansion of all dimensions of the torus is blocked by winding strings, and see how likely it is to discover another cross-section of the same universe, in which a certain number of winding strings have annihilated and a certain number of dimensions have expanded to much bigger sizes.)

4. Acknowledgments

I would like to thank the organizers of the Conference on Modern Physics of Compact Stars and Relativistic Gravity in Yerevan, Armenia (September 18-21, 2013), and specifically Armen Sedrakian. I would also like to thank Jarah Evslin, Chethan Krishnan, Karapet Mkrtchyan and Levon Pogosian for stimulating communication during the conference, and Ben Craps and Anatoly Konechny for collaboration on the research reviewed here. This work has been supported in part by Ratchadaphisek Sompote Endowment Fund.

References

[1] R. H. Brandenberger and C. Vafa, Nucl. Phys. B **316** (1989) 391.
[2] T. Battefeld and S. Watson, Rev. Mod. Phys. **78** (2006) 435, arXiv:hep-th/0510022.
[3] R. H. Brandenberger, arXiv:0808.0746 [hep-th];
Class. Quant. Grav. **28** (2011) 204005, arXiv:1105.3247 [hep-th].
[4] M. Sakellariadou, Nucl. Phys. B **468** (1996) 319, arXiv:hep-th/9511075.
[5] R. Danos, A. R. Frey and A. Mazumdar, Phys. Rev. D **70** (2004) 106010, arXiv:hep-th/0409162.
[6] R. Easther, B. R. Greene, M. G. Jackson and D. N. Kabat, JCAP **0502** (2005) 009, arXiv:hep-th/0409121.
[7] B. Greene, D. Kabat and S. Marnerides, Phys. Rev. D **82** (2010) 043528, arXiv:0908.0955 [hep-th];
Phys. Rev. D **88** (2013) 043527 arXiv:1212.2115 [hep-th].
[8] O. Evnin, Fortschr. Phys. **60** (2012) 998-1004, arXiv:1201.6606 [hep-th]
[9] R. Rajaraman, *Solitons and instantons*, North Holland, 1982.
[10] N. H. Christ and T. D. Lee, Phys. Rev. **D12** (1975) 1606.
[11] V. Periwal and O. Tafjord, Phys. Rev. D **54** (1996) 3690, arXiv:hep-th/9603156.
[12] S. Hirano and Y. Kazama, Nucl. Phys. B **499** (1997) 495, arXiv:hep-th/9612064.
[13] B. Craps, O. Evnin and S. Nakamura, JHEP **0612** (2006) 081, arXiv:hep-th/0609216.
[14] B. Craps, O. Evnin and A. Konechny, JHEP **1310** (2013) 177, arXiv:1308.4016 [hep-th].
[15] A. Cohen, G. Moore, P. Nelson, and J. Polchinski, Nucl. Phys. B **281** (1987) 127;
Nucl. Phys. B **267** (1986) 143.
[16] J. B. Hartle and S. W. Hawking, Phys. Rev. D **28** (1983) 2960.
[17] J. J. Halliwell, in Jerusalem 1989, Proceedings, Quantum cosmology and baby universes, p. 159-243, arXiv:0909.2566 [gr-qc].
[18] C. J. Isham, in proceedings of *Salamanca 1992*, pp. 157-287, arXiv:gr-qc/9210011.
[19] D. Marolf, Phys. Rev. D **53** (1996) 6979, arXiv:gr-qc/9602019.
[20] J. Polchinski, *String theory* (CUP, 1998) v. 1.
[21] B. Craps, O. Evnin and S. Nakamura, JHEP **0701** (2007) 050, arXiv:hep-th/0608123.

Journal of Physics: Conference Series **496** (2014) 012023 doi:10.1088/1742-6596/496/1/012023

Giant monopoles as a dark matter candidate

Jarah Markar Evslin

TPCSF, Institute of High Energy Physics, CAS, YuQuanLu 19B, 100049 Beijing, China

E-mail: jarah@ihep.ac.cn

Abstract. The most prominent successes of weakly interacting massive particle (WIMP) models of dark matter are at comoving scales of megaparsecs. At kiloparsec scales they face challenges in explaining the density profiles, abundances and phase space distributions of satellite galaxies in our local group. An alternate dark matter candidate, the giant 't Hooft-Polyakov monopole, is proposed which may share the large distance successes of WIMPs while evading their short distance problems. These are classical field theory solutions of a dark sector including a nonabelian gauge field, an adjoint scalar field and fundamental fermions. In such models each halo consists of a single monopole characterized by a conserved integer charge.

1. Successes of WIMP models

The most successful dark matter model is the cold dark matter (CDM) WIMP. These are nonrelativistic particles which interact gravitationally at long distances and via weak interactions at short distances. While experimental searches for their weak interactions have yielded no convincing signals [1], the consequences of their gravitational interactions are regularly verified.

Perhaps the most famous of these confirmations comes from the bullet cluster: Two clusters of galaxies have passed through each other, leaving most of their luminous matter in the form of a hot plasma at the collision point. Via gravitational lensing it was shown in Ref. [2] that most of the matter continued past the collision point, separating from the plasma. The fact that the dark matter separated from the visible matter means that it has its own inertia. This observation eliminates the most elegant modified gravity explanations of dark matter. In addition, the fact that the dark matter halos of the clusters passed through each other yields a strong upper bound on the dark matter self-interaction cross section [3]. It has been claimed [4] that this observation rules out a large class of models in which, like the model presented below, dark matter consists of classical field theory solutions such as the scalar condensate model of Ref. [5].

Once structure formation begins, CDM WIMPs are nonrelativistic and only interact gravitationally. Based on numerical simulations it was claimed in Ref. [6] that, as a result of these two facts, the shape of a pure dark matter halo assumes a universal form which is well-approximated by the NFW profile. Observations of galaxy clusters confirm that the NFW profile indeed provides an excellent approximation to the dark matter density profile [7].

The CMB power spectrum observed up to multipoles of about $l = 2000$ [8] is just what would arise from an initial power law spectrum of adiabatic fluctuations in a mixture of baryonic matter, photons and a nonrelativistic fluid in a universe with a cosmological constant and massless neutrinos. Around $l = 2000$, combining CMB data with other data, the interpretation becomes more controversial [9] and above $l = 3000$ the primordial fluctuations have been erased by Silk damping [10]. Therefore CMB observations demonstrate that up to redshifts of about

10^4 and at angular scales down to about $l = 2000$ dark matter behaves as a perfect fluid with an equation of state of about $w = 0$.

The above evidence for dark matter is subject to various assumptions and unknown systematics. For example, since the power spectrum of primordial perturbations is entirely unknown there is no way to estimate the error caused by its deviation from a power law. While the bullet cluster seems to be a convincing demonstration of the existence of particulate dark matter, it is worth noting that in the case of the merger Abell 520 the dark matter coincides with the plasma. If the dark matter was stopped by a self-interaction, the cross section for this process would need to be appreciably larger than the upper bound established using the bullet cluster [11]. Similarly the dark matter profile of a galaxy cluster depends upon its unknown formation history. We have chosen not to discuss type Ia supernovae as the standardizability of these candles is an empirical observation, still not supported by a calculation, and so again the errors cannot be estimated. Indeed, assuming ΛCDM the cosmic distance ladder based on standardizable candles [12, 13] is now in about 3σ disagreement with CMB observations [8].

However one cosmological probe is independent of any known systematics or cosmological assumptions, the baryonic acoustic oscillation peak (BAO) [14]. This is the location of an isolated peak in the position space two point function of the matter distribution. As the matter moves nonrelativistically, the location of this peak in comoving coordinates is time independent. Thus, while the size itself depends on the cosmological model, once the size is measured at any redshift, it will be the same in comoving coordinates at all redshifts. Furthermore, the location of the peak is isotropic in the sense that the peak of the matter distribution in the radial and angular directions at the same redshift agree. Thus at each redshift one can measure two quantities, the angular and radial BAO size. The knowledge that each scale is the same in comoving coordinates then allows one to robustly determine the expansion of the universe as a function of time. Using Einstein's equations this rate of expansion yields the pressure and density of the universe as a function of time. If one further assumes that the universe consists of various components whose stress energy tensors are separately conserved and that dark matter corresponds to a nonrelativistic fluid ($w = 0$) then one obtains the average dark matter density at each time.

The dark matter density so obtained, going back to about $z = 2.3$ [15], is consistent with that obtained from the CMB [8] corresponding to redshifts going back to 10^4. This implies that the quantity of dark matter in a sufficiently large comoving volume should have been roughly constant since redshift $z = 10^4$. While this result is inconsistent with modified gravity models of dark matter, it is consistent with models of stable particles or other such compact objects.

2. Challenges faced by WIMPs

The successes of WIMPs have one thing in common. They are all at large distance scales, comoving scales of Mpc. What about short distance predictions?

One robust prediction of CDM WIMPs is that the Milky Way should have at least of order 10^4 satellite dark matter halos [16, 17]. Only about 25 such satellites have so far been observed, in particular the lightest ones seem to be missing. In itself this is not a contradiction, current lensing probes are not yet sufficiently sensitive to observe these halos. It may be that, since the lightest halos lead to shallow gravitational potential wells, the primordial gas in these wells was blown away by ultraviolet radiation from reionization [18, 19] or cosmic ray pressure [20] before it could condense into stars, or that the supernovas of the first generation of stars blew out of all the gas so that no further generations formed [21]. Whether these mechanisms really would have suppressed star formation in these light halos depends on many unknowns, such as the ultraviolet and cosmic ray fluxes in the distant past and the efficiency with which supernovae can transfer their energy to the gas. Even if these mechanisms may evade star formation in small satellites and so explain the nondetection of light halos, there also appear to be missing heavy halos [22] for which such mechanisms are of no avail [23]. One can eliminate these heavy halos from simulations by reducing the mass of the Milky Way [24, 25] but this leads to tension with theoretical models of the mass distribution of the Milky Way [26]. Also such a low Milky Way mass would imply that the dwarf galaxy Leo I [27] and the Magellanic clouds [28] are gravitationally unbound, which makes their observed proximity to the Milky Way unlikely.

Pure dark matter simulations yield a second prediction, that the density profile of a dark matter halo diverges at small radii [6]. Such divergent halos are said to be cusped. Recent simulations [29] suggest that the power law divergence be replaced by a more complicated Einasto form [30], but in practice the diminishing divergence occurs at such small radii that it will not have any effect on the discussion below. Observed halos do not appear cusped, but they are not pure dark matter, they also contain stars and often gas. Could the baryonic physics of stars eliminate the cusps? A number of simulations have indicated that in principle the outflow of gas from a supernova may turn a cusped profile into a cored profile [31, 32]. However this transformation only works in the supernovae is able to efficiently transfer its energy, which requires these models to use a very high threshold gas density for star formation [33]. There has been considerable debate concerning whether such a threshold is reasonable.

While it may be that baryonic physics can transform the cusps into cores in some dark matter halos, clearly such a transformation is impossible if the baryon density is insufficient. The main proponents of this mechanism [34] find, as summarized in their Eq. (1), that it can totally remove cusps in galaxies with more than $10^9\ M_\odot$ of stars whereas the central density diverges at least as $1/r$ if the mass is beneath $4 \times 10^6\ M_\odot$. Thus smaller galaxies, such as the majority of the dwarf spheroidal galaxies in our local group, will unavoidably have cusped dark matter halos if dark matter is made of WIMPs. Are these cusps in contradiction with observations? These light galaxies are not rotating, so their rotation curves can not be used to determine their dark matter halo profiles. The stars are dispersion supported, so in principle the halo profiles can be determined from the Jeans equation [35]. However, as a result of a degeneracy [36] this is not sufficient to determine the halo density. Instead, if a few basic assumptions are included [37], the Jeans equation yields the amount of mass within the half-light radius [38] of each stellar population [39]. Applying this method to galaxies with multiple stellar populations one can obtain the densities at distinct radii. This has been done for the Fornax [39] and Sculptor [40] dwarfs and in both cases the result is consistent with a cored profile and not a cusped profile, although the stellar mass of Sculptor is only $2 \times 10^6\ M_\odot$ and so the results of [34] suggest a cusped profile. Furthermore, assuming it not to be recently accreted, the existence of old substructure in the Fornax [41], Ursa Minor [42] and Sextans [43] dwarfs is inconsistent with a cusped profile, as the resulting tidal force would both delocalize the substructure and also pull it towards the center of the halo.

Recently WIMP models have faced new challenges arising from the concentration in phase space of dwarf satellites of the Milky Way [44] and Andromeda [45] as well as the planes of Ref. [46] and the filamentary structure of [47] and the abundance of pairs in Refs. [48, 49]. In CDM WIMP simulations such concentrations do not appear, although perhaps they may be caused by recent mergers [50].

3. Proposal: giant monopoles

The standard model contains no dark matter candidates which are consistent with the tight bounds on MACHOs. As a result, any explanation of dark matter must introduce either new and exotic couplings or else new fields. We choose the later path, introducing a dark SU(2) gauge field, a scalar Higgs field which transforms in the adjoint representation of this dark SU(2) and also an undetermined number of fundamental fermions. These fields are coupled via the same gauge-invariant and renormalizable terms that enter in the standard model. These fields all inhabit a dark sector, they are in addition to the usual standard model fields. They will interact

Journal of Physics: Conference Series **496** (2014) 012023 doi:10.1088/1742-6596/496/1/012023

with the standard model fields gravitationally, although renormalizability and gauge invariance also allow the Higgs portal interaction of Ref. [51].

This dark sector admits a stable classical field theory solution, the 't Hooft-Polyakov monopole. While it is known that a small fraction of dark matter consists of active, massive neutrinos, following Ref. [52], we will propose that the rest of the dark matter in our universe consists of these monopoles. Each monopole, in a steady state and up to a discrete choice that we will discuss below, is completely characterized by a single integer, its charge Q. We will propose that each dark matter halo consists of a *single* monopole of some charge Q, thus recovering the observational fact that galaxies tend to inhabit one parameter families. This observation has consistently posed a challenge to WIMP models which naturally yield two parameters, corresponding intuitively to a mass and a temperature. Note that, since halos are extremely large, so are the monopoles. The monopoles will have diameters measured in parsecs and masses measured in solar masses, in contrast with the more familiar and smaller GUT monopoles. In the rest of this talk we investigate the consequences of this proposal.

4. Giant monopoles at large scales

Does the giant monopole model reproduce the successes of WIMPs at large scales? At scales much larger than the monopole, the monopoles can be treated as point particles. As we will discuss later, at long distances only their gravitational interactions will be significant. Therefore their behavior will be indistinguishable from WIMPs. In particular, the profiles of galaxy clusters will be the same as will the relative BAO scales.

The overall BAO scale is determined by the speed of sound of the primordial plasma, which is unchanged in our scenario, and by the age of the universe at recombination. What is the age of the universe at recombination in this model? First of all we can calculate the time when the monopoles form using the usual Kibble mechanism. The monopoles form when the scalar Higgs field unfreezes, which occurs when the Hubble length grows beyond r_1, the inverse tachyonic mass of the Higgs field. The scale r_1 is just the radius of the core of a dark matter halo which is about 1 kpc, so monopoles form when the Hubble radius is about one inverse kpc. This is before matter-radiation equality, so before the monopoles have formed the dark matter, in either WIMP or monopole models, is an insignificant part of the total energy budget of the universe. At matter-radiation equality, when the contribution of dark matter to the energy density of the universe becomes relevant, the monopoles have already formed and so, as the monopoles are stable, it is reasonable that the mass in each comoving volume is then fixed as in a WIMP cosmology. In this case the energy density of monopoles before recombination is the same as that of WIMPs in the standard cosmological model, and so recombination happens at the same moment. This implies that the overall scale of the BAO feature will be the same as in WIMP cosmologies, in agreement with observations. Note that, had the cores of galaxies today been larger by a factor of ten, the monopoles would have formed too late and this model would have been falsified.

What about the CMB? As the monopoles have formed before matter-radiation equality, they may contribute the evolution of the power spectrum. Once they are fully formed, they will have the correct nonrelativistic equation of state $w = 0$, although at early times w begins at -1. Were they a fluid? Evolving back the density of dark matter today one finds about 10^2 monopoles in each $l = 2000$ volume [52], just enough for the fluid approximation at these multipoles.

The monopoles are particles with inertia and so, as is observed in the bullet cluster, will tend to continue straight through a collision unless scattered. The bullet cluster bound on the cross section [3] of about 1 cm^2/g, for a $10^{12} M_\odot$ halo corresponds to the cross section of a black disk of radius equal to about 5 kpc. While we have not yet computed the cross sections for our solutions, the dark force interactions will be stronger than the gravitational interactions by only about 5 orders of magnitude and so it is plausible that this bound is satisfied.

5. Giant monopoles at small scales

We have seen that it is plausible that giant monopoles share the same large scale successes as WIMPs, although this is contingent on details of their formation before recombination. The real strength of the monopole models is at small scales, where they effortlessly solve the challenges described in Sec. 2. First of all, the missing satellite problem. The missing light satellite problem is essentially the observation that satellite galaxies are observed to have a minimum mass, in contradiction with the universal expectation from WIMP simulations. On the other hand, Dirac has shown that monopole solutions always have a minimum mass, corresponding to the $Q = 1$ monopole. Below we will fix the parameters of our theory such that the $Q = 1$ monopole has the observed minimum mass, and thus the missing halo problem will be solved.

What about the cusp problem? The profile of a 't Hooft-Polyakov monopole in the presence of gravity, at least in the case $Q = 1$, is well known. There is a central core, of radius r_1 with a reasonably constant density and a nearly zero Higgs and gauge field. Thus the core problem is automatically solved. At the radius r_1 the Higgs field turns on, as r_1 is proportional to the inverse Higgs mass, breaking the SU(2) gauge symmetry to U(1). Let r_2 be the inverse W boson mass, where the W boson is not the standard model W boson but the gauge field which has been Higgsed. Then, between r_1 and r_2 the gauge fields are essentially zero but the Higgs field enjoys a topologically nontrivial winding which gives this region a density proportional to $1/r^2$. This automatically yields the observed flat rotation curves. At higher radii the gauge fields also become nontrivial.

And the phase space correlations? The dark forces in these models are more attractive than gravity in some regimes, and so one expects more binding and so more correlations. Whether this increase agrees with observations will be studied in the near future.

What are the parameters of the model? Dwarf spheroidal galaxies (dSphs) are among the purest concentrations of dark matter in the universe. By using Refs. [38, 37] to fix to their masses inside of their half-light radii and assigning each dSph a charge Q we have fit the Higgs VEV to be $v \sim 10^{14}$ GeV, the quartic Higgs selfcoupling to be $\lambda \sim 10^{-96}$ and the gauge coupling to be less than 10^{-48}. The value of v is tantalizing, the inputs were astrophysical scales and the output is the leptogenesis scale.

6. Challenge

Unfortunately these monopoles repel and so the charge $Q > 1$ halos are unstable. This may rule out our model. Then again, protons repel but visible matter is mostly made of protons, as the repulsion at small distances is canceled by neutrons and at large distances is screened by electrons. The monopoles only repel at long distances. So what are the analogs of the electrons? Electrons carry the opposite charge from protons but cannot annihilate with protons as they carry a flavor quantum number and the lightest state for a decay product, the neutron, is too massive for the decay to be kinematically allowed. Similarly such a flavor quantum number for the monopoles is an automatic consequence of our fermionic couplings [53]. The masses of the various flavors of monopoles can be adjusted by choosing the Yukawa couplings. We propose to include light antimonopoles of a different flavor which screen the long distance repulsion of our monopoles. If such a screening cannot be made to work, our proposal will be excluded.

Acknowledgments

Many of the results presented in this talk were obtained in collaboration with Sven Bjarke Gudnason. JE is supported by the Chinese Academy of Sciences Fellowship for Young International Scientists grant number 2010Y2JA01.

References

[1] Akerib D S,*et al.* [LUX Collaboration], First results from the LUX dark matter experiment at the Sanford Underground Research Facility 2013 *Preprint* arXiv:1310.8214
[2] Clowe D, Gonzalez A and Markevitch M 2004, *Astrophys. J.* **604** 596
[3] Randall S W, Markevitch M, Clowe D, Gonzalez A H, Bradac M 2008, *Astrophys. J.* **679** 1173
[4] Slepian Z, Goodman J 2011, *MNRAS* **427** 839
[5] Matos T, Guzman F S 2000, *CQG* **17** L9
[6] Navarro J F, Frenk C S and White S D M 1997, *Astrophys. J.* **490** 493
[7] Lin Y T, Mohr J J and Stanford S A 2004, *Astrophys. J.* **610** 745
[8] Ade P A R *et al.*, Planck 2013 results. XVI. Cosmological parameters *Preprint* arXiv:1303.5076
[9] Hou Z, *et al.*, Constraints on Cosmology from the Cosmic Microwave Background Power Spectrum of the 2500-square degree SPT-SZ Survey *Preprint* arXiv:1212.6267
[10] Silk J 1968, *Astrophys. J.* **151** 459
[11] Mahdavi A, Hoekstra H, Babul A, Balam D D, Capak P L 2007, *Astrophys. J.* **668** 806
[12] Riess A G, Macri L, Casertano S, *et al.* 2011, *Astrophys.J.* **730** 119
[13] Freedman W L, Madore B F, Scowcroft V, *et al.* 2012, *Astrophys.J.* **758** 24
[14] Bashinsky S, Bertschinger E 2001 *Phys. Rev. Lett.* **87** 1301
[15] Busca N G, *et al.* 2013, *Astron.Astrophys.* **552** A96
[16] Klypin A A, Kravtsov A V, Valenzuela O, *et al.* 1999 *Astrophys. J.* **522** 82
[17] Moore B, Ghigna S, Governato F, *et al.* 1999 *Astrophys. J.* **524** L19
[18] Couchman H M P, Rees M. J 1986, *Mon. Not. Roy. Astron. Soc.* **221** 53
[19] Efstathiou G 1992, *Mon. Not. Roy. Astron. Soc.* **256** 43P
[20] Wadepuhl, M. & Springel, V. 2010, *Mon. Not. Roy. Astron. Soc.* **410** 1975
[21] Larson R B 1974, *Mon. Not. Roy. Astron. Soc.* **169** 229
[22] Boylan-Kolchin M, Bullock J S, Kaplinghat M 2012 *Mon. Not. Roy. Astron. Soc.* **422** 1203
[23] Garrison-Kimmel, *et al. Mon. Not. Roy. Astron. Soc.* **433** 3539
[24] Vera-Ciro C A, Helmi A, Starkenburg E, Breddels M A, Not too big, not too small: the dark halos of the dwarf spheroidals in the Milky Way 2012 *Preprint* arXiv:1202.6061
[25] Wang J, Frenk C S, Navarro J F, *et al.* 2012 *Mon. Not. Roy. Astron. Soc.* **424** 2715
[26] McMillan P J 2011 *Mon. Not. Roy. Astron. Soc.* **414** 2446
[27] Li Y-S, White S D M 2008 *Mon. Not. Roy. Astron. Soc.* **384** 1459
[28] Besla G., *et al.* 2007 *Astrophys. J.* **668** 949
[29] Springer V 2008 *Mon. Not. Roy. Astron. Soc.* **391** 1685
[30] Einasto J 1965 *Trudy Inst. Astroz. Alma-Ata* **51** 87
[31] Maschensko S, Couchman H M P, Wadsley J 2006 *Nature* **442** 539
[32] Governato F, *et al.* 2010 *Nature* **463** 203
[33] Ceverino D,Klypin A 2009 *Astrophys. J.* **695** 292
[34] Governato F, *et al.* 2012 *Mon. Not. Roy. Astron. Soc.* **422** 1231
[35] Jeans J H *Mon. Not. Roy. Astron. Soc.* **76** 70
[36] Binney J, Mamon G A 1982 *Mon. Not. Roy. Astron. Soc.* **200** 361
[37] Wolf J, Martinez G D, Bullock J S, *et al.* 2010, *Mon. Not. Roy. Astron. Soc.* **406**, 1220
[38] Walker M G, Mateo M, Olszewski E W, *et al.* 2009, *Astrophys. J.* **704** 1274 [Erratum-ibid. **710** (2010) 886]
[39] Walker M G, Penarrubia, J 2011, *Astrophys. J.* **742** 20
[40] Battaglia G, *et al.* 2008 *Astrophys. J.* **681** L13
[41] Goerdt T, *et al.* 2006 *Mon. Not. Roy. Astron. Soc.* **368** 1073
[42] Kleyna J T, *et al.* 2003 *Astrophys. J.* **588**, L21 [Erratum-ibid. **589** (2003) L59]
[43] Lora V, *el al.* 2013 *Astrophys. J.* **777**, 65
[44] Kroupa P, Theis C, Boily C M 2005 *A&A* **431** 517
[45] Ibata R A, *et al.* 2013 *Nature* **493** 62
[46] Pawlowski M S, Kroupa P, Jerjen H 2013 *Mon. Not. Roy. Astron. Soc.* **435** 1928
[47] Bellazzini M, *et al.*, Dwarfs walking in a row. The filamentary nature of the NGC3109 association 2013, *Preprint* arXiv:1310.6365
[48] Fattahi A, *et al.* 2013, *Mon. Not. Roy. Astron. Soc.* **431** L73
[49] Evslin J, Binary Satellite Galaxies 2013, *Preprint* arXiv:1310.6896
[50] Belokurov V., Galactic Archaeology. The dwarfs that survived and perished 2013, *Preprint* arXiv:1307.0041
[51] Baek S, Ko P, Park Wan-Il, Hidden sector monopole, vector dark matter and dark radiation with Higgs portal 2013, *Preprint* arXiv:1311.1035
[52] Evslin J, Gudnason S B, Dwarf Galaxy Sized Monopoles as Dark Matter? 2012 *Preprint* arXiv:1202.0560
[53] Jackiw R. & Rebbi J F 1976, *Phys. Rev.* D **13** 3398

Journal of Physics: Conference Series **496** (2014) 012024 doi:10.1088/1742-6596/496/1/012024

Aspects of Transplanckian Scattering

Hovhannes R. Grigoryan
Center for Cosmology and Particle Physics
Department of Physics, New York University
4 Washington Place, New York, NY 10003, USA

E-mail: hg41@nyu.edu

Abstract. We attempt to address the well known problem of quantum gravity - the information paradox, through the study of transplanckian collisions, that may proceed with the production of intermediate black hole states. We apply the S-matrix approach to study $2 \to 2$ scattering of particles with the center of mass energy much greater than the Planck mass, and investigate the dependence of the scattering amplitude on the impact parameter. It is expected that the main properties of the S-matrix (unitarity, analyticity and crossing) may provide sufficiently strong constrain for the scattering amplitude to unravel non-trivial information about the underlying properties of quantum gravity. However, we argue that our present understanding is far from complete and additional novel ideas are required to resolve this puzzle.

1. Introduction

Although we understand how to construct consistent quantum field theory describing electromagnetic, strong and weak interactions, we still don't know how to unite gravity with the quantum mechanics. One of the reasons is that in gravity, violation of unitarity occurs not only for short scales but also at long distances, not linked to the UV-behavior of gravity. Another major difference of gravity from the rest of the Standard Model interactions is that gravity generates an energy dependent scale – Schwarzschild radius. Although a lot of effort has been directed towards the resolution of this puzzle (see, e.g., [1] and references therein), there is no universally accepted understanding of even the simplest possible process involving $2 \to 2$ exclusive scattering that proceeds with the black hole production and its subsequent evaporation. Clearly, better understanding of some of these questions will allow us to address one of the fundamental problems of quantum gravity - the information paradox. This problem is one of the obstacles, preventing the unification of gravity with quantum mechanics.

In an attempt to reconcile gravity (GR) with quantum mechanics (QM) one faces multiple difficulties. Three of these problems are:

A) GR is not renormalizable;

B) unitarity violation (information paradox);

C) problem in defining gauge invariant observables in GR.

It seems, B) and C) suggest that we might want to abandon the notion of locality in order to find consistent theory of QG. Therefore, to study the problems of quantum gravity a framework is required that preserves causality without the need of demanding locality.

Although, string theory might seem to provide such a framework, it has been argued [2] that "stringy" effects are not very essential in the regime of interest relevant for the resolution of the information paradox. On the other hand, it is expected that nonperturbative dual formulations such as AdS/CFT should resolve the problem [3].

Taking into account the problems with locality and renormalizability, along with the lack of knowledge of the exact theory of QG, the attempt has been made to understand this problem by employing the S-matrix formalism. Certainly, the existence of such a gravitational S-matrix is a very strong assumption that can only be tested a posteriori.

It appears that the main properties of the S-matrix – unitarity, analyticity and crossing, are in no contradiction with non locality. The goal of this approach is to understand what the S-matrix description along with the expected properties of classical gravity can teach us about the underlying theory of quantum gravity. However, one should keep in mind that the S-matrix description has a lot of uncertainties. For instance, we don't know what are the precise quantum numbers describing asymptotic states, and if these states are complete. Moreover, to avoid problems with IR divergences, the extension of the S-matrix to inclusive amplitudes has to be made, much in the same way as in QED [7].

In what follows, we will describe how to approach one of the simplest QG problems, involving the exclusive collision of two light particles at the center of mass energies much larger than the Planck scale. We consider scattering at different impact parameters. Studying the eikonal approximation, we argue that there exists a parameter, which is closely related to the average number of gravitons involved in the collision, and using it, explore the large-N limit of gravitational scattering. We then comment on the idea of fractionation, and think that this is a new constructive idea that is worth exploring. We also present the best available ansatz for the amplitude that describes the production of intermediate black hole state, and discuss its implications and limitations. Finally, we formulate the questions that are of immediate relevance, when one attempts to study the problem of transplanckian scattering.

2. Transplanckian Scattering

The relevant length scales involved in the two particle scattering (in $D = 4$) with the center of mass (c.o.m.) energy $\sqrt{s} \gg M_P$, where M_P is the Planck mass, are:

i) $\ell_P \equiv \sqrt{\hbar G_N}$, quantum Planck's length scale (since it contains $\hbar$);

ii) $R = 2G_N\sqrt{s}$, classical characteristic Schwarzschild radius;

iii) b - impact parameter of scattering;

iv) $\ell_s = \sqrt{\hbar\alpha'}$, string length scale.

As usual, to classify different regimes of scattering, or to establish a perturbative expansion, it is instructive to define the dimensionless ratios made from these length scales, which are:

a. $(\ell_P/b)^2 = \hbar G_N/b^2 \equiv \alpha(b)$. We expect b to be related to the typical (exchanged) graviton's wave length, λ, in which case α will become the dimensionless graviton selfcoupling.

b. $(R/\ell_P)^2 = 4G_N s/\hbar \equiv N(s)$. For example, in [9], $N(s)$ itself plays the role of the dimensionless coupling. We will show below in what sense N is related to the number of gravitons.

c. $(R/b)^2 = \alpha N$. While α and N are quantum, αN is a classical parameter. For $b \leq R$ or $\alpha N \geq 1$, collision produces a trapped surface, therefore a black hole [8].

Consider the regime, where the string length is the shortest of the scales (that is $\ell_s \ll \max\{R, b\}$), then for $b \gg R$, we have a small angle scattering (single graviton exchange dominates the amplitude). Elastic amplitude grows too fast with energy, and therefore, breaks unitarity. To maintain unitarity the full sum over ladder and cross-ladder diagrams is required (eikonal regime). Decreasing b farther, while keeping $b > R$, we will enter a regime, where intermediate graviton exchange diagrams among different legs of the ladder (H-diagrams) will start to contribute. Simplest H-diagram $\sim G_N s(R/b)^2$, will be described by $(R/b)^2$ classical expansion [13]. We have to add all H-type of diagrams, which in certain respect is a classical problem. For $b \sim R$, we have a large angle scattering. Finally, for $b \leq R$, collision produces a trapped surface, signalling the presence of a black hole (BH) [8].

3. N as the Number of Gravitons

For $b \gg R, \ell_P$, the single graviton exchange dominates the scattering amplitude, $A_0(s,t) \approx -8\pi G_N s^2/t$. The latter is just the leading Born amplitude. When the impact parameter becomes smaller, but still much larger than R, the scattering enters the eikonal regime, in which case [16]:

$$iA_{eik}(s,t) = 2s \int d^2x_\perp \, e^{-i\mathbf{q}_\perp \mathbf{x}_\perp} \left(e^{i\chi(x_\perp,s)} - 1 \right) , \tag{1}$$

$$\chi(x_\perp, s) = \frac{1}{2s} \int \frac{d^2k_\perp}{(2\pi)^2} \, e^{i\mathbf{k}_\perp \mathbf{x}_\perp} \, A_0(s, -k_\perp^2) , \tag{2}$$

where $|\mathbf{x}_\perp| = b$, and $\chi \sim N(s)\ln(b\lambda_{IR})$ is an eikonal phase with λ_{IR} being an IR cutoff. Soft graviton divergences can be eliminated in case the number of dimensions is $D \geq 5$, (see, e.g. [10] in case of $D > 5$). However, this issue will not be relevant for our discussion.

The eikonal amplitude starts to dominate the Born amplitude, when $\chi \geq 1$. In $D > 5$ dimensions, $\chi \sim G_D s/b^{D-4}$, therefore, there is a strong dependence on the impact parameter, while in $D = 4$, the dependence is only logarithmic. This suggests that in $D = 4$, the value of $N(s)$ determines the approximation that we have to employ. In particular, for $N \leq 1$, the Born approximation can be sufficient, while in case $N(s) \gg 1$, eikonal approximation has to be used (if also $R \ll b$).

The eikonal approximation is described by the sum of Feynman ladder diagrams with multiple graviton exchanges. The sum of ladder and crossed-ladder diagrams at n-loop order can be written as follows [9]:

$$iA_n(s,q) = \frac{2s}{(n+1)!} \int \left[\prod_{j=1}^{n+1} \frac{d^2k_\perp}{(2\pi)^2} \frac{iA_0(s,-k_j^2)}{2s} \right] (2\pi)^2 \delta^{(2)} \left(\sum_j k_j - q_\perp \right) . \tag{3}$$

It can be shown that for $n \sim N(s)$, we have: $A_{eik} = \sum_i A_i \approx A_N$. That is the eikonal sum is dominated by the N-ladder graviton diagram [9]. Indeed, the integrand in (1) has a phase function $f(b) \equiv -iqb + i\chi(b,s)$, saddle point of which is at $b \sim N(s)/q$. On the other hand, if we Taylor expand $e^{i\chi}$ in the eikonal amplitude, the phase function would become: $f_n(b) \approx -iqb + n\ln(\chi)$, saddle point of which is at $b \sim n/q$.

Therefore, even for $R \ll b$, the (ladder and cross-laddered) diagrams that give the dominant contribution to the scattering amplitude are characterized by N exchanged gravitons. It is also clear that in either the Born or eikonal regime, $\alpha \ll 1/N$.

3.1. *Meaning of Imaginary N*

The scattering amplitude of one particle, seen as a test body, in the background metric of the other particle, an Aichelburg-Sexl metric [17], was computed by 't Hooft in [18]. The same result can be obtained applying eikonal approximation, see, e.g. [19].

Now, consider small angle scattering (when $\theta^2 \sim -t/s \ll 1$) for which the important contribution comes from $b \approx R/\theta \gg R$. In this case, the small-angle amplitude can be computed by just extending the previous result up to $b = 0$. As a result:

$$iA_{eik}(s,t) = \frac{2i\pi sN}{-t} \frac{\Gamma(1 - iN/4)}{\Gamma(1 + iN/4)} \left(\frac{4\lambda_{IR}^2}{-t} \right)^{-iN/4} , \tag{4}$$

where the first fraction is just the Born amplitude. (We also assumed $m \ll \sqrt{s}$, where m is the mass of the scatteried particle.) The complex N-poles of this amplitude are analogous to those in the Coulomb scattering, due to the $1/r$ potential. Therefore, 't Hooft poles originate from the singularity at $b = 0$, that we attempted to ignore, without consistently treating the $b < R$ region. More about the nature of these poles can be found in [19].

3.2. Fractionation

It is claimed in [9] that the transplanckian scattering in gravity probes long-distance physics. This suggests that the short-distance singularities in the loop amplitudes are not crucial in studying such collisions, however, unitarity is. Also, in transplanckian scattering the exchanged gravitons seems to redistribute democratically the momentum transfer. This idea, is called *fractionation*, and is partly motivated by the saddle point approximation made above.

However, we think that even thought the idea of fractionation is plausible, some of the details must be clarified. In particular, since the eikonal approximation is essentially a random walk in the scattering angle, it can be shown that the average value of the momentum transfer in a single graviton, for a dominant diagram with N ladders, must be $q/\sqrt{N}$, instead of q/N, expected in [9]. This suggests that the average graviton wavelength in the scattering process must be: $\lambda \sim b \sim \sqrt{N}/q \geq \sqrt{N}\ell_P$. Since $b = \ell_P/\sqrt{\alpha}$, when $R \sim b$, we have $\alpha \sim 1/N$, therefore, $b \sim \sqrt{N}\ell_P$. Taking into account that $b \sim \sqrt{N}/q$, we can verify that when $R \sim b$, we have $q \sim M_P$. In other words, in the regime of black hole formation we expect $s \sim M_P^2$ and $-t/s \sim \alpha N \sim 1$.

Combining ideas of classicalization and fractionation one can deduce that in any transplanckian scattering the gravitons, can not have wavelengths shorter than ℓ_P, as a result they are getting multiplied in quantity and redistribute the large momentum transfer. This principle can be used to calculate the entropy of the collision process, when $N \gg 1$. In this case, the entropy would be related to the number of distributions of momenta k_i among N gravitons, such that $k_i \leq M_P$ and $\sum_{i=1}^{N} k_i = q$.

When b approaches R, the diagrams with intermediate graviton exchanges among gravitons themselves in the ladder (H-diagrams) may become important. Moreover, one can wonder about the possible Reggeization of the gravitons by considering vertical ladder diagrams. However, due to fractionation, one expects two vertical gravitons in the H-diagram to be soft, making the contribution of the vertical ladder suppressed. Therefore, one would expect that the Regge regime will be irrelevant [9], unless $b \sim R$.

3.3. The Black Hole Ansatz in the Large N limit

Assuming that the S-matrix obeys unitarity condition from the beginning, the partial-wave expansion of the scattering amplitude (in D dimensions) can be written as follows [10, 14]:

$$A(s,t) = \psi_h s^{2-D/2} \sum_{\ell=0}^{\infty} (\ell + h) C_\ell^h (1 + 2t/s) f_\ell(s) \ , \tag{5}$$

$$f_\ell(s) = \frac{1}{2i}\left(e^{2i\delta_\ell(s) - 2\beta_\ell(s)} - 1\right) \ , \tag{6}$$

where $h = (D-3)/2$, $\psi_h \equiv 2^{4h+3}\pi^h \Gamma(h)$, δ_ℓ is the phase shift, and β_ℓ is an absorption parameter ($C_\ell^h(x)$ are Gegenbauer polynomials that in case of $D = 4$ coincide with the Legendre polynomials). As one can check, partial-wave amplitudes satisfy unitarity condition:

$$\mathrm{Im}\, f_\ell \geq |f_\ell|^2 \ , \tag{7}$$

where $\ell \geq 0$ and is an integer. Also, notice that in terms of our dimensionless parameters the angular momentum is: $\ell \sim \sqrt{s}b \sim N/\sqrt{\alpha N}$.

In the eikonal regime, and when $D = 4$, one has: $\delta_\ell^{eik}(s) \sim N(s)$ and $\beta_\ell^{eik}(s) = 0$. The contribution from the graviton bremsstrahlung of soft gravitons can be estimated to give the following correction to the absorptive part: $\beta_\ell^{soft} \sim \alpha N^2$.

According to Giddings and Srednicki [10], the entropy of the produced BH is $S_{BH}(s,b) \sim G_N s/\hbar \sim N(s)$, in which case, for $R \leq b$ or $\ell \leq N$, they obtained:

$$\delta_\ell^{GS}(s) \sim \pi S_{BH} \ , \quad \beta_\ell^{GS} \sim S_{BH} \ . \tag{8}$$

As a result, the elastic and absorptive cross sections can be shown to be:

$$\sigma_{el} \approx \sigma_{abs} \sim \sum_{\ell=0}^{N} (\ell + h) \sim N^2 \ . \tag{9}$$

To my knowledge this result (with later modifications in [1]) seems to be the best guess for such an amplitude. However, I think, a lot of work has to be done to obtain a more realistic result. In particular, the following questions must be addressed.

4. Reformulating Old Questions

What is the number of BH states? Consider multiparticle collision with the c.o.m. energy E that results to the production of BH states. Assume that the initial state is $|E, a\rangle_{in}$, where a run through the number of states that can be produced from the collapse of matter of energy E, therefore, a takes $\exp\{E^{3/2}\}$ integer values. Indeed, in D-dimensions, $E \sim R^{D-1}T^D$, $S \sim (TR)^{D-1}$ and we need $R > R_S$, where T is the temperature, R is the size of the sphere inside of which matter is confined, and $R_S = (G_D E)^{1/(D-3)}$.. Since not all of the energy goes into the production of BH, the produced state will be a mixture of BH state and radiation. The numbers of states corresponding to BH of mass M and radiation of energy $E - M$ are: $\exp\{M^2\}$ and $\exp\{E^2\}$, respectively. This is so, since the typical radiation quanta have energies $\sim 1/R$, then given the radiation energy $E - M$, the entropy would be of order $\sim R(E - M) \sim E^2$. Therefore, the number of states from the collapse of matter of energy E is smaller than the number of states of the BH, suggesting that not all BH states can be accessed by the collision, since $E/2 < M < E$. This raises an important question: What is the number of BH states accessed by the collision? Recently, similar issues were discussed in [15].

Can BH be resonances of quantum gravity? It may seem that the answer is positive, since the width of the BH state is $\Gamma \sim 1/R$, and moreover, $\Gamma/M \sim 1/S_{BH} \ll 1$, suggesting that the resonances might even be sharp. On the other hand, no matter how narrow the resonances are, if the distance between the subsequent resonances is smaller than their widths, resonances can not be distinguished. In this case, the application of the Levinson's theorem, that led to the result in [10] might be under question. Since we don't have a solid ground to address this problem, we can discuss this question in the frameworks of certain models. As an example, consider a model of the BH, where it is assumed that the area is quantized [20]. In this case, it was shown that BH states indeed can be considered as resonances. However, I think that this assumption is not universal if not incorrect. The reason for such a suspicion becomes clear when we try to extend this idea to AdS BH, with subsequent application of the AdS/CFT correspondence. Indeed, the area quantization of the AdS BH in 5D implies $M_n \sim n^{3/2}$ and $\Delta M_n \sim \sqrt{n}$, where n is a large natural number. This suggests that the dual CFT side, should contain such widely separated states with huge degeneracy $\sim \exp\{\#n\}$. Neither this spectral pattern nor this degeneracy is expected to be present in CFT.

Should we modify the unitarity condition? If we only consider channels that (classically) are doomed to undergo gravitational collapse, then there is no reason to assume that the S-matrix has a non-scattering term. This assumption leads to a modification of the optical theorem. We proposed a model [21] without such a non-scattering term in the S-matrix. This model is assumed to describe interactions of particle and BH states. It is based on the appropriate generalization of the well-known Breit-Wigner formula for resonant scattering. The other main ingredient is that couplings between BH and other states are governed by random complex numbers obeying certain sum-rules. These assumptions lead to the emergence of the Hawking-like thermal properties.

Finally, why is it justified to consider only processes that proceed with intermediate BH formation, and ignore the infinite number of diagrams that do not have intermediate BH states?

It may be that the Hawking's original arguments that lead to the information paradox should still work, if we ignore diagrams (with trivial topology) that do not contain intermediate BH states.

Needless to say, there are many more questions that are left to be answered, however, to our view, the questions above require immediate attention, if one attempts to understand simple aspects of quantum gravity.

Acknowledgments

This work is supported by NSF grant PHY-0758032. We would like to thank the organizers of the conference "The Modern Physics of Compact Stars and Relativistic Gravity," where this talk was presented.

References

[1] S. B. Giddings and R. A. Porto, Phys. Rev. D **81**, 025002 (2010).
[2] S. B. Giddings, Phys. Rev. D **74**, 106006 (2006); S. B. Giddings, D. J. Gross and A. Maharana, Phys. Rev. D **77**, 046001 (2008).
[3] J. Polchinski, hep-th/9901076.
[4] S. B. Giddings, Phys. Rev. D **61**, 106008 (2000).
[5] T. Banks and W. Fischler, hep-th/9906038.
[6] S. B. Giddings, arXiv:1105.2036 [hep-th].
[7] S. Weinberg, Phys. Rev. **140**, B516 (1965).
[8] D. M. Eardley and S. B. Giddings, Phys. Rev. D **66**, 044011 (2002).
[9] S. B. Giddings, M. Schmidt-Sommerfeld and J. R. Andersen, Phys. Rev. D **82**, 104022 (2010).
[10] S. B. Giddings and M. Srednicki, Phys. Rev. D **77**, 085025 (2008).
[11] D. Amati, M. Ciafaloni and G. Veneziano, JHEP **0802**, 049 (2008).
[12] H. L. Verlinde and E. P. Verlinde, Nucl. Phys. B **371**, 246 (1992).
[13] D. Amati, M. Ciafaloni and G. Veneziano, Nucl. Phys. B **403**, 707 (1993).
[14] M. Soldate, Phys. Lett. B **186**, 321 (1987).
[15] S. B. Giddings, arXiv:1308.3488 [hep-th].
[16] D. Amati, M. Ciafaloni and G. Veneziano, Nucl. Phys. B **347**, 550 (1990).
[17] P. C. Aichelburg and R. U. Sexl, Gen. Rel. Grav. **2**, 303 (1971).
[18] G. 't Hooft, Phys. Lett. B **198**, 61 (1987).
[19] D. N. Kabat and M. Ortiz, Nucl. Phys. B **388**, 570 (1992).
[20] J. D. Bekenstein and V. F. Mukhanov, Phys. Lett. B **360**, 7 (1995).
[21] H. R. Grigoryan and G. Veneziano, work in progress.

Journal of Physics: Conference Series **496** (2014) 012025 doi:10.1088/1742-6596/496/1/012025

Primordial magnetism in CMB polarization

Levon Pogosian

Department of Physics, Simon Fraser University, Burnaby, BC, V5A 1S6, Canada

E-mail: levon@sfu.ca

Abstract. A large scale B-mode signal in the CMB polarization would constitute a smoking gun of Inflation and is the main target of several ongoing and upcoming experiments. In this contribution, I consider distinguishing features of another potential source of primordial B-modes – magnetic fields. In particular, the Faraday Rotation of CMB polarization provides a distinctive signature of cosmic magnetic fields through the characteristic frequency dependence and the mode-coupling correlations of the CMB variables. I discuss constraints on primordial magnetism that can be expected from future CMB experiments, taking into account the obstruction caused by the magnetic field of the Milky Way.

1. Introduction

The unexplained origin of the galactic and extragalactic magnetic fields is one of the long standing problems of cosmology and astrophysics. While in the case of mature galaxies magnetic fields perhaps could be generated through a dynamo mechanism [1], explaining their presence in young protogalaxies and, possibly, in the intergalactic space [2], is more challenging. It is possible that there is yet unknown astrophysical mechanism that can generate magnetic fields on large scales and high redshifts. Another possibility is that magnetic fields existed before astrophysical structures began to form [3]. A primordial magnetic field (PMF) could be produced in the aftermath of cosmic phase transitions [4, 5] or in specially designed inflationary scenarios [6, 7]. Measurements of cosmic microwave background (CMB) temperature and polarization could decisively prove their primordial origin if they contained magnetic signatures present at the time of last scattering. A discovery of PMF would have profound implications for our understanding of the early universe, with critical insights into fundamental problems such as the matter-antimatter asymmetry [8].

A stochastic PMF influences CMB observables in several ways. Magnetic stress-energy perturbs the metric which leads to CMB anisotropies, while the Lorentz force deflects moving electrons and protons coupled to photons. Recently, it has been suggested [9] that small-scale fields can appreciably alter the recombination history and, consequently, the distance to last scattering, because of the enhanced small scale baryonic inhomogeneities. Here we focus on another signature of PMF – the Faraday Rotation (FR) of CMB polarization.

FR produces a B mode type polarization with a characteristic spectrum [10, 11] as well as non-trivial 4-point correlations of the CMB temperature and polarization. In [12, 13, 14], we examined detectability of PMF using different correlators and evaluated their relative merits. We found that a Planck-like experiment can detect scale-invariant PMF of a several nano-Gauss (nG) strength using the FR diagnostic, while realistic future experiments at the same frequency can detect 10^{-10} Gauss. This is comparable or better than other CMB probes of PMF, and using multiple frequency channels can further improve on these prospects [14].

2. B-modes from Faraday Rotation

At a given direction $\hat{\mathbf{n}}$ on the sky, CMB is characterized by its intensity and two additional Stokes parameters, Q and U, quantifying its linear polarization. While $Q(\hat{\mathbf{n}})$ and $U(\hat{\mathbf{n}})$ are the quantities that experiments directly measure, their values depend on the choice of the coordinate axes. Instead, it has become customary to interpret polarization maps by separating them into parity-even and parity-odd patterns, or the so-called E and B modes [15, 16]. Existence of intensity fluctuations at last scattering implies generation of E modes, which by now have been observed and found to be consistent with the spectrum of temperature anisotropies. On the other hand, B modes would not be generated at last scattering unless there were gravitational waves or other sources of metric perturbations with parity-odd components such as cosmic defects [17] or magnetic fields [18]. Weak lensing (WL) of CMB photons by the large scale structures along the line of sight distorts polarization patterns generated at last scattering and converts some of the E mode into B modes, which is expected to be measured with upcoming CMB experiments.

A primordial magnetic field present at and just after last scattering will Faraday-rotate the plane of polarization of the CMB photons. The rotation angle along $\hat{\mathbf{n}}$ is given by

$$\alpha(\hat{\mathbf{n}}) = \frac{3}{16\pi^2 e}\lambda_0^2 \int \dot{\tau}\, \mathbf{B}\cdot d\mathbf{l}\ , \tag{1}$$

where $\dot{\tau} \equiv n_e \sigma_T a$ is the differential optical depth, n_e is the line of sight free electron density, σ_T is the Thomson scattering cross-section, a is the scale factor, λ_0 is the observed wavelength of the radiation, $\mathbf{B}$ is the "comoving" magnetic field, and $d\mathbf{l}$ is the comoving length element along the photon trajectory.

A statistically homogeneous, isotropic and Gaussian distributed stochastic magnetic field can be characterized by a two-point correlation function in Fourier space

$$\langle b_i(\mathbf{k}) b_j(\mathbf{k}')\rangle = (2\pi)^3 \delta^{(3)}(\mathbf{k}+\mathbf{k}')[(\delta_{ij} - \hat{k}_i\hat{k}_j)S(k) \tag{2}$$

where $S(k)$ is the symmetric magnetic power spectrum, and where we omit the anti-symmetric contribution that quantifies the amount of magnetic helicity because only $S(k)$ contributes to the CMB observables evaluated in this paper. The shape of $S(k)$ depends on the mechanism responsible for production of PMF and generally can be taken to be a power law up to a certain dissipation scale, $S(k) \propto k^{2n-3}$ for $0 < k < k_{\rm diss}$, and zero at higher k. The dissipation scale, $k_{\rm diss}$, should, in principle, be dependent on the amplitude and the shape of the magnetic fields spectrum. We assume that $k_{\rm diss}$ is determined by damping into Alfven waves [19, 20] and can be related to $B_{\rm eff}$ as

$$\frac{k_{\rm diss}}{1{\rm Mpc}^{-1}} \approx 1.4\ h^{1/2} \left(\frac{10^{-7}{\rm Gauss}}{B_{\rm eff}}\right)\ , \tag{3}$$

where $B_{\rm eff}$ is defined as the effective homogeneous field strength that would have the same total magnetic energy density. $B_{\rm eff}$ is related to the fraction of magnetic energy density to the total radiation density, $\Omega_{B\gamma}$, via [11]

$$B_{\rm eff} = 3.25 \times 10^{-6} \sqrt{\Omega_{B\gamma}}\ {\rm Gauss}\ . \tag{4}$$

The generation of CMB polarization and the FR happen concurrently during the epoch of last scattering. However, as we have shown in [11], assuming an instantaneous last scattering, i.e. that E modes were produce first and subsequently rotated by PMF, results in relatively minor

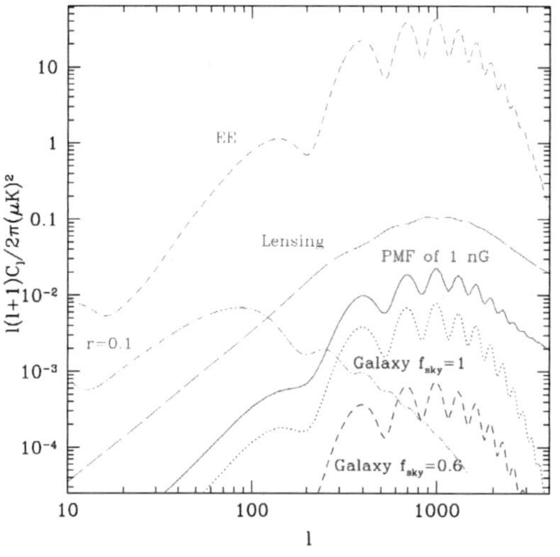

Figure 1. The CMB B-mode spectrum from Faraday rotation sourced at 30 GHz by a scale-invariant primordial magnetic field of 1 nG strength (red solid), by the full sky galactic magnetic field (blue dot), and by the galactic field with Planck's sky mask ($f_{\rm sky} = 0.6$). The black short-dash line is the input E-mode spectrum, the black dash-dot line is the contribution from inflationary gravitational waves with $r = 0.1$, while the black long-dash line is the expected contribution from gravitational lensing by large scale structure.

inaccuracies. In this approximation, the relation between the spherical expansion coefficients of the E, B and α fields can be written as

$$B_{lm} = \sum_{LM} \sum_{l_1 m_1} \alpha_{LM} E_{l_1 m_1} \mathcal{M}^{LM}_{l_1 m_1} , \tag{5}$$

where $\mathcal{M}^{LM}_{l_1 m_1}$ is defined in terms Wigner 3-j symbols [21]. We note that B modes from WL can also be schematically written as (5) but with a different mixing matrix $\mathcal{M}^{LM}_{l_1 m_1}$. Importantly, the mixing matrix for WL has a parity opposite to that of FR so that the two rotations are orthogonal to each other, making it possible to reconstruct them separately.

In Fig. 1 we show the B-mode auto-correlation spectra due to FR by a scale-invariant stochastic magnetic fields of 1 nG strength. Also shown is the contribution from FR by the galactic MF, the E mode auto-correlation spectrum which acts as a source for the FR B modes, as well as B modes from inflationary gravitational waves with $r = 0.1$, and the expected contribution from WL.

The FR induced B mode spectra have certain characteristic features. In the case of the nearly scale invariant magnetic spectrum, the spectrum is oscillatory. In fact, the shape of the B-mode spectrum mimics that of the E-mode, except for the lack of exponential damping on small scales. This is because E modes are suppressed by the Silk damping, while PMF can remain correlated on small scales. The damping of the FR induced B-mode power is due to averaging over many random rotations along the line of sight. This translates into a $1/l$ suppression of the angular spectrum, i.e. asymptotically we have $l^2 C_l^{BB} \propto l^{2n-1}$ at large l.

Name - freq (GHz)	$f_{\rm sky}$ ($f^{\rm opt}_{\rm sky}$)	FWHM (arcmin)	Δ_P(μK-arcmin)	$B_{\rm eff}$ (2σ, nG)
Planck LFI - 30	0.6	33	240	16
Planck HFI - 100	0.7	9.7	106	23
Polarbear - 90	0.024^a	6.7	7.6	3.3 (3.0)
QUIET II - 40	0.04^a	23	1.7	0.46 (0.26)(0.25)
CMBPOL - 30	0.6	26	19	0.56 (0.55)(0.51)
CMBPOL - 45	0.7	17	8.25	0.38 (0.35)(0.29)
CMBPOL - 70	0.7	11	4.23	0.39 (0.32)(0.26)
CMBPOL - 100	0.7	8	3.22	0.52 (0.4)(0.34)
Suborbital - 30	0.1	1.3	3	0.09 (0.07)(0.05)
Suborbital - 90	0.1	1.3	3	0.63 (0.45)
Space - 30	0.6 (0.2)	4	1.4	0.06 (0.04)(0.02)
Space - 90	0.7 (0.4)	4	1.4	0.26 (0.15)(0.12)

Table 1. The expected 2σ bound in nano-Gauss on the strength of a scale-invariant PMF. The numbers in the first parenthesis are for the case when 99% of the lensing contribution to the B-mode spectrum is subtracted, while the number the in the second parenthesis is after the addition subtraction of the 90% of the galactic FR contribution. Note that for full sky experiments there is an optimal sky cut ($f^{\rm opt}_{\rm sky}$) that gives the best bounds on the PMF. (^{a} based on 0.1 of RM sky)

3. Detectability of the primordial magnetic field

Spatially dependent FR produces additional non-Gaussian signatures in the CMB polarization. Namely, a particular realization of the FR distortion field correlates the respective Legendre coefficients E_{lm} and $B_{l'm'}$. In fact, as shown in [21], it is possible to reconstruct the distortion field $\alpha(\hat{\mathbf{n}})$ from specially constructed linear combinations of products $E_{lm}B_{l'm'}$. The additional correlations induced by FR also manifest themselves as connected 4-point functions of the CMB, which, in turn, provide a measurement of the distortion spectrum $C_L^{\alpha\alpha}$ [22, 23]. In principle, one can construct four estimators of the distortion spectrum, based on products of two CMB fields one of which contains polarization: TE, EE, TB, and EB. Of these four, the first two receive a large contribution to their variance from the usual scalar adiabatic Gaussian perturbations which makes it harder to find the FR signal. In [12, 13, 14], we studied the last two, i.e we considered estimators based on 4-point correlations $\langle EBEB\rangle$ and $\langle TBTB\rangle$. We have fully accounted for the contamination by weak lensing, which contributes to the variance, but whose contribution to the 4-point correlations is orthogonal to that of FR. We also account for the RM from our own galaxy, using the map provide by [29].

We will consider several sets of experimental parameters corresponding to ongoing, future and hypothetical experiments. Namely, we consider Planck's 30 GHz LFI and 100 GHz HFI channels based on actual performance characteristics [24], POLARBEAR [25] at 90 GHz with parameters compiled in [26], QUIET Phase II [27] at 40 GHz using the parameters compiled in [26], 30, 45, 70 and 100 GHz channels of a proposed CMBPol satellite [28], as well as an optimistic hypothetical sub-orbital and space-based experiments at 30 and 90 GHz. The assumed sky coverage ($f_{\rm sky}$), resolution ($\Theta_{\rm FWHM}$), and instrumental noise (Δ_P) parameters are listed in Table 1.

In Table 1 we present our forecasts for 2σ bounds on $B_{\rm eff}$ that can be expected from various ongoing and future CMB experiments. For each experiment, we check the effect of de-lensing as well as further partial subtraction of the galactic rotation measure. The optimal sky cuts are indicated when relevant. One of the interesting facts one can extract from this table is that very competitive bounds can be placed by POLARBEAR and QUIET, while even $\mathcal{O}(10^{-11}\rm G)$ fields can be constrained in principle by future experiments.

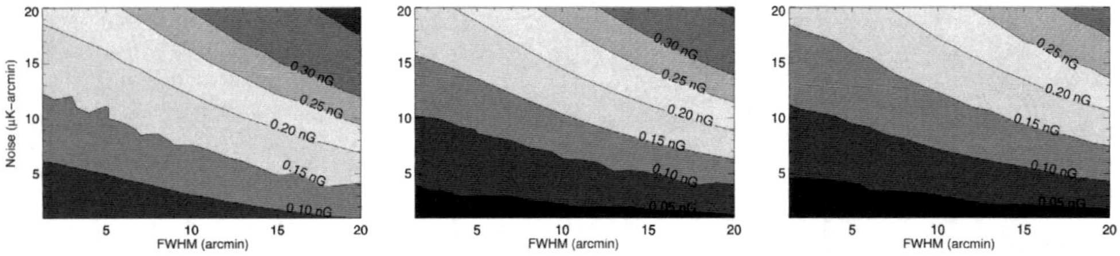

Figure 2. Contour plots of the 2σ bound on $B_{\rm eff}$ (in nG) in the instrumental noise – CMB resolution plane, with the frequency channel taken to be 30 GHz. The left panel shows the case with no de-lensing and without subtracting the Milky Way RM. The middle panel is with de-lensing, while the right panel is with de-lensing and with subtraction of the galactic FR. It is assumed that up to 0.6 of the sky is available and an optimal sky cut is found in each case.

To show the dependence of the PMF bound on the instrumental noise and resolution, we plot contours of constant 2σ bounds on $B_{\rm eff}$ in Fig. 2. The three panels show the cases with and without de-lensing and with an additional partial subtraction of the galactic RM. An optimal sky cut was used for each set of parameters.

4. Summary

We have studied the detectability of a scale-invariant PMF due to FR of the CMB using polarization correlators. We have found that the galactic RM is not a serious barrier to observe a scale-invariant PMF and that the EB quadratic correlator is in general more powerful than the BB correlator.

We have also studied the detectability of the PMF as a function of the sky coverage. We find that suborbital experiments can be almost as effective as space borne experiments, and the obstruction caused by the galaxy is relatively weak if the observed patch is near the poles. Also, as the mode-coupling correlations of CMB are mostly sourced by the largest scale features (low L) of the rotation measure, a full sky CMB map is not necessary to access the PMF on the largest scales. Cross-correlating polarization maps at multiple frequencies with comparable sensitivity to FR can further boost the significance of detection [14].

Acknowledgments

I would like to thank my collaborators Soma De, Tanmay Vachaspati and Amit Yadav. I also want to thank the organizers and the sponsors of the "The Modern Physics of Compact Stars and Relativistic Gravity" workshop in Yerevan, commemorating the 100th birthday of Gurgen Serobi Sahakyan, the founder of the Theoretical Physics Chair at the Department of Physics of Yerevan State University and a prominent armenian scientist. Being a graduate of that Chair, I was honoured to be given the opportunity to present at this meeting.

[1] L. M. Widrow, Rev. Mod. Phys. **74**, 775 (2002) [astro-ph/0207240].
[2] A. Neronov and I. Vovk, Science **328**, 73 (2010) [arXiv:1006.3504 [astro-ph.HE]].
[3] D. Grasso and H. R. Rubinstein, Phys. Rept. **348**, 163 (2001) [astro-ph/0009061].
[4] T. Vachaspati, Phys. Lett. B **265**, 258 (1991).
[5] J. M. Cornwall, Phys. Rev. D **56**, 6146 (1997) [hep-th/9704022].
[6] M. S. Turner and L. M. Widrow, Phys. Rev. D **37**, 2743 (1988).
[7] B. Ratra, Astrophys. J. **391**, L1 (1992).
[8] T. Vachaspati, Phys. Rev. Lett. **87**, 251302 (2001) [astro-ph/0101261].
[9] K. Jedamzik and T. Abel, arXiv:1108.2517 [astro-ph.CO].

[10] A. Kosowsky, T. Kahniashvili, G. Lavrelashvili and B. Ratra, Phys. Rev. D **71**, 043006 (2005) [astro-ph/0409767].
[11] L. Pogosian, A. P. S. Yadav, Y. -F. Ng and T. Vachaspati, Phys. Rev. D **84**, 043530 (2011) [Erratum-ibid. D **84**, 089903 (2011)] [arXiv:1106.1438 [astro-ph.CO]].
[12] A. P. S. Yadav, L. Pogosian and T. Vachaspati, arXiv:1207.3356 [astro-ph.CO].
[13] S. De, L. Pogosian and T. Vachaspati, Phys. Rev. D **88**, 063527 (2013) [arXiv:1305.7225 [astro-ph.CO]].
[14] L. Pogosian, arXiv:1311.2926 [astro-ph.CO].
[15] M. Kamionkowski, A. Kosowsky and A. Stebbins, Phys. Rev. Lett. **78**, 2058 (1997) [astro-ph/9609132].
[16] U. Seljak and M. Zaldarriaga, Phys. Rev. Lett. **78**, 2054 (1997) [astro-ph/9609169].
[17] U. Seljak, U. -L. Pen and N. Turok, Phys. Rev. Lett. **79**, 1615 (1997) [astro-ph/9704231].
[18] T. R. Seshadri and K. Subramanian, Phys. Rev. Lett. **87**, 101301 (2001) [astro-ph/0012056].
[19] K. Jedamzik, V. Katalinic and A. V. Olinto, Phys. Rev. D **57**, 3264 (1998) [astro-ph/9606080].
[20] K. Subramanian and J. D. Barrow, Phys. Rev. D **58**, 083502 (1998) [astro-ph/9712083].
[21] M. Kamionkowski, Phys. Rev. Lett. **102**, 111302 (2009) [arXiv:0810.1286 [astro-ph]].
[22] A. P. S. Yadav, R. Biswas, 1, M. Su and M. Zaldarriaga, Phys. Rev. D **79**, 123009 (2009) [arXiv:0902.4466 [astro-ph.CO]].
[23] V. Gluscevic, M. Kamionkowski and A. Cooray, Phys. Rev. D **80**, 023510 (2009) [arXiv:0905.1687 [astro-ph.CO]].
[24] P. A. R. Ade *et al.* [Planck Collaboration], arXiv:1303.5062 [astro-ph.CO].
[25] B. Keating, S. Moyerman, D. Boettger, J. Edwards, G. Fuller, F. Matsuda, N. Miller and H. Paar *et al.*, arXiv:1110.2101 [astro-ph.CO].
[26] Y. -Z. Ma, W. Zhao and M. L. Brown, JCAP **1010**, 007 (2010) [arXiv:1007.2396 [astro-ph.CO]].
[27] D. Samtleben [QUIET Collaboration], Nuovo Cim. B **122**, 1353 (2007) [arXiv:0802.2657 [astro-ph]].
[28] D. Baumann *et al.* [CMBPol Study Team Collaboration], AIP Conf. Proc. **1141**, 10 (2009) [arXiv:0811.3919 [astro-ph]].
[29] N. Oppermann, H. Junklewitz, G. Robbers, M. R. Bell, T. A. Ensslin, A. Bonafede, R. Braun and J. C. Brown *et al.*, arXiv:1111.6186 [astro-ph.GA].

2nd International Symposium on the Modern Physics of Compact Stars and Relativistic Gravity IOP Publishing
Journal of Physics: Conference Series **496** (2014) 012026 doi:10.1088/1742-6596/496/1/012026

An almost Einsteinian theory of gravitation

L Sh Grigoryan[1]

Institute of Applied Prob lems in Physics, 25 Hr. Nersessian str., 0014 Yerevan, Armenia

E-mail: levonshg@mail.ru

Abstract. A new version of relativistic theory of gravitation is proposed, in which the GR is supplemented with two postulates. Owing to these postulates it turns possible to introduce into consideration a covariant energy-momentum tensor of gravitational field. Additionally the influence of Universe on the gravitating system under consideration is taken into account. The equations describing the gravitation field in the framework of proposed tensor mono-metric theory of gravitation are derived. In its simplest version the proposed theory contains nine free parameters. At a particular choice of these parameters the theory is reduced to GR with a cosmological term.

1. Introduction

Yet after almost 100 years since GR creation: (a) some important conclusions of GR (e.g., the black holes, gravitational waves) have not been directly confirmed by observational data, (b) there still exist some unresolved problems in GR (e.g., the lack in the theory of the covariant energy-momentum tensor of the gravitational field), and some important phenomena have been discovered (e.g., accelerated expansion of Universe and the associated problem of "dark energy"), that had not found any consistent and non-contradictory explanation in the frameworks of GR. It is the desire to bypass the above difficulties that justifies the efforts for development of more accurate, improved or, possibly, the modified GR.

There is always several directions of search. We indicate only a few variants of relativistic theories of gravitation discussed in the science literature:

(a) Scalar-tensor theories of gravitation, in which the gravitation «constant» $G(x^{\mu})$ is a function of space-time coordinates x^{μ} (see, e.g., [1,2]),

(b) Bimetric theories of gravitation, in which along with the principal metric g_{ik} there is another "background" metric γ_{ik}. In these theories the operation of covariant differentiation with respect to the background metric γ_{ik} plays a crucial role (see, e.g., [3-9]).

(c) The theories in which: (i) the contributions of the summands that are nonlinear in Riemann tensor R_{iklm}, its convolutions and covariant derivatives are taken into account, (ii) the space-time is multi-dimensional and other theories of gravitation [10].

In this paper a modified version of GR is proposed, in which *the GR is supplemented with two postulates*.

[1] To whom any correspondence should be addressed.

2. The Tensor Mono-metric Theory of Gravitation (TMTG)

2.1. The principal provisions of the proposed theory

As in GR, in the proposed Tensor Mono-metric Theory of Gravitation (TMTG) (a) *the space-time is described by one metric tensor* g^{ik}. Then as in GR, g^{ik} describes the gravitational field as well as also the effects connected with the choice of non-inertial reference system. Thus, as in GR (b) only a part of g^{ik} describes the gravitational field. If *one denotes this part of metric tensor* as φ^{ik}, then

$$g^{ik} = g_*^{ik} + \varphi^{ik}. \tag{1}$$

The first summand in (1) *is not related to the gravitational field* and is obtained from g^{ik}, when in the space-time there is no gravitating system in question: $\varphi^{ik} = 0$ (g_*^{ik} is defined, e.g., by initial conditions of the problem under consideration and the choice of the reference frame). And, at last, as in GR, (c) the influence of gravitation on the matter and on non-gravitational fields is described by metrical tensor g^{ik}.

In TMTG it is additionally assumed in comparison with GR that (A) φ^{ik} is a covariant symmetrical second rank tensor. So, g_*^{ik} is also a covariant tensor that does not change at variation of gravitational field action with respect to φ^{ik}. In TMTG it is also assumed that (B) *the effect of gravitational field* on the matter, non-gravitational fields and *on itself is produced in the same way* (via metrical tensor g^{ik}). For this reason the covariant density of gravitational field Lagrangian L_g must depend not only on φ^{ik} and covariant derivatives of this tensor, but also on g^{ik}:

$$L_g = L_g(g^{\alpha\beta}, \varphi^{\mu\nu}, \varphi^{\mu\nu}{}_{;\lambda}). \tag{2}$$

It follows from the aforesaid that the total action of gravitating system is

$$S = S_g + S_M^* = c^{-1}\int \left(L_g + L_M^*\right)\sqrt{-g}\,d^4x, \tag{3}$$

where L_M^* is the covariant density of the Lagrangian of the matter and non-gravitation fields, c is the velocity of light.

2.2. Equations of the gravitational field

First let us find the variation of the first term in (3):

$$c\delta S_g = \int \delta(L_g\sqrt{-g})d^4x = 0.5\int(-G_{(ik)}\delta\varphi^{ik} + T_{(ik)}^{g*}\delta g^{ik})\sqrt{-g}\,d^4x, \tag{4}$$

where $G_{(ik)}$ tensor is obtained by means of variation of S_g only with respect to φ^{ik}, and $T_{(ik)}^{g*}$ tensor – by variation of S_g only with respect to g^{ik}, the brackets in (ik) implying the operation of symmetrization in *i* and *k* indices. In particular,

$$c\delta S_g = 0.5\int T_{(ik)}^{g*}\delta g^{ik}\sqrt{-g}\,d^4x \tag{5}$$

if φ^{ik} is not varied, and the metrical tensor g^{ik} is varied independently of φ^{ik}.

The equations of gravitational field are obtained based on the principle of least action:

$$\delta S = \delta S_g + \delta S_M^* = 0. \tag{6}$$

Subjected to variation is the gravitational field, i.e., the quantities φ^{ik}. At the substitution of (4) to (6) it follows in view of the equality $\delta g^{ik} = \delta\varphi^{ik}$ (see (1)) that

$$G_{(ik)} = T_{(ik)}^{M*} + T_{(ik)}^{g*}. \tag{7}$$

These are the equations of gravitational field, - the main equations of TMTG. Here $T_{ik}^{M_*}$ is the covariant energy-momentum tensor of the matter and of non-gravitational fields determined by the equality

$$c\delta S_M^* = 0.5\int T_{(ik)}^{M_*}\delta g^{ik}\sqrt{-g}d^4x \tag{8}$$

(see [11]).

Comparing (5) and (8) one ascertains that the *covariant symmetrical tensor* $T_{(ik)}^{g_*}$ is determined just as the energy-momentum tensor of the matter and of non-gravitating fields is determined in GR. In this sense $T_{(ik)}^{g_*}$ is a *covariant tensor of the energy-momentum of the gravitational field.*

2.3 Covariant conservation laws

The action S_g is a scalar and hence its variation (4) must be zero:

$$\delta S_g = 0, \tag{9}$$

in case of transformation of coordinates

$$x'^i = x^i + \zeta^i(x^\mu), \tag{10}$$

where ζ^i are small values. Here

$$\delta g^{ik}(x^\mu) = \zeta^{i;k}(x^\mu) + \zeta^{k;i}(x^\mu)$$

$$\delta\varphi^{ik}(x^\mu) = \varphi^{in}\partial\zeta^k / \partial x^n + \varphi^{nk}\partial\zeta^i / \partial x^n - \zeta^n\partial\varphi^{ik} / \partial x^n \tag{11}$$

in the linear approximation in ζ^i. At the substitution of these expressions to (4) it follows from (9) in view of the arbitrariness of ζ^i that

$$T_{(im);n}^{g_*} g^{mn}\sqrt{-g} = \partial G_{(im)}\phi^{mn}\sqrt{-g} / \partial x^n + 0.5\sqrt{-g}G_{(mn)}\partial\phi^{mn} / \partial x^i. \tag{12}$$

In GR based on condition

$$\delta S_M^* = 0 \tag{13}$$

and transformation of coordinates (10) the covariant conservation law

$$T_{(im);n}^{M_*} g^{mn} = 0 \tag{14}$$

is analogously deduced for matter and non-gravitating fields [11].

The conservation law (12) is not reduced to equality $T_{(im);n}^{g_*} g^{mn} = 0$ and in this sense it differs from an analogous law (14). The difference is due to the fact that the gravitational field is the *only field that interacts with matter and other (non-gravitational) fields by means of metrical tensor* g^{ik}. It should be added here that an equation

$$(T_{(im)}^{g_*} - G_{(im)})_{;n} g^{mn} = 0 \tag{15}$$

follows from (7) and (14). Using this equation one can rewrite the differential conservation law (12) in an explicitly covariant form.

Really, let us consider the space-time when there is no gravitating system under consideration. In this space-time (the “initial” space-time) the square of linear element

$$ds_0^2 = g_{(0)ik}dx^i dx^k. \tag{16}$$

Here we shall identify the contravariant components $g_{(0)}^{ik}$ of the metrical tensor $g_{(0)ik}$ with g_*^{ik}:

$$g_{(0)}^{ik} = g_*^{ik}, \tag{17}$$

and denote the operation of covariant differentiation in such a space-time by the vertical bar (as opposed to the symbol “;”, designating the operation of covariant differentiation in the space-time with metrical tensor g_{ik}). Using the introduced operation of covariant differentiation in the "initial" space-time, as well as (15), we can write the differential conservation law (12) in the following clearly covariant form:

$$[\eta(T^{M*}_{(im)}+T^{g*}_{(im)})]_{|n}g_*^{mn}=0, \tag{18}$$

where $\eta=\sqrt{g/g_{(0)}}$ is a scalar quantity, and $g_{(0)}$ is the determinant of metrical tensor $g_{(0)ik}$. Thus, the differential conservation law that follows from the invariance of action S_g with respect to the transformation of coordinates x^μ, may be represented in the form of covariant conservation law for the total energy-momentum tensor of the matter and all fields (including the gravitational field).

3. Influence of the Universe

The total density of the Lagrangian of matter and non-gravitational fields is

$$L^*_M = L_M(g^{\alpha\beta},q,\partial q/x^\lambda)+L_{UN}(g^{\alpha\beta},p,\partial p/\partial x^\lambda), \tag{19}$$

where the density of Lagrangian L_M describes the contribution of gravitating system under consideration, and the density of Lagrangian L_{UN} describes the contribution of the "remaining part of Universe". L_{UN} is a certain function of quantities p describing the state of this "remaining part of Universe", and of time and space derivatives of p. As the metrical tensor $g^{\alpha\beta}$ enters into the equations of motion

$$(\frac{\partial}{\partial x^\lambda}\frac{\partial}{\partial p_{,\lambda}}-\frac{\partial}{\partial p})L_{UN}\sqrt{-g}=0 \tag{20}$$

the solutions of these equations must depend not only on space-time coordinates of x^λ, but also on $g^{\alpha\beta}$:

$$p=p(g^{\alpha\beta},x^\lambda). \tag{21}$$

After substitution of these expressions to L_{UN} one finds

$$L^*_M = L_M(g^{\alpha\beta},q,\partial q/x^\lambda)+L_{UN}[g^{\alpha\beta},x^\lambda]. \tag{22}$$

According to this equality the covariant energy-momentum tensor of the matter and of non-gravitational fields splits into two summands:

$$T^{M*}_{(ik)}=T^M_{(ik)}+W_{(ik)}, \tag{23}$$

where

$$\delta\int L_{UN}[g^{\alpha\beta},x^\lambda]\sqrt{-g}d\Omega=0.5\int W_{(ik)}\delta g^{ik}\sqrt{-g}d\Omega. \tag{24}$$

Here $T^M_{(ik)}$ is the energy-momentum tensor of matter and non-gravitational fields for the gravitating system under consideration, and $W_{(ik)}$ is the tensor describing the contribution of the "remaining part of Universe". As a result one may rewrite the field equations (7) of TMTG in the following form:

$$G_{(ik)}=T^M_{(ik)}+T^{g*}_{(ik)}+W_{(ik)}. \tag{25}$$

4. The simplest version of the theory

Now consider the "simplest" version of TMTG, in which the density of Lagrangian of the gravitational field (2) is square-law dependent on $\varphi^{\mu\nu}$ and $\varphi^{\mu\nu}{}_{;\lambda}$:

$$L_g=(a_1\varphi^{im;r}\varphi^{kn;s}+a_2\varphi^{im;r}\varphi^{ks;n})g_{ik}g_{mn}g_{rs}+a_3\varphi_{,n}\varphi^{,n}+a_4\varphi_{,n}\varphi^{nr}{}_{;r}+$$

$$+a_5\varphi^{mr}{}_{;r}\varphi^{ns}{}_{;s}g_{mn}+a_6\varphi^2+a_7\varphi^{im}\varphi^{kn}g_{ik}g_{mn}, \qquad (26)$$

where $\varphi = g_{\alpha\beta}\varphi^{\alpha\beta}$, and the indices are raised and descended with the help of metrical tensor g^{ik} and of its covariant components g_{ik}. In (26) there are seven free parameters a_i.

Generally speaking, the density of Lagrangian $L_{UN}[g^{\alpha\beta}, x^{\lambda}]$ is determined by the tensor R_{tklm} of space-time curvature, by its convolutions and covariant derivatives. In the simplest case

$$L_{UN} = b_1 + b_2 R, \qquad (27)$$

where b_1 and b_2 are the free parameters of the theory. Here it follows from (24) that

$$W_{(ik)} = -b_1 g_{ik} + 2b_2(R_{ik} - 0.5g_{ik}R). \qquad (28)$$

Consequently the equations of gravitational field (25) assume the form

$$G_{(ik)} = T^{M}_{(ik)} + T^{g_*}_{(ik)} - b_1 g_{ik} + 2b_2(R_{ik} - 0.5g_{ik}R). \qquad (29)$$

The final explicit expressions for $G_{(ik)}$ and $T^{g_*}_{(ik)}$ are not given here for brevity.

4.1. Partial choice of the values of free parameters of TMTG

In case of a partial choice of the values of free parameters of theory:

$$a_1 = a_2 = a_3 = a_4 = a_5 = a_6 = a_7 = 0, \qquad b_1 = \frac{c^4}{8\pi G}\Lambda, \qquad b_2 = -\frac{c^4}{16\pi G} \qquad (30)$$

(29) is reduced to the Einstein equations with the cosmological term:

$$R_{ik} - 0.5g_{ik}R + g_{ik}\Lambda = 8\pi G T^{M}_{(ik)} / c^4 \qquad (31)$$

(G is the gravitation constant and Λ is the cosmological constant).

4.2. Comparison with observation data

In case of (30) the problem of matching TMTG with observation data is reduced to the problem of matching GR with these data. When a_i, b_1 and b_2 insignificantly differ from (30) one can consider TMTG as an "almost Einsteinian theory of gravitation". The problem consists only in the following: "To what extent these insignificant differences from (30) are essential for astrophysical applications of TMTG?" The question is not idle since in (26) a_6 and a_7 determine the energy of gravitational field that is constant in time and in space. The difference of a_6 and a_7 from zero may be crucial for models of Universe within the frameworks of TMTG.

In future it is supposed to determine the PPN-parameters of TMTG to afford the determination of the limits of permissible values of a_i, b_1 and b_2.

4.3. Comparison with bimetric theories of gravitation

The secondary ("background") metric γ_{ik} of the space-time mentioned in Introduction, plays an important part in bimetric theories of gravitation. For instance, the determination of the action of gravitational field is based in these theories on the operation of covariant differentiation with respect to the background metric γ_{ik}. In (18) we had also introduced the second ("initial") metrical tensor $g^{ik}_{(0)}$. However, in contrast with the bimetric theories, the operation of covariant differentiation with respect to an "initial" metrical tensor $g^{ik}_{(0)}$ in the action of gravitational field is not generally employed (see (2),(4)).

Journal of Physics: Conference Series **496** (2014) 012026 doi:10.1088/1742-6596/496/1/012026

5. Conclusions

In the present work a modified version of GR, named the Tensor Mono-metric Theory of Gravitation (TMTG), is proposed, in which the GR is supplemented with two postulates. It is first assumed that the part of metrical tensor describing the gravitational field, is a *second rank covariant tensor.* Second, it is assumed that the gravitational field acts on the matter, on non-gravitational fields and on itself *in a similar way* (through the metrical tensor). This permitted an introduction of a second rank *covariant* symmetrical *energy-momentum tensor for the gravitational field* that is defined in exactly the same way as the energy-momentum tensor of the matter and non-gravitational fields is defined in GR. And, at last, in TMTG an allowance is made for the influence of the "remaining parts of Universe" on the gravitating system under consideration.

Further, the equations describing the gravitational field in the framework of TMTG were derived and then the differential law of conservation, stemming from an invariance of gravitational field action with respect to transformation of four-dimensional coordinates of space-time, have been obtained. This conservation law is represented in the covariant form as the conservation law for the *total energy-momentum tensor of the matter and of all fields including the gravitational field.*

In the "simplest" version of TMTG it contains nine free parameters. At a particular choice of these parameters the theory is reduced to GR with a cosmological term. Two of nine parameters specify the energy of constant gravitational field (constant in time and space), and, hence, may be crucial for models of Universe in TMTG.

References

[1] Sahakian G S 1974 *Equilibrium Configurations of Degenerate Gaseous Masses* (Wiley, New York), Chapters IX,X
[2] Weinberg S 1975 *Gravitation and Cosmology* (Mir, Moscow) (in Russian)
[3] Rosen N 1973 *Gen. Rel. Grav.* **4** 435
[4] Rosen N 1974 *Ann. Phys. N.Y.* **84** 455
[5] Rosen J and Rosen N 1977 *Astrophys. J.* **212** 605
[6] Sahakian G S, Sarkissian A V, Khachatrian B V and Choubarian E V 1978 *Astrofizika* **14** 489
[7] Sarkissian A V, Khachatrian B V and Choubarian E V 1979 *Astrofizika* **15** 506
[8] Avakian R M and Grigorian L Sh 1988 *Astrophys. Space Sci.* **146** 183
[9] Grigorian L Sh and Saharian A A 1990 *Astrophys. Space Sci.* **167** 271
[10] Will C M 1985 *Theory and Experiment in Gravitational Physics* (Energoatomizdat, Moscow) (in Russian)
[11] Landau L D and Lifshitz E M 1980 *The Classical Theory of Field* (Oxford: Pergamon Press)

2nd International Symposium on the Modern Physics of Compact Stars and Relativistic Gravity IOP Publishing
Journal of Physics: Conference Series **496** (2014) 012027 doi:10.1088/1742-6596/496/1/012027

Casimir densities for a conducting plate in de Sitter spacetime

A. S. Kotanjyan, A. A. Saharian, H. A. Nersisyan

Department of Physics, Yerevan State University, 1 Alex Manoogian Street, 0025 Yerevan, Armenia

E-mail: anna.kotanjyan@ysu.am, saharian@ysu.am

Abstract. Two-point functions for the electromagnetic field in background of $(D + 1)$-dimensional dS spacetime are evaluated assuming that the field is prepared in the Bunch-Davies vacuum state. By using these functions, the vacuum expectation values (VEVs) of the field squared and the energy-momentum tensor are investigated in the geometry of a conducting plate. The VEVs are explicitly decomposed into the boundary-free and plate-induced parts. For points outside of the plate the renormalization is needed for the first parts only. Because of the maximal symmetry of the background spacetime and of the Bunch-Davies vacuum state, the boundary-free parts do not depend on spacetime coordinates, whereas the plate-induced parts are functions of the proper distance of the observation point from the plate. The plate-induced part in the VEV of the energy-momentum tensor vanishes for $D = 3$ which is a direct consequence of a conformal invariance of the electromagnetic field for this spatial dimension. For $D > 3$, in addition to the diagonal components, the vacuum energy-momentum tensor has nonzero off-diagonal component which describes energy flux along the direction normal to the plate.

1. Introduction

De Sitter (dS) spacetime is among the most popular backgrounds in the gravitational physics. There are several reasons for this. First of all, dS spacetime is the maximally symmetric solution of Einstein's equations with a positive cosmological constant and due to this maximal symmetry numerous physical problems are exactly solvable on its background. Further, dS spacetime is of key importance in the inflationary cosmology scenario [1], in accordance of which in the early stages of the cosmological expansion our Universe passed through a phase with the geometry well approximated by dS spacetime. During an inflationary epoch, quantum fluctuations in the inflaton field introduce inhomogeneities which play a central role in the generation of cosmic structures from inflation. Recent astronomical observations of high redshift supernovae, galaxy clusters and cosmic microwave background have indicated that at the present epoch the Universe is accelerating and can be approximated by a world with a positive cosmological constant [2]. If the Universe were to accelerate indefinitely, the Standard Cosmology would lead to an asymptotic dS Universe. It is therefore important to investigate physical effects in dS spacetime for understanding the early Universe and its future.

In particular, the investigation of quantum field theoretical effects in dS background has attracted a great deal of attention. The interaction of a fluctuating quantum field with the background gravitational field leads to the vacuum polarization. Another type of vacuum

Journal of Physics: Conference Series **496** (2014) 012027 doi:10.1088/1742-6596/496/1/012027

polarization arises by the imposition of boundary conditions on the field operator. These conditions may arise because of the presence of boundaries having different physical nature: macroscopic bodies in QED, extended topological defects, horizons and branes in higher-dimensional models. Boundary conditions, imposed on the field operator, modify the zero-point modes of a quantized field and, as a result, the vacuum expectation values of physical observables are changed. In particular, forces arise acting on constraining boundaries. This is the well-known Casimir effect [3]. The nature of the Casimir forces depend on the quantum field, the type of spacetime manifold, the boundary geometry, and the specific boundary conditions imposed on the field.

In this paper, based on [4], we consider the combined effects of the background gravitational field and of boundaries on the properties of the electromagnetic vacuum. The background geometry is the dS spacetime and as a boundary we will consider a single plate on which the field obeys perfectly conducting boundary conditions. The Casimir effect in dS spacetime for a massive scalar field with general curvature coupling parameter obeying Robin boundary condition has been recently considered in [5, 6] and [7] for flat and spherical boundaries respectively (see also [8, 9] and [10] for special cases of conformally and minimally coupled massless fields). The case of the electromagnetic filed in background of Friedmann-Robertson-Walker cosmological models with power-law scale factors is discussed in [11]. By taking into account that the higher-dimensional models play an important role in high-energy physics, in particular, in string theories and in supergravity, we assume the general value of spatial dimension D.

2. Electromagnetic two-point functions

We consider the change in the properties of the electromagnetic vacuum induced by the presence of a perfectly conducting plate in the background of $(D+1)$-dimensional dS spacetime. The geometry is described by the line element:

$$ds^2 = dt^2 - e^{2t/\alpha} \sum\nolimits_{l=1}^{D} \left(dz^l\right)^2, \tag{1}$$

where the parameter α is related to the positive cosmological constant Λ by the formula $\alpha^2 = D(D-1)/(2\Lambda)$. In terms of conformal time τ, defined as $\tau = -\alpha e^{t/\alpha}$, $-\infty < \tau < 0$, the metric tensor takes a conformally flat form:

$$g_{ik} = (\alpha/\tau)^2 \, diag\,(1, -1, ..., -1)\,. \tag{2}$$

We assume that the plate is placed at $z^D = 0$. On the plate the field obeys the boundary condition $n^{\nu_1}\,{}^*F_{\nu_1\cdots\nu_{D-1}} = 0$, with the tensor ${}^*F_{\nu_1\cdots\nu_{D-1}}$ dual to electromagnetic field tensor $F_{\mu\nu}$, and with n^μ being the normal to the plate. Here we consider the region $z^D > 0$.

All properties of the quantum vacuum are encoded in two-point functions. In the presence of the plate, the two-point function of the electromagnetic field tensor is decomposed as:

$$\langle F_l(x)F_m(x')\rangle = \langle F_l(x)F_m(x')\rangle_0 + \langle F_l(x)F_m(x')\rangle_{\mathrm{b}}\,, \tag{3}$$

where $\langle F_l(x)F_m(x')\rangle_0$ is the corresponding function in the boundary-free dS spacetime and the part $\langle F_l(x)F_m(x')\rangle_{\mathrm{b}}$ is induced by the plate. Assuming that the electromagnetic field is prepared

Journal of Physics: Conference Series **496** (2014) 012027 doi:10.1088/1742-6596/496/1/012027

in the Bunch-Davies vacuum state, for the boundary-free part of the two-point functions we have

$$\begin{aligned}
\left\langle F_{0l}(x)\, F_{0m}(x')\right\rangle_0 &= \frac{(\eta\eta')^{-2}}{2B_D\alpha^{D-3}}\left[(\delta_{lp}\delta_{mq}-\delta_{lm}\delta_{pq})\frac{\Delta z^p\Delta z^q}{2\eta\eta'}\partial_z+(D-1)\delta_{lm}\;\; G_D(z),\right.\\
\left\langle F_{pl}(x)\, F_{0m}(x')\right\rangle_0 &= \frac{(\eta\eta')^{-2}}{B_D\alpha^{D-3}}\delta_{[pm}\delta_{l]q}\frac{\Delta z^q}{\eta'}\left[2+(z-\frac{\eta+\eta'}{2\eta})\partial_z\;\; F_D(z),\right.\\
\left\langle F_{pl}(x)\, F_{qm}(x')\right\rangle_0 &= \frac{(\eta\eta')^{-2}}{B_D\alpha^{D-3}}\left(\delta_{[pr}\delta_{l][q}\delta_{m]s}\frac{\Delta z^r\Delta z^s}{\eta\eta'}\partial_z+2\delta_{[pq}\delta_{l]m}\right)F_D(z),
\end{aligned} \tag{4}$$

where $l,m,p,q=1,2,\ldots,D$, $\mathbf{z}=(z^1,\ldots,z^D)$ and

$$B_D=(4\pi)^{(D-1)/2}\Gamma\left((D+3)/2\right),\; z=1+\frac{(\Delta\eta)^2-|\Delta\mathbf{z}|^2}{4\eta\eta'},\;\; \Delta\mathbf{z}=\mathbf{z}-\mathbf{z}'. \tag{5}$$

In (4), we have defined the functions

$$\begin{aligned}
F_D(z) &= \Gamma(D)\,F\left(D,2;\frac{D+3}{2};z\right),\\
G_D(z) &= 2\Gamma(D-1)\,F\left(D-1,3;\frac{D+3}{2};z\right),
\end{aligned} \tag{6}$$

with $F(a,b,c;z)$ being the hypergeometric function. The expression for the two-point function $\langle F_{0m}(x)F_{pl}(x')\rangle$ is obtained from the expression (4) for $\langle F_{pl}(x)F_{0m}(x')\rangle$ by changing the sign and by the interchange $\eta\rightleftarrows\eta'$.

The plate-induced part in (3) are related to the boundary-free functions by the formulas

$$\begin{aligned}
\left\langle F_{pl}(x)F_{qm}(x')\right\rangle_{\rm b} &= -\left\langle F_{pl}(x)F_{qm}(x'_-)\right\rangle_0,\\
\left\langle F_{pl}(x)F_{Dm}(x')\right\rangle_{\rm b} &= \left\langle F_{pl}(x)F_{Dm}(x'_-)\right\rangle_0,
\end{aligned} \tag{7}$$

with $p,l=0,1,...,D,\;\; q,m=0,1,...,D-1$, and x'_- is the image of x' with respect to the plate: $x'_-=\left(\tau',z^{1\prime},...,z^{D-1\prime},-z^{D\prime}\right)$.

For $D=3$ for the functions (6) one has $F_3(z)=G_3(z)=2(z-1)^{-2}$. In this case the two-point functions (4) are related to the corresponding functions in Minkowski spacetime by the standard conformal transformation. The latter property is a consequence of the fact that the electromagnetic field is conformally invariant for $D=3$.

3. Casimir densities

First let us consider the VEV of the electric field squared. The boundary-induced part in this VEV is obtained from the two-point function $\langle F_{0l}(x)F_{0m}(x')\rangle_{\rm b}$ in the coincidence limit:

$$\langle E^2\rangle_{\rm b}=\frac{D-1}{2B_D\alpha^{D+1}}\left[2(1-y)\partial_y-D+2\right]G_D(y), \tag{8}$$

where and in what forllows we use the notation

$$y=1-(z^D/\eta)^2. \tag{9}$$

At proper distances from the plate much smaller than the dS curvature scale, $z^D/\eta\ll 1$, to the leading order we have

$$\langle E^2\rangle_{\rm b}\approx\frac{3\,(D-1)\,\Gamma((D+1)/2)}{2(4\pi)^{(D-1)/2}(\alpha z^D/\eta)^{D+1}}. \tag{10}$$

In this limit, the boundary-induced part dominates in the VEV of the electric field squared. At large distances from the plate one has the asymptotics:

$$\langle E^2\rangle_{\rm b} \approx \frac{2^{3-D}\Gamma(D+1)\Gamma(2-D/2)}{(4\pi)^{D/2}\alpha^{D+1}(z^D/\eta)^{2(D-1)}},\ D<4,\ \langle E^2\rangle_{\rm b} \approx 3\frac{\ln(2z^D/\eta \ -5/8}{\pi^2\alpha^5(z^D/\eta)^6},\ D=4,$$
$$\langle E^2\rangle_{\rm b} \approx \frac{(D-1)(8-D)\Gamma(D/2-2)}{2^{4-D}(4\pi)^{D/2}\alpha^{D+1}(z^D/\eta)^6},\ D>4. \tag{11}$$

For $D=8$ the leading term vanishes. For $3\leqslant D\leqslant 8$ the plate-induced part $\langle E^2\rangle_{\rm b}$ is positive everywhere. For $D\geqslant 9$, it is positive near the plate and negative at large distances (see figure 1 below).

Another important characteristic of the vacuum state is the VEV of the energy-momentum tensor. It describes the local structure of the vacuum state and acts as the source of gravity in the quasiclassical Einstein equations. Therefore, the VEV of the energy-momentum tensor plays an important role in modelling self-consistent dynamics involving the gravitational field.

Having the two-point functions we can evaluate the VEV of the energy-momentum tensor by making use of the formula

$$\langle T_\mu^\nu\rangle = -\frac{1}{4\pi}\left\langle F_\mu^{.\beta}(x)F_{.\beta}^\nu(x)\right\rangle + \frac{\delta_\mu^\nu}{16\pi}\left\langle F_{\beta\sigma}F^{\beta\sigma}\right\rangle. \tag{12}$$

Of course, the expression in the right-hand side is divergent and some regularization with the subsequent renormalization procedure is needed. The important thing here is that we have already decomposed the two-point functions. This allows us to have a similar decomposition of the vacuum energy-momentum tensor:

$$\langle T_\mu^\nu\rangle = \langle T_\mu^\nu\rangle_0 + \langle T_\mu^\nu\rangle_{\rm b}, \tag{13}$$

with the boundary-free, $\langle T_\mu^\nu\rangle_0$, and boundary-induced, $\langle T_\mu^\nu\rangle_{\rm b}$, parts. For points outside the plate the local geometry is not changed by the presence of the plate and the divergences in $\langle T_\mu^\nu\rangle$ and $\langle T_\mu^\nu\rangle_0$ are the same. Hence, with the decomposition (13), for the points outside the plate the renormalization of $\langle T_\mu^\nu\rangle$ is reduced to the one for boundary-free dS spacetime. Because of the maximal symmetry of de Sitter spacetime and the Bunch-Davies vacuum state, the corresponding VEV of the energy-momentum tensor is proportional to the metric tensor: $\langle T_\mu^\nu\rangle_0 = \mathrm{const}\cdot\delta_\mu^\nu$.

Our main interest here is the boundary-induced part in the VEV. By using (3), (4) and (7), from (12) for the diagonal components of vacuum energy-momentum tensor one finds (no summation over l)

$$\begin{aligned}
\langle T_0^0\rangle_{\rm b} &= \frac{\alpha^{-D-1}}{A_D}\{[2(1-y)\partial_y - D+2]G(y) - [2(1-y)\partial_y + D-4]F_D(y)\},\\
\left\langle T_l^l\right\rangle_{\rm b} &= -\frac{\alpha^{-D-1}}{A_D}\frac{D-3}{D-1}\left\{\left[2(1-y)\partial_y - (D-4)\frac{D-3}{D-1}\ \ G(y)\right.\right.\\
&\quad\left.+\left[2(1-y)\partial_y + (D-4)\frac{D-1}{D-3} - 4\ \ F_D(y)\right\},\ l=1,\dots,D-1,\\
\langle T_D^D\rangle_{\rm b} &= \frac{\alpha^{-D-1}}{A_D}[2(1-y)\partial_y - D][F_D(y) - G_D(y)],
\end{aligned} \tag{14}$$

with the notation

$$A_D = (4\pi)^{(D+1)/2}(D+1)\Gamma\left(\frac{D-1}{2}\right). \tag{15}$$

In addition to the diagonal components there is a non-zero off-diagonal component:

$$\langle T_0^D \rangle_{\rm b} = 4\frac{\alpha^{-D-1}}{A_D}\frac{x^D}{\eta}\left[(1-y)\,\partial_y - 2\right] F_D\left(y\right), \tag{16}$$

which describes the energy flux along the direction normal to the plate. By using the expressions given above for the functions $F_D(y)$ and $G_D(y)$, we can see that the boundary-induced part in the VEV vanishes for $D=3$. Again, this property is a consequence of the conformal invariance of the electromagnetic field in $D=3$.

As it is seen from the expressions given above, the VEVs of the field squared and of the energy-momentum tensor depend on the coordinates z^D and η in the combination z^D/η. This ratio is the proper distance from the plate, $\alpha z^D/\eta$, measured in the units of the curvature radius α. This property is a consequence of the maximal symmetry of the background spacetime and of the Bunch-Davies vacuum state.

Simple expressions for the VEVs of the components of the energy-momentum tensor are obtained at small and large distances from the plate. At proper distances smaller than the dS curvature radius, $z^D/\eta \ll 1$, to the leading order one has (no summation over $l=0,\ldots,D-1$):

$$\langle T_l^l \rangle_{\rm b} \approx -\frac{\eta}{z^D}\langle T_0^D \rangle_{\rm b} \approx \frac{D-1}{(z^D/\eta)^2}\langle T_D^D \rangle_{\rm b} \approx -\frac{(D-3)\,(D-1)\Gamma((D+1)/2)}{2(4\pi)^{(D+1)/2}(\alpha z^D/\eta)^{D+1}}. \tag{17}$$

In this region, all diagonal components are negative for $D \geqslant 4$ and one has $|\langle T_0^0 \rangle_{\rm b}| \gg |\langle T_D^D \rangle_{\rm b}|$. At large proper distances from the plate, compared with the dS curvature radius, $z^D/\eta \gg 1$, and for $D>4$ for the leading terms in the corresponding asymptotic expansions we get (no summation over $l=1,\ldots,D$)

$$\begin{aligned}\langle T_0^0 \rangle_{\rm b} &\approx \frac{D}{D-4}\langle T_l^l \rangle_b \approx -\frac{2^{D-4}D(D-1)\Gamma(D/2-1)}{(4\pi)^{D/2+1}\alpha^{D+1}(z^D/\eta)^4},\\ \langle T_0^D \rangle_{\rm b} &\approx \frac{2^{D-2}(D-1)\Gamma(D/2-1)}{(4\pi)^{D/2+1}\alpha^{D+1}(z^D/\eta)^5}.\end{aligned} \tag{18}$$

For $D=4$ the asymptotic expressions for the components $\langle T_0^0 \rangle_{\rm b}$ and $\langle T_0^D \rangle_{\rm b}$ remain the same, whereas for the stresses one has (no summation over $l=1,\ldots,D$)

$$\langle T_l^l \rangle_{\rm b} = -\frac{3\alpha^{-5}\ln(z^D/\eta)}{16\pi^3(z^D/\eta)^6}. \tag{19}$$

As it is seen, at large distances the stresses are isotropic. From (18) it follows that the decay of the VEVs, as functions of the proper distance from the plate, at large distances is power-law. For massive scalar field with relatively large value of the mass (compared with the dS energy scale) the decay is oscillatory [5]. This effect is qualitatively new feature induced by the curvature of the background spacetime.

On the left panel of figure 1 we have plotted the plate-induced part in the VEV of the electric field squared as a function of z^D/η for various values of D (numbers near the curves). Note that $D=9$ corresponds to the critical dimension in string theories. On the right panel the boundary-induced parts in the components of the energy-momentum tensor, $\langle T_l^l \rangle_{\rm b}$, are plotted versus the distance from the plate. The numbers near the curves correspond to the value of the index l. The dashed curve corresponds to the energy flux, $\langle T_0^D \rangle_{\rm b}$. As it is seen, the corresponding energy flux is directed from the plate.

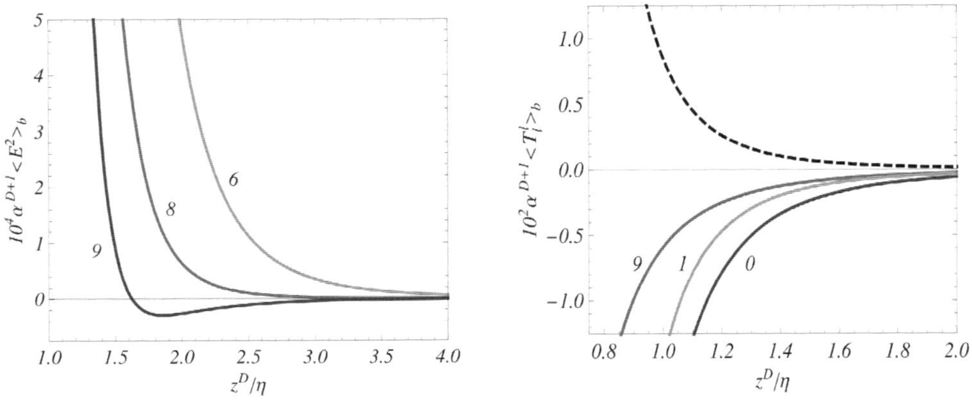

Figure 1. The plate-induced parts in the VEVs of the field squared (left panel) and of the energy-momentum tensor (right panel) as functions of z^D/η. On the right panel the numbers near the curves are the corresponding values of D and for the right panel $D = 9$.

4. Conclusion

Two-point function for the electromagnetic field tensor is evaluated in $(D + 1)$-dimensional de Sitter spacetime in the presence of a conducting plate. We have assumed that the field is prepared in the Bunch-Davies vacuum state. The VEVs of the field squared and of the energy-momentum tensor are decomposed into boundary-free and plate-induced parts. Because of the maximal symmetry of the background spacetime and of Bunch-Davies vacuum state, the boundary-free parts do not depend on spacetime coordinates. The diagonal components of plate-induced part in the vacuum energy-momentum tensor are negative and, hence, the effective pressures are positive. The VEV of the energy-momentum tensor contains a nonzero off-diagonal component which is positive and corresponds to the energy flux directed from the plate. The results obtained above can be used for the investigation of the vacuum characteristics in the geometry of two parallel plates, including the Casimir forces acting on the plates.

The work was supported by State Committee Science MES RA, within the frame of the research project No. SCS 13-1C040.

References

[1] Linde A D 1990 *Particle Physics and Inflationary Cosmology* (Switzerland: Harwood Academic Publishers)
[2] Komatsu E et al 3009 Astrophys. J. Suppl. Ser. **180** 330
[3] Bordag M Klimchitskaya G L Mohideen U and Mostepanenko V M 2009 *Advances in the Casimir Effect* (Oxford: Oxford University Press)
[4] Saharian A A Kotanjyan A S and Nersisyan H A 2013 Phys. Lett. B in press
[5] Saharian A A and Vardanyan T A 2009 Class. Quantum Grav. **26** 195004
[6] Elizalde E Saharian A A and Vardanyan T A 2010 Phys. Rev. D **81** 124003
[7] Milton K A and Saharian A A 2012 Phys. Rev. D **85** 064005
[8] Setare M R and Mansouri R 2001 Class. Quantum Grav. **18** 2331
[9] Setare M R 2001 Class.Quantum Grav. **18** 4823.
[10] Burda P 2011 JETP Lett. **93** 632
[11] Bellucci S and Saharian A A 2013 Phys. Rev D **88** 064034

Journal of Physics: Conference Series **496** (2014) 012028 doi:10.1088/1742-6596/496/1/012028

Brane-world black holes with post-Newtonian parameter: astrophysical signatures

Roldão da Rocha

Centro de Matemática, Computação e Cognição, Universidade Federal do ABC 09210-170, Santo André, SP, Brazil

E-mail: roldao.rocha@ufabc.edu.br

Arpine Piloyan

Yerevan State University, Faculty of Physics, Alex Manoogian 1, Yerevan 0025, Armenia

E-mail: arpine.piloyan@ysu.am

Abstract. The existence of the many unanswered questions in fundamental physics, in particular, in astrophysics allows for a great variety of theories to remain viable candidates for becoming the correct theory at energies not accessible in current experiments. One special class of these type of theories is the class of extra-dimensional brane-world models. Besides answering many fundamental problems, for instance, the hierarchy problem, they may produce testable predictions. In this work, we find and investigate brane-world induced black string horizon corrections, when the black hole solution has a post-Newtonian parameter. For suitable choices of such a parameter, the Hawking radiation on the brane is precluded, and the Hawking radiation in the bulk causes the black hole to slightly recoil into the bulk, which modifies the black hole apparent horizon. It has an impact on quasars luminosity and, therefore, might be detected and measured.

1. Introduction

Dark Matter and Dark Energy hypotheses are the basis of the answers to many open questions in contemporary fundamental physics. There exist a large number of theories that are aimed at the explanation of physics at scales not accessible in current terrestrial experiments as well as in astrophysics. The brane-world scenarios are a class of successful theories which, among other things, provide a solution of the hierarchy problem. The underlying idea is that, unlike matter fields, which live on a four-dimensional brane, the gravity lives in a five-dimensional space and, therefore, leaks into the bulk which greatly reduces its strength as measured on the brane.

There exists a high level of degeneracy among the candidate theories and observational tests are needed in order to make progress towards the correct theory. In this work, we propose an observational test to distinguish the brane-world scenarios from other classes of theories of gravity.

The black hole horizon is sensitive to the details of the extra-dimensional theories. This might impact astrophysical models of quasars, which invoke accretion of matter onto a black hole. The masses of black hole and the accretion onto them are detectable through the radiation caused by the accretion process, which also depends on the distribution of the mass in the galaxy

Journal of Physics: Conference Series **496** (2014) 012028 doi:10.1088/1742-6596/496/1/012028

hosting the black hole. Here, we present the Casadio-Fabbri-Mazzacurati black hole solutions, that possess a post-Newtonian parameter, and study how these solutions can be extended to the bulk, thus generalizing the concept of black string. For suitable values of the post-Newtonian parameter, the Hawking radiation on the brane is suppressed, but not the Hawking radiation in the bulk. It causes the black hole to slightly recoil into the bulk, which makes the apparent horizon on the brane to be modified.

In Section 2, we use the Gauss-Codazzi formalism to project the 5D space-time quantities on the 4D brane, taking into account the brane tension and junction conditions. The Taylor expansion of metric along the Gaussian coordinate perpendicular to the brane is carried out, providing the bulk metric (and also the black string warped horizon) near the brane. In the Section 3, we present a complete model and illustrate main results in figures. Section 4 is an explanation of how brane-world induced effects might be calculated and how they change quasars luminosity. Our results and conclusions are summarized in Sections 5 and 6.

2. Projection of 5D model on the 4D brane: the Gauss-Codazzi formalism and junction conditions

We are working in 5D space-time where y is a Gaussian coordinate orthogonal to the brane so the brane is defined by $y = 0$. The 5D metric is $\widehat{g}_{AB}dx^A dx^B = g_{\mu\nu}(x^\alpha, y)\, dx^\mu dx^\nu + dy^2$, where $g_{\mu\nu}$ is the metric on the brane. n^A is an orthonormal vector field normal to the brane so that the metrics on the brane and in the bulk are related as follows:

$$g_{\mu\nu} = \widehat{g}_{\mu\nu} - n_\mu n_\nu. \tag{1}$$

The Einstein's field equations in 5D read

$$G^{(5)}_{AB} = -\Lambda_5 g^{(5)}_{AB} + k_5^2 T^{(5)}_{AB}. \tag{2}$$

The sub-and superscripts '(5)' indicate that the quantity is 5-dimensional.

As the phenomenology is accessible only on the brane, we need to project this equation onto the brane. Besides the projection, we need to take into account junction conditions

$$g^-_{\mu\nu} - g^+_{\mu\nu} = 0, \tag{3}$$

$$K^-_{\mu\nu} - K^+_{\mu\nu} = -k_5^2 (T^{brane}_{\mu\nu} - \frac{1}{3} T^{brane}), \tag{4}$$

where T^{brane} is the energy-momentum tensor on the brane which is defined as

$$T^{brane}_{\mu\nu} = T_{\mu\nu} - \lambda g_{\mu\nu}, \tag{5}$$

where λ is the brane tension, and $K_{\mu\nu}$ is the extrinsic curvature. As a result, one obtains the known junction condition for the Lie derivative of the metric

$$K_{\mu\nu} = -\frac{1}{2} k_5^2 (T_{\mu\nu} + \frac{1}{3}(\lambda - T) g_{\mu\nu}). \tag{6}$$

For more details on the derivation of the above equation see Ref. [1]. Using the junction conditions after the projection, one obtains the field equations on the brane (see again Ref. [1])

$$G_{\mu\nu} = -\Lambda g_{\mu\nu} + k^2 T_{\mu\nu} + 6\frac{k^2}{\lambda} S_{\mu\nu} - \epsilon_{\mu\nu} + 4\frac{k^2}{\lambda} F_{\mu\nu}. \tag{7}$$

Here $\epsilon_{\mu\nu}$ is a Weyl tensor, which provides the wave equations for Kaluza-Klein modes if we apply the covariant derivative to the field equation and take into account Bianchi identity and energy conservation on the brane. Like the Weyl tensor, $S_{\mu\nu}$ is also a high energy term and contributes to the wave equation. $S_{\mu\nu}$ is defined as

$$S_{\mu\nu} = \frac{1}{12}TT_{\mu\nu} - \frac{1}{4}T_{\mu\alpha}T^{\mu}_{\alpha} + \frac{1}{24}[3T_{\alpha\beta}T^{\alpha\beta} - T^2]. \tag{8}$$

The last term in (7) vanishes if we take into account that there is no exchange between bulk and brane. Indeed, if one applies the Bianchi identity and energy conservation on the brane, one obtains $F_{\mu\nu} = 0$ with $F_{\mu\nu}$ defined as

$$F_{\mu\nu} = T^{(5)}_{AB}g^{A}_{\mu}g^{B}_{\nu} + \left(T^{(5)}_{AB}n^{A}n^{B} - \frac{1}{4}T^{(5)}\right). \tag{9}$$

Interestingly, after the projection, new terms appear proportional to the square of the energy-momentum tensor. Junction condition and brane tension relate bulk with the brane, and so the brane holds information about the bulk. Therefore, we can expect that brane-world effects and corrections can appear in high energy processes due to S. For more details on Eq. (7) able see Ref. [1]. In the next section we will provide techniques of our calculation which is based on Taylor expansion.

3. Taylor expansion of metric

The field equation can not be exactly solved, so we have to apply different approaches to find some approximate solution. One successful and simple technique is an expansion in Taylor series. Thus, by using Lie derivatives and the definition of the extrinsic curvature $K_{\mu\nu}$, the equations above and the junction condition provide the bulk metric near the brane in terms of brane metric exclusively. This is achieved carrying out a Taylor expansion of the metric along the extra dimension y in Gaussian coordinates.

This approach is discussed in several works (see Refs. [1, 2, 3]). We present just the final result from Ref. [3]. The bulk metric obtained from the metric on the brane, up to the fourth order along the Gaussian coordinate y, reads

$$\begin{aligned}
g_{\mu\nu}(x^\alpha, y) &= g_{\mu\nu} - \kappa_5^2\left[T_{\mu\nu} + \frac{1}{3}(\lambda - T)g_{\mu\nu}\right]|y| \\
&+ \left[\frac{1}{2}\kappa_5^4\left(T_{\mu\alpha}T^{\alpha}{}_{\nu} + \frac{2}{3}(\lambda - T)T_{\mu\nu}\right) - 2\mathcal{E}_{\mu\nu} + \frac{1}{3}\left(\frac{1}{6}\kappa_5^4(\lambda - T)^2 - \Lambda_5\right)g_{\mu\nu}\right]\frac{y^2}{2!} \\
&+ \Bigg[2K_{\mu\beta}K^{\beta}{}_{\alpha}K^{\alpha}{}_{\nu} - \mathcal{E}_{(\mu|\alpha}K^{\alpha}{}_{|\nu)} - \nabla^{\rho}\mathcal{B}_{\rho(\mu\nu)} + \frac{1}{6}\Lambda_5 g_{\mu\nu}K + K^{\alpha\beta}R_{\mu\alpha\nu\beta} - K\mathcal{E}_{\mu\nu} \\
&\qquad +3K^{\alpha}{}_{(\mu}\mathcal{E}_{\nu)\alpha} + K_{\mu\alpha}K_{\nu\beta}K^{\alpha\beta} - K^2K_{\mu\nu}\Bigg]\frac{|y|^3}{3!} \\
&+ \Bigg[\frac{\Lambda_5}{6}\left(R - \frac{\Lambda_5}{3} + K^2\right)g_{\mu\nu} + \left(\frac{K^2}{3} - \Lambda_5\right)K_{\mu\alpha}K^{\alpha}{}_{\nu} + (R - \Lambda_5 + 2K^2)\mathcal{E}_{\mu\nu} \\
&+ \left(K^{\alpha}{}_{\tau}K^{\tau\beta} + \mathcal{E}^{\alpha\beta} + KK^{\alpha\beta}\right)R_{\mu\alpha\nu\beta} + K^2\,K\,K_{\mu\nu} - \frac{1}{6}\Lambda_5 R_{\mu\nu} + 2K_{\mu\beta}K^{\beta}{}_{\rho}K^{\rho}{}_{\alpha}K^{\alpha}{}_{\nu} \\
&+ \frac{7}{2}KK^{\alpha}{}_{\mu}\mathcal{E}_{\nu\alpha} - \frac{7}{6}K^{\sigma\beta}K_{\mu}{}^{\alpha}R_{\nu\sigma\alpha\beta} + \mathcal{E}_{\mu\alpha}\left(\frac{1}{2}KK^{\alpha}{}_{\nu} - 3K^{\alpha}{}_{\sigma}K^{\sigma}{}_{\nu}\right) \\
&- \frac{13}{2}K_{\mu\beta}\mathcal{E}^{\beta}{}_{\alpha}K^{\alpha}{}_{\nu} - 4K^{\alpha\beta}R_{\mu\nu\gamma\alpha}K^{\gamma}{}_{\beta} - K_{\mu\alpha}K_{\nu\beta}\mathcal{E}^{\alpha\beta}\Bigg]\frac{y^4}{4!} + \cdots
\end{aligned} \tag{10}$$

where $C \equiv C^{\mu}{}_{\mu}$ and $C^2 \equiv C_{\rho\sigma}C^{\rho\sigma}$, for any tensor C that has rank 2, and $g_{\mu\nu}(x^\alpha, 0) \equiv g_{\mu\nu}$ is the metric on the brane. Given the 5D Weyl tensor $C_{\mu\nu\sigma\rho}$ one commonly defines the electric $\mathcal{E}_{\mu\nu} = C_{\mu\nu\sigma\rho}n^{\sigma}n^{\rho}$ and the magnetic $\mathcal{B}_{\mu\nu\alpha} = g_{\mu}^{\rho}g_{\nu}^{\sigma}C_{\rho\sigma\alpha\beta}n^{\beta}$ Weyl tensor components.

We need a metric for vacuum on the brane which reads:

$$\begin{aligned} g_{\mu\nu}(x,y) &= g_{\mu\nu} - \frac{1}{3}\kappa_5^2\lambda g_{\mu\nu}\,|y| + \left[\frac{1}{6}\left(\frac{1}{6}\kappa_5^4\lambda^2 - \Lambda_5\right) g_{\mu\nu} - \mathcal{E}_{\mu\nu}\right] y^2 \\ & -\frac{1}{6}\left[\left(\frac{193}{36}\lambda^3\kappa_5^6 + \frac{5}{3}\Lambda_5\kappa_5^2\lambda\right) g_{\mu\nu} + \kappa_5^2 R_{\mu\nu}\right]\frac{|y|^3}{3!} + \\ & + \left[\frac{1}{6}\Lambda_5\left(\left(R - \frac{1}{3}\Lambda_5 - \frac{1}{18}\lambda^2\kappa_5^4\right) + \frac{7}{324}\lambda^4\kappa_5^8\right) g_{\mu\nu} + \left(R - \Lambda_5 + \frac{19}{36}\lambda^2\kappa_5^4\right)\mathcal{E}_{\mu\nu}\right. \\ & \left. +\frac{1}{6}\left(\frac{37}{36}\lambda^2\kappa_5^4 - \Lambda_5\right) R_{\mu\nu} + \mathcal{E}^{\alpha\beta}\, R_{\mu\alpha\nu\beta}\right]\frac{y^4}{4!} + \cdots \end{aligned} \tag{11}$$

Therefore, in order to proceed, $g_{\mu\nu}$ must be chosen. There are solutions that can be obtained on the brane, which differ from Schwarzschild one. These are the so called Casadio-Fabbri-Mazzacurati (CFM) brane-world solutions [4, 5]. In general, a static spherically symmetric metric can be written as follows

$$g_{\mu\nu}dx^\mu dx^\nu = -N(r)dt^2 + A(r)dr^2 + r^2 d\Omega^2. \tag{12}$$

CFM brane-world solutions have an additional post-Newtonian parameter β. Although for solar system scales such parameter satisfies the relation $|\beta - 1| \sim 10^{-4}$ [6], for extremal black holes this is not the case, as we shall see below. By A certain choice of β the black hole Hawking radiation can be suppressed on the brane so that we can analyze pure brane-world effect. Indeed for some values of β the Hawking radiation, although precluded on the brane, is not zero in the bulk, which makes the black hole to recoil into the bulk. Even more, for these solutions, Hawking radiation is altered and the evaporation of mini-black holes is slowed, so they can be detected at the LHC.

The CFM I brane-world solution reads [4]

$$N(r) = 1 - \frac{2GM}{c^2 r} \quad and \quad A(r) = \frac{1 - \frac{3GM}{2c^2 r}}{\left(1 - \frac{2GM}{c^2 r}\right)\left(1 - \frac{GM}{2c^2 r}(4\beta - 1)\right)}. \tag{13}$$

Hawking radiation for this CFM I brane-world solution is suppressed if $\beta = 5/4$.

The CFM II brane-world solution is given by [5]

$$N(r) = 1 - \frac{2GM}{c^2 r} + \frac{2G^2M^2}{c^4 r^2}(\beta - 1) \quad and \quad A(r) = \frac{1 - \frac{3GM}{2c^2 r}}{\left(1 - \frac{2GM}{c^2 r}\right)\left(1 - \frac{GM}{2c^2 r}(4\beta - 1)\right)}. \tag{14}$$

Now we need to choose $\beta = 3/2$ since for this value of β, Hawking radiation is suppressed and gives an opportunity to demonstrate pure brane-world effects. In the following Section, we show an imprint of brane-world solutions on the black-hole accretion rate, i.e. on the quasar luminosity, which is the physical quantity that is observed.

4. Accretion and brane-world effects

Accretion depends on the black hole mass and horizon and its rate is proportional to the derivative of mass so that for the luminosity L can be written as

$$L(\ell) = \eta(\ell)\dot{M}c^2, \tag{15}$$

where ℓ is the bulk curvature radius parameter. And accretion efficiency η is commonly given by $\eta = \frac{GM}{6c^2 R}$, where R denotes the black hole horizon. Thus, for both the CFM I and the CFM

II brane-world solutions, the Taylor expansion of Eq. (11) provides the black string horizon that equals the black hole horizon when it slightly recoils from the brane into the bulk, which is denoted by $R(y) = \sqrt{g_{\theta\theta}(x^\alpha, y)}$ and is given by Eq. (11). Then, we compare the luminosity with the Schwarzschild case R_S. Indeed, $R(y)$ is the black hole horizon when the black hole recoils from the brane. Therefore, the result for quasar luminosity correction can be given by the following form

$$\Delta L = \frac{GM}{6c^2}\left(R(y)^{-1} - R_S^{-1}\right)\dot{M}c^2 = \frac{1}{12}\left(\frac{R_S}{\sqrt{g_{\theta\theta}(x^\alpha, y)}} - 1\right)\dot{M}c^2. \quad (16)$$

All the analysis depends upon the previous expressions and we are ready to provide our main results in figures in the next Section.

5. Results

As a result, we illustrate how the luminosity varies as the black hole recoils into the extra dimension y. The most important and distinguishable feature of quasar luminosity in brane-world scenarios is its dependence on the black hole mass. In Fig. 1, we show plots of normalized correction of luminosity versus the size of extra dimension y, for different masses. We see that for masses lower than $M = 10^6 M_\odot$, the relative correction is a decreasing function of the extra dimension. Here $M_\odot$ denotes the mass of the sun. For the second model CFM II for all cases, the normalized correction of luminosity is always a decreasing function for all black hole masses, see Fig. 2. The correction is normalized by solar luminosity $L_\odot \sim 3.91 \times 10^{33}\mathrm{erg\,s^{-1}}$. For typical supermassive black hole of mass $M \approx 10^9 M_\odot$, the luminosity correction can be improved by two orders if we keep in the expansion up to the fourth order terms (instead of restricting to second order). Therefore, now the correction is $\Delta L \sim 2.1 \times 10^{-3}\ L_\odot$. In previous works it was found that $\Delta L \sim 1.0 \times 10^{-5}\ L_\odot$. Now we can investigate the dependence of the luminosity on the black hole mass as well as the extra dimension. Fig. 3 illustrates that the black hole mass does not influence the normalized correction of the quasar luminosity, as the CFM I black hole recoils into

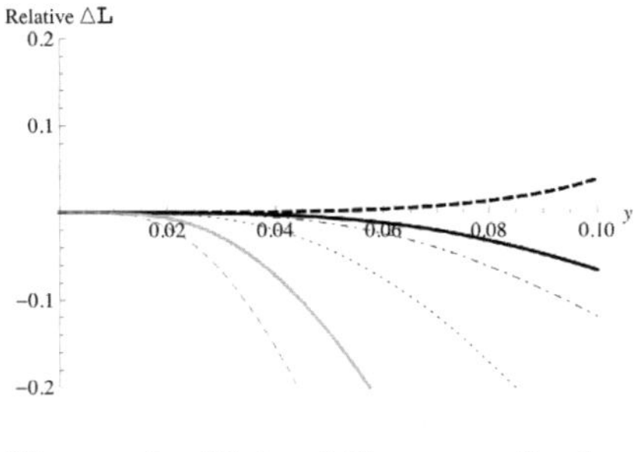

Figure 1. Plots of the normalized correction to luminosity $\Delta L/\dot{M}c^2$ in a CFM type I model as a function of the extra dimension y for different values of masses $M/M_\odot = 10^7$ (dashed black line) 10^6 (continuous black line), 10^5 (dash-dotted line), 10^4 (light-gray thick line), 10^3 (gray dashed line), 10^2 (dark-gray line).

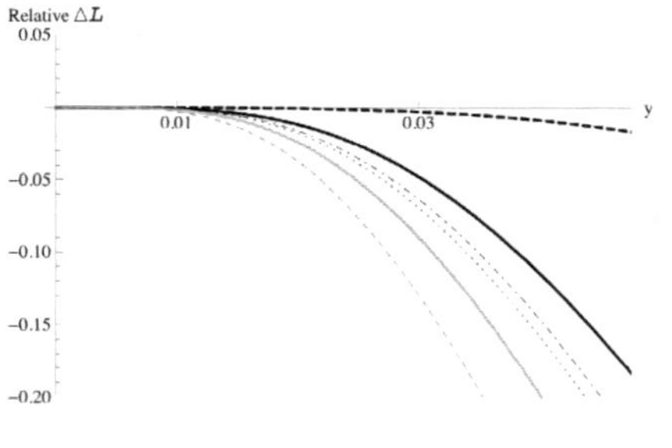

Figure 2. The same as in Fig. 1 but for the CFM II model. The labeling of curves is as follows: $M/M_\odot = 10^7$ - dashed black line, 10^6 - continuous black line, 10^5 - dash-dotted line, 10^4 - dark dotted line, 10^3 - light-gray thick line, for 10^2 - gray dashed line.

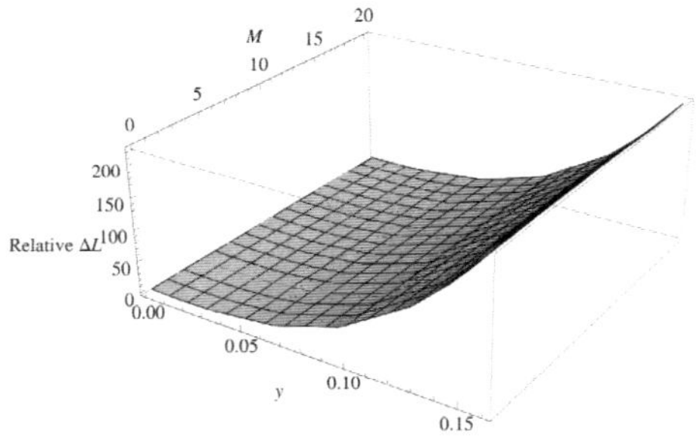

Figure 3. The normalized correction of luminosity $\Delta L/\dot{M}c^2$ in CFM type I model as function of the extra dimension y and the black hole mass $\ln M$.

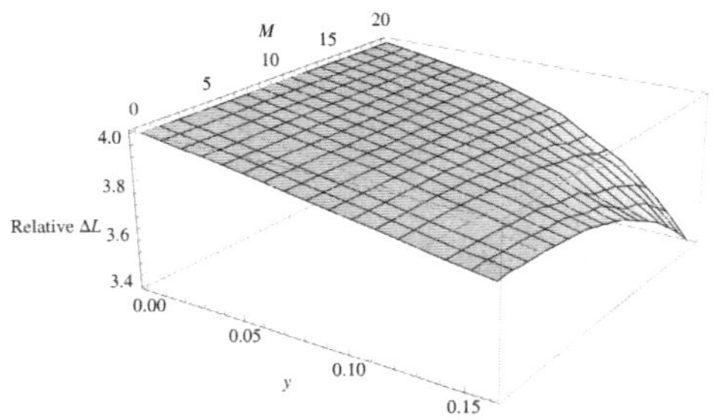

Figure 4. Same as in Fig. 3, but for CFM type II model.

the bulk. For all values for the mass in the range analysed, the relative luminosity is essentially the same. On the another hand, the CFM II black hole has a completely distinct behavior, for the same range of the black hole mass. As it can be seen at Fig. 4, as the mass varies in the range illustrated, the normalized correction of the quasar luminosity is as abrupt as the mass increases. Indeed, it is in full compliance to Eq.(14) describing the CFM II metric, that differs from the CFM I metric (13) solely by the post-Newtonian correction for the Schwarzschild term $N(r)$ in (13). As such term contains the a term M^2, the profile in Fig. 4 is expected.

6. Conclusion

In this work we carried out an analysis of possible signatures of extra-dimensional theories. Our analysis shows that there could be variations in quasar luminosity in different brane-world scenarios. These variations provide a promising observational test of these intriguing theories. The corrections to the quasar luminosity could be three-orders of magnitude and may be detectable in future observations. While these corrections emerge at all the wavelengths of radiation, they should be still detectable for quasars with their narrow range of radiation wavelengths. In particular, the corrections to the luminosity of a quasar (which is modeled as a black hole combined with the canonical model of accretion) show a new feature: the luminosity is a function of black hole's mass, see Figs. 1 and 2. We find that even if the size of the extra-dimension is very small, the correction is detectable for the both models (CFM I and CFM II) considered in this study. The work on other models, which feature non-constant accretion rate and time dependent mass of black hole is in progress.

Acknowledgments

R. da Rocha is grateful for the support by the CNPq grants 303027/2012-6 and 480482/2012-8.

References

[1] Maartens R and Koyama K 2010 *Living Rev. Relativity* **13** 5
[2] da Rocha R and da Silva J H 2012 *Phys. Rev. D* **85** 046009
[3] da Rocha R, Piloyan A, Kuerten A M and Coimbra-Araujo C H 2013 *Class. Quantum Grav.* **30** 045014
[4] Casadio R, Fabbri A and Mazzacurati L 2002 *Phys. Rev. D* **65** 084040
[5] Casadio R and Mazzacurati L 2003 *Mod. Phys. Lett. A* **18** 651–660
[6] Williams J G, Turyshev S G and Boggs D H 2004 *Phys. Rev. Lett.* **93** 261101

Quantum effects from boundaries in de Sitter and anti-de Sitter spacetimes

A. A. Saharian

Department of Physics, Yerevan State University, 1 Alex Manoogian Street, 0025 Yerevan, Armenia

E-mail: saharian@ysu.am

Abstract. We discuss the features of the Casimir effect for two parallel plates in background of de Sitter and anti-de Sitter spacetimes. A massive scalar field with general curvature coupling parameter obeying Robin boundary conditions on the plates is considered. The corresponding results are compared with those in Minkowski spacetime.

1. Introduction

The Casimir effect (for a review see [1]) is now known to be common to systems of very different kinds, involving fluctuating quantities on which external boundary conditions are imposed. It can have important implications on all scales, from subnuclear to cosmological. An interesting topic in the investigations of the Casimir effect is its explicit dependence on the geometry of background spacetime. The relevant information is encoded in the vacuum fluctuations spectrum and analytic solutions can be found for highly symmetric geometries only. In the present paper, we will discuss the Casimir stresses on parallel plates for a scalar field in background of maximally symmetric spaces with zero, positive and negative curvature corresponding to Minkowski, de Sitter (dS) and anti-de Sitter (AdS) spacetimes, respectively.

2. Casimir forces in Minkowski spacetime

Consider a scalar field $\varphi(x)$ with the curvature coupling parameter ξ. The corresponding field equation has the form

$$(\nabla_l \nabla^l + m^2 + \xi \mathcal{R})\varphi = 0, \tag{1}$$

where ∇_l is the covariant derivative operator and $\mathcal{R}$ is the Ricci scalar of the background spacetime. The values of the curvature coupling parameter $\xi = 0$ and $\xi = \xi_D \equiv (D-1)/4D$, with D being the number of spatial dimensions, correspond to the most important special cases of minimally and conformally coupled fields. The background geometry is described by the line element

$$ds^2 = R^2 \eta_{ik} dz^i dz^k = R^2 [dt^2 - \sum\nolimits_{i=1}^{D} (dz^i)^2], \tag{2}$$

where η_{ik} is the Minkowskian metric tensor. We will consider special cases $R = 1$ (Minkowski spacetime), $R = \alpha/|t|$ (dS spacetime) and $R = \alpha/z^D$ (AdS spacetime).

Our main interest is the Casimir forces in the geometry of two infinite, parallel plates located at $z^D = a_j$, $j = 1, 2$. On the plates the field obeys Robin boundary conditions (BCs)

$$(1 + \beta_j n^l_{(j)} \partial_l)\varphi(x) = 0, \quad z^D = a_j, \tag{3}$$

with $n^l_{(j)}$ being the inward-pointing unit normal to the boundary at $z^D = a_j$ (with respect to the region under consideration). Robin type conditions are an extension of Dirichlet ($\beta_j = 0$) and Neumann ($\beta_j = \infty$) BCs and appear in a variety of situations, including the considerations of vacuum effects for a confined charged scalar field in external fields, spinor and gauge field theories, quantum gravity and supergravity. The Robin BCs may, in some geometries, be useful for depicting the finite penetration of the field into the boundary with the 'skin-depth' parameter related to the Robin coefficient. For the region between the plates one has $n^l_{(j)} = (-1)^{j-1}\delta^l_D/R_j$, where $R_j = R|_{z^D=a_j}$. In this case the BCs are written in the form $(1 + (-1)^{j-1} b_j \partial_D)\varphi(x) = 0$, with $b_j = \beta_j/R_j$. In the discussion below we assume that $b_j = \text{const}$. The imposition of BCs leads to a modification of the spectrum for vacuum fluctuations of quantum fields and, as a result, to the changes in the vacuum expectation values (VEVs) for physical quantities. In particular, forces arise acting on constraining boundaries.

First we consider the case of the Minkowski spacetime corresponding to $R = 1$ in (2). The VEV of the energy-momentum tensor, $\langle T^k_i \rangle = \langle 0|T^k_i|0\rangle$, has been investigated in [2]. The normal force acting per unit surface of the plates, p, is determined by the VEV of the normal stress $\langle T^D_D \rangle$: $p = -\langle T^D_D \rangle$. This stress is uniform in the region between the plates and vanishes in the regions $z^D < 0$ and $z^D > 0$ (note that, in general, this is not the case for the energy density and parallel stresses). The Casimir force per unit area (the vacuum effective pressure) is given by

$$p = -\frac{2(4\pi)^{-D/2}}{\Gamma(D/2)} \int_m^\infty dt \, \frac{t^2(t^2 - m^2)^{D/2-1}}{\frac{\beta_1 t-1}{\beta_1 t+1}\frac{\beta_2 t-1}{\beta_2 t+1} e^{2at} - 1}. \tag{4}$$

The force is attractive/repulsive for negative/positive values of p. Note that the Casimir force is the same for both plates and it does not depend on the curvature coupling parameter. The vacuum densities and the Casimir forces for a massive scalar field, subject to Robin boundary conditions on two codimension-one parallel plates, located on a $(D+1)$-dimensional background spacetime with an arbitrary internal space have been investigated in [3]. For Dirichlet and Neumann scalars the forces are attractive: $p_{\rm D} = p_{\rm N} < 0$. For Dirichlet BC on one plate and the Neumann one on the other, the force is repulsive: $p_{\rm DN} < 0$. For a massless field one gets

$$p_{\rm D} = p_0 = -\frac{D\zeta_R(D+1)}{(4\pi)^{(D+1)/2} a^{D+1}} \Gamma\left(\frac{D+1}{2}\right), \; p_{\rm DN} = -(1 - 2^{-D})p_0, \tag{5}$$

with $\zeta_R(x)$ being the Riemann zeta function.

For small separations between the plates, $ma \ll 1$, $a \ll |\beta_j|$, to the leading order one has $p \approx p_0$. The same is the case for Dirichlet BCs on both plates. For Dirichlet BC on one plate and non-Dirichlet BC on the other one has $p \approx p_{\rm DN}$. At large separations, $ma \gg 1$, $a \gg |\beta_j|$, in the leading order we get $p \approx -m^{D/2+1}(4\pi a)^{-D/2}e^{-2ma}$ for a massive field and $p \approx p_0$ for a massless field. The corresponding forces are attractive. For the Neumann BC on one plate and the non-Neumann BC on the other, at large distances we have $p \approx m^{D/2+1}(4\pi a)^{-D/2}e^{-2ma}$ and $p \approx -(1 - 2^{-D})p_0$, for massive and massless fields, respectively. In this case the forces are repulsive. As we see, for Dirichlet BC on one plate and non-Dirichlet BC on the other the forces are repulsive at small separations and attractive at large separations. At some intermediate separation the force vanishes and this equilibrium position is stable. For Neumann BC on one plate and non-Neumann BC on the other the forces are attractive at small separations and repulsive at large separations. In this case the equilibrium position is unstable.

The results given above can be generalized for non-local BCs on the parallel plates [4]:

$$n^l_{(j)}\partial_l\varphi(x) + \int d\mathbf{z}'_{\parallel}\, f_j(|\mathbf{z}_{\parallel} - \mathbf{z}'_{\parallel}|)\varphi(x') = 0,\ z^D = a_j, \tag{6}$$

where $\mathbf{z}_{\parallel} = (z^1, \ldots, z^{D-1})$. These BCs state that the normal derivative at a given point depends on the values of the field at other points on the boundary. The kernel functions f_j determine the properties of the boundaries. The vacuum force acting per unit surface of the plate:

$$p = -\frac{2^{2-D}\pi^{-(D+1)/2}}{\Gamma((D-1)/2)}\int_0^\infty du \int_{\sqrt{u^2+m^2}}^\infty \frac{u^{D-2}t^2dt}{\sqrt{t^2-u^2-m^2}}\left[\frac{(t-F_1(u))(t+F_2(u))}{(t+F_1(u))(t-F_2(u))}e^{2at}-1\right]^{-1}, \tag{7}$$

where

$$F_j(x) = \frac{(2\pi)^{(D-1)/2}}{x^{(D-3)/2}}\int_0^\infty du\, u^{(D-1)/2} f_j(u) J_{(D-3)/2}(ux), \tag{8}$$

and $J_\nu(z)$ is the Bessel function. Examples of functions $f_j(z)$ have been considered in [12].

3. Casimir Forces in dS spacetime

In this section we consider the Casimir forces in dS spacetime. This spacetime is among the most popular backgrounds in gravitational physics. It is the maximally symmetric solution of Einstein's equation with a positive cosmological constant and due to this high symmetry numerous physical problems are exactly solvable on this background. A better understanding of physical effects in this background could serve as a handle to deal with more complicated geometries. De Sitter spacetime also plays an important role in most inflationary models. More recently astronomical observations of high redshift supernovae, galaxy clusters and cosmic microwave background indicate that at the present epoch the universe is accelerating and can be well approximated by a world with a positive cosmological constant.

In the case of dS spacetime described in inflationary coordinates, the line element is given by (2) with the function $R = \alpha/|t|$, where t is the conformal time coordinate, $-\infty < t < 0$. The latter is related to the synchronous time t_s by $t = -\alpha e^{-t_s/\alpha}$ and $R = e^{t_s/\alpha}$. For the Ricci scalar one has $\mathcal{R} = D(D+1)/\alpha^2$. In the construction of a quantum field theory in a fixed classical gravitational background, the choice of the vacuum state is among the most important steps. dS spacetime is a maximally symmetric, and it is natural to choose a vacuum state having the same symmetry. In fact, there exists a one-parameter family of maximally symmetric quantum states. Here we will assume that the field is prepared in the dS-invariant Bunch-Davies vacuum state [5]. Among the set of dS-invariant quantum states, the Bunch-Davies vacuum is the only one for which the ultraviolet behavior of the two-point functions is the same as in Minkowski spacetime.

The VEVs of the field squared and the energy-momentum tensor in the geometry of parallel plates in background of dS spacetime have been investigated in [6, 7]. The vacuum force acting per unit surface of the plate at $z^D = a_j$ is determined by the ^D_D-component of the vacuum energy-momentum tensor evaluated at this point. For the region between the plates, the corresponding effective pressures are presented as $p^{(j)} = p_1^{(j)} + p_{(\text{int})}^{(j)}$, $j = 1, 2$. The first term, $p_1^{(j)}$, is the pressure for a single plate at $z^D = a_j$, when the second plate is absent. This term is divergent due to surface divergences in the subtracted VEVs and needs additional renormalization. The term $p_{(\text{int})}^{(j)}$ is the pressure induced by the second plate, and can be termed as an interaction force. This contribution is finite for all nonzero distances between the plates. In the regions $z^D < a_1$ and $z^D > a_2$ we have $p^{(j)} = p_1^{(j)}$. As a result, the contributions to the vacuum force coming from the term $p_1^{(j)}$ are the same from the left and from the right sides of the plate, so there is no net contribution to the effective force.

The interaction force on the plate at $z^D = a_j$ is given by the expression [7]:

$$p_{(\mathrm{int})}^{(j)} = -\frac{8(4\pi)^{-(D+1)/2}}{\Gamma((D-1)/2)\alpha^{D+1}} \int_0^\infty dy\, y^{1-D} \int_y^\infty dx\, \frac{x^2(x^2-y^2)^{(D-3)/2}}{\frac{b_1 x/\eta-1}{b_1 x/\eta+1}\frac{b_2 x/\eta-1}{b_2 x/\eta+1}e^{2ax/\eta}-1} \times \left[\frac{2\,(b_j/\eta)^2\, G_\nu(y)}{(b_j x/\eta)^2-1} + F_\nu(y)\right], \tag{9}$$

where $\eta = |t|$, and we introduced the notations

$$F_\nu(y) = y^D\left[I_\nu(y) + I_{-\nu}(y)\right] K_\nu(y),\ \nu = \left[D^2/4 - D(D+1)\xi - m^2\alpha^2\right]^{1/2},$$
$$G_\nu(y) = \left\{(\xi - 1/4)\, y^2\partial_y^2 + \left[\xi(2-D) + (D-1)/4\right] y\partial_y - \xi D\right\} F_\nu(y). \tag{10}$$

In (10), $I_\nu(y)$ and $K_\nu(y)$ are the modified Bessel functions. The time dependence of the forces appears in the form a/η and b_j/η. Note that the ratio a/η is the proper distance between the plates measured in units of dS curvature radius α. Unlike to the Minkowskian case, for $\beta_1 \neq \beta_2$ the Casimir forces acting on the left and on the right plates are different. For large values of α, to the leading order, the corresponding result for the geometry of two parallel plates in Minkowski spacetime is obtained.

In the special cases of Dirichlet and of Neumann BCs one finds:

$$p_{(\mathrm{int})}^{(\mathrm{D})} = -\frac{4\alpha^{-D-1}}{(2\pi)^{D/2+1}} \sum_{n=1}^{\infty} \int_0^\infty dy\, y F_\nu(y) \left[(D-1) f_{D/2}(yu_n) + f_{D/2-1}(yu_n)\right], \tag{11}$$

$$p_{(\mathrm{int})}^{(\mathrm{N})} = p_{(\mathrm{int})}^{(\mathrm{D})} - \frac{8\alpha^{-D-1}}{(2\pi)^{D/2+1}} \sum_{n=1}^{\infty} \int_0^\infty dy\, \frac{G_\nu(y)}{y} f_{D/2-1}(yu_n),\ u_n = 2na/\eta, \tag{12}$$

where $f_\mu(x) = K_\mu(x)/x^\mu$. For $0 \leqslant \nu < 1$ the integrand in the expression for $p_{(\mathrm{int})}^{(\mathrm{D})}$ is positive which corresponds to an attractive force for all separations.

In the limit of small proper distances between the plates, $a/\eta \ll 1$, the effects induced by the curvature of dS spacetime are subdominant and to leading order we find the result for the Minkowski spacetime, given in the previous section, with the replacement $a \to \alpha a/\eta$.

In considering the large distance asymptotics, corresponding to $a/\eta \gg 1$, the cases of real and imaginary ν must be studied separately. For positive values of ν, one has $p_{(\mathrm{int})}^{(j)} \propto (a/\eta)^{2\nu-D-2}$ for non-Neumann BC on the plate at $z^D = a_j$ ($|b_j| < \infty$) and $p_{(\mathrm{int})}^{(j)} \propto (a/\eta)^{2\nu-D}$ for Neumann BC ($b_j = \infty$). In the case of non-Neumann BC we have assumed that $|b_j|/a \ll 1$. For positive values of ν, at large distances the ratio of the Casimir forces acting on the plate with Neumann and non-Neumann BCs is of the order $(a/\eta)^2$. In neither of these cases does the leading order term in the force depend on the specific value of Robin coefficient in the BC on the second plate. For Dirichlet BC on the plate at $z^D = a_j$ ($b_j = 0$), at large separations the Casimir force acting on that plate is repulsive (attractive) for Neumann (non-Neumann) BCs on the other plate. The nature of the force acting on the plate with Neumann BC can be either repulsive or attractive, in function of the curvature coupling parameter and of the field mass. For minimally and conformally coupled massive scalar fields the corresponding force is attractive (repulsive) for Neumann (non-Neumann) BC on the second plate. For imaginary ν, the leading order terms at large separations between the plates are in the form $p_{(\mathrm{int})}^{(j)} \propto (a/\eta)^{-D-2}\cos[2|\nu|\ln(2a/\eta) + \phi_{(j)}]$ for $|b_j| < \infty$, and $p_{(\mathrm{int})}^{(j)} \propto (a/\eta)^{-D}\cos[2|\nu|\ln(2a/\eta) + \phi_{(j)}^{\mathrm{N}}]$ for $b_j = \infty$ (Neumann BC). In this case the decay of the vacuum forces is oscillatory.

From the discussion given above it follows that for proper distances between the plates larger than the curvature radius of the dS spacetime, $\alpha a/\eta \gtrsim \alpha$, the gravitational field essentially changes the behavior of the Casimir forces compared with the case of the plates in Minkowski spacetime. The forces may become repulsive at large separations between the plates. Recall that, for the geometry of parallel plates on the background of Minkowski spacetime, the only case with repulsive Casimir forces at large distances corresponds to Neumann BC on one plate and non-Neumann BC on the other. A remarkable feature of the influence of the gravitational field is the oscillatory behavior of the Casimir forces at large distances, which appears in the case of imaginary ν. In this case, the values of the plate distance yielding zero Casimir force correspond to equilibrium positions. Among them, the positions with negative derivative of the force with respect to the distance are locally stable. At large separations between the plates the decay of the Casimir forces as functions of the distance is power-law for both cases of massive and massless fields. Recall that, in Minkowski spacetime the corresponding Casimir forces are exponentially suppressed by the factor $\exp(-2ma)$ for a massive filed.

4. AdS spacetime

Now let us turn to the AdS spacetime as a background geometry. Much of the earlier interest in this geometry was motivated by questions of principal nature, mainly related with the quantization of fields on curved backgrounds. Further interest in this subject arose from the discovery that the AdS spacetime generically arises as a ground state in extended supergravity and string theories, what is again potentially most important. In recent developments of the topic, the AdS geometry is an arena for two classes of models. The first is the AdS/CFT correspondence, which relates string theories or supergravity in the AdS bulk with a conformal field theory living on its boundary. The second class of models with the AdS spacetime as background geometry is a realization of the braneworld scenario with large extra dimensions and provides a solution to the hierarchy problem which arises between the gravitational and electroweak mass scales.

For AdS spacetime described in Poincare coordinates the line element is given by (2) with $R = \alpha/z^D$ and for the Ricci scalar one has $\mathcal{R} = -D(D+1)/\alpha^2$. The coordinate z^D is related to the coordinate y, measuring the proper distance from the plates, by $z^D = \alpha e^{y/\alpha}$. The VEVs of the field squared and the energy-momentum tensor in the geometry of two parallel plates have been investigated in [8] (see also [9] for the case of a conformally coupled massless field). The corresponding effective pressures on the plates can be presented as a sum of two terms: $p^{(j)} = p_1^{(j)} + p_{(\mathrm{int})}^{(j)}$, $j = 1, 2$. The first term on the right is the pressure for a single plate at $z^D = a_j$ when the second plate is absent. This term is divergent due to the surface divergences in the VEVs. For the interaction force per unit surface of the plate at $z^D = a_j$ one has:

$$p_{(\mathrm{int})}^{(j)} = \frac{\alpha^{-D-1}a_j^D}{(4\pi)^{D/2}\Gamma(D/2)}\int_0^\infty dx\, x^{D-1}C_{j\mu}(xa_1, xa_2) \times \frac{(x^2a_j^2 - \mu^2 + 2m^2\alpha^2)B_j^2 - D(4\xi - 1)A_jB_j - A_j^2}{\bar{K}_\mu^{(1)}(xa_1)\bar{I}_\mu^{(2)}(xa_2) - \bar{K}_\mu^{(2)}(xa_2)\bar{I}_\mu^{(1)}(xa_1)}. \quad (13)$$

where $\mu = \sqrt{D^2/4 - D(D+1)\xi + m^2\alpha^2}$ and we have introduced the functions

$$C_{1\mu}(u,v) = \bar{K}_\mu^{(2)}(v)/\bar{K}_\mu^{(1)}(u),\ C_{2\mu}(u,v) = \bar{I}_\mu^{(1)}(u)/\bar{I}_\mu^{(2)}(v), \quad (14)$$

The barred notations for $j = 1, 2$ are defined as

$$\bar{F}^{(j)}(x) = A_jF(x) + B_jxF'(x),\ A_j = 1 + (-1)^{j-1}(\beta_j/a_j)D/2,\ B_j = (-1)^{j-1}\beta_j/a_j. \quad (15)$$

The parameter μ must be real to ensure stability [10]. For given β_j/a_j, $j = 1, 2$, the interaction forces depend on a_1 and a_2 in the form of the ratio a_2/a_1, which is related to the proper distance between the plates by $a = \alpha \ln(a_2/a_1)$. As it is seen from (13), the forces acting on the plates are not symmetric under the interchange of the brane indices. It can be seen that $p_{(\text{int})}^{(j)}$ is negative for Dirichlet scalar and for a scalar with $A_1 = A_2 = 0$. The corresponding interaction forces are attractive for all values of the interplate distance.

At small separations compared with the AdS curvature radius, $a \ll \alpha$, the leading terms in the asymptotic expansion of the forces concides with the expressions for the plates in the Minkowski bulk. At large separations, $a \gg \alpha$ (this limit is realized in the Randall-Sundrum model), one has $p_{(\text{int})}^{(1)} \propto e^{-(D+2\mu)a/\alpha}$ and $p_{(\text{int})}^{(2)} \propto e^{-2\mu a/\alpha}$. Note that in AdS spacetime the exponential suppression of the forces with the separation between the plates takes place for a massless fields as well. This is in contrast with the cases of Minkowski and dS spacetimes. In dependence of the values for the coefficients in the boundary conditions, the corresponding forces can be either attractive or repulsive. The expressions of the forces for an untwisted scalar in the $(D + 1)$-dimensional Randall-Sundrum braneworld model are obtained from (13) with an additional factor $1/2$ and with

$$A_1/B_1 = [D(1-4\xi) - \alpha c_1]/2, \quad A_2/B_2 = [D(1-4\xi) + \alpha c_2]/2, \tag{16}$$

with c_j being the brane mass term on the brane $z^D = a_j$. For a twisted scalar one has Dirchlet BCs on both branes.

The VEV of the surface energy-momentum tensor in the geometry of two parallel plates on AdS bulk is evaluated in [11]. It has been shown that in the Randall-Sundrum braneworld model, for the interbrane distances solving the hierarchy problem, the cosmological constant generated on the visible brane is of the right order of magnitude with the value suggested by the cosmological observations. The Casimir densities and forces in AdS spacetime with a warped internal compact space have been discussed in [12, 13].

5. Conclusion

From the analysis carried out above, it follows that the curvature of the background spacetime decisively influences the behavior of the Casimir forces at distances larger than the curvature scale. As we have seen, in dS spacetime the decay of the forces at large separations between the plates is power-law. This is quite remarkable and clearly in contrast with the corresponding features of the same problem in Minkowski and AdS spacetimes.

The work was supported by State Committee Science MES RA, within the frame of the research project No. SCS 13-1C040.

References

[1] Bordag M Klimchitskaya G L Mohidden U and Mostepanenko V M 2009 *Advances in the Casimir Effect* (Oxford: Oxford University Press)
[2] Romeo A and Saharian A A 2002 J. Phys. A: Math. Gen. **35** 1297
[3] Elizalde E Odintsov S D and Saharian A A 2009 Phys. Rev. D **79** 065023
[4] Saharian A and Esposito G 2006 J. Phys. A: Math. Gen. **39** 5233
[5] Bunch T S and Davies P C W 1978 Proc. R. Soc. London A **360** 117
[6] Saharian A A and Vardanyan T A 2009 Class. Quantum Grav. **26** 195004
[7] Elizalde E Saharian A A and Vardanyan T A 2010 Phys. Rev. D **81** 124003
[8] Saharian A A 2005 Nucl. Phys. B **712** 196
[9] Saharian A A and Setare M R 2003 Phys. Lett. B **552** 119
[10] Breitenlohner P and Freedman D Z 1982 Ann. Phys. (N.Y.) **144** 249
[11] Saharian A A 2004 Phys. Rev. D **70** 064026
[12] Saharian A A 2006 Phys. Rev. D **73** 044012
[13] Saharian A A 2006 Phys. Rev. D **73** 064019

Journal of Physics: Conference Series **496** (2014) 012030 doi:10.1088/1742-6596/496/1/012030

On reduction of t e general t ree body ewtonian problem and t e cur ed geometry

A S Gevorkyan[1,2]

[1]Institute for Informatics and Automation Problems NAS of Armenia, 1 P. Sevak street, 0014 Yerevan, Republic of Armenia
[2] LIT JINR, 141980 Moscow region, Dubna, Russia

E-mail: g_ashot@sci.am

Abstract. In the framework of an idea of separation of rotational and vibrational motions, we have examined the problem of reducing the general three-body problem. The class of differentiable functions allowing transformation of the $6D$ Euclidean space to the $6D$ conformal-Euclidean space is defined. Using this fact the general classical three-body problem is formulated as a problem of geodesic flows on the energy hypersurface of the bodies system. It is shown that when the total potential depends on relative distances between the bodies, three from six ordinary differential equations of second order describing the non-integrable hamiltonian system are integrated exactly, thus allowing reducing the initial system in the phase space to the autonomous system of the 6th order. In the result of reducing of the initial Newtonian problem the geometry of reduced problem becomes curved. The latter gives us new ideas related to the problem of geometrization of physics as well as new possibilities for study of different physical problems.

1. Introduction

The general three-body classical problem concerns the question of understanding the motions of three arbitrary point masses traveling in space according to Newton's laws of mechanics. Many works on analytical mechanics, celestial mechanics, stellar and molecular dynamics (see for example [1, 2, 3, 4, 5]) are devoted to the study of this problem. For solution of the general problem different approaches based on series expansions methods have been proposed; however, due to the poor convergence of these expansions they are often used and are useful only for solving of particular problems where the system of three-bodies is in a stable bound state [2]. Moreover, the three-body problem is a typical example of a dynamical system where on the large scales of the phase space we observe all features of a complex motion including the bifurcation and chaos. That makes the numerical simulation method a basic way of research of the mentioned problems.

The main aim of this work is to find new opportunities for separation of the internal and external motions in the general classical three-body problem. The latter will have a key importance for reducing the dimensionality of the dynamical problem that will allow to develop an effective algorithm for numerical simulation.

Journal of Physics: Conference Series **496** (2014) 012030 doi:10.1088/1742-6596/496/1/012030

2. The classical three-body system in the laboratory frame

The classical Hamiltonian of the three-body system after Jacobi and mass-scale [6] transformations can be written in the form (see also [7]):

$$\mathrm{H}(\mathbf{r};\mathbf{p}) = \frac{\mathbf{p}^2}{2\mu_0} + \mathrm{V}(\mathbf{r}), \tag{1}$$

where $\mathbf{r} = \boldsymbol{r} \oplus \boldsymbol{R} \in R^6$ and $\mathbf{p} \in R^6$ are correspondingly the position vector and the momentum of the effective mass (imaginary point) $\mu_0 = [m_1 m_2 m_3/(m_1 + m_2 + m_3)]^{1/2}$, ($m_1, m_2$ and m_3 masses of the bodies). Note that $\boldsymbol{r}$ stands for the distance between the bodies 2 and 3, while $\boldsymbol{R}$ is the distance between the body 1 and the center of mass of the pair (2,3). In addition, the total potential $\mathrm{V}(\mathbf{r})$ depends on the distances between the bodies, that means that the interaction potential in fact depends on three variables.

Let us consider the following system of coordinates:

$$\rho_1 = r = ||\boldsymbol{r}||, \quad \rho_2 = R = ||\boldsymbol{R}||, \quad \rho_3 = \theta, \quad \rho_4 = \Theta, \quad \rho_5 = \Phi, \quad \rho_6 = \Psi, \tag{2}$$

where the first set of three coordinates; $\{\bar{\rho}\} = (\rho_1, \rho_2, \rho_3)$ describes a position of an imaginary point on the plane formed by bodies triangle, while; $\Theta \in (-\pi, +\pi]$, $\Phi = (-\pi, +\pi]$ and $\Psi \in [0, \pi]$ are Euler angles describing the rotation of the plane in $3D$ space. The kinetic energy of three-body system in these variables has the form (see also [8]):

$$T = \frac{1}{2\mu_0}\{\dot{\boldsymbol{r}}^2 + \dot{\boldsymbol{R}}^2\} = \frac{1}{2\mu_0}\Big\{\dot{r}^2 + r^2[\boldsymbol{\omega}\times\mathbf{k}]^2 + (\dot{\boldsymbol{R}} + [\boldsymbol{\omega}\times\boldsymbol{R}])^2\Big\}, \tag{3}$$

where the direction of unit vector $\mathbf{k}$ in the moving reference frame $\{\varrho\}$ is defined by expression: $\boldsymbol{R}\boldsymbol{R}^{-1} = \pm\mathbf{k}$. Below the vector $\mathbf{k} = (0,0,1)$ is directed towards a positive direction of the axis OZ, while the angular velocity $\boldsymbol{\omega}$ describes the rotation of the frame $\{\varrho\}$ with respect to laboratory system. By simple calculations in the expression (3) we can find:

$$T = \frac{1}{2\mu_0}\Big\{\dot{r}^2 + \dot{R}^2 + R^2\dot{\theta}^2 + Ar^2 + BR^2\Big\}, \tag{4}$$

where the following notations are introduced:

$$A = (\dot{\Theta}^2 + \dot{\Phi}^2\sin^2\Theta) = \omega_X^2 + \omega_Y^2, \qquad B = (\omega_X\cos\theta - \omega_Z\sin\theta)^2.$$

Let us note that in deriving the expression (4) we have used the definition of the moving system $\{\varrho\}$ by the requirement that unit vector $\boldsymbol{\gamma}$ lies in the plane OXZ forming angle θ to OZ, i.e. $\boldsymbol{\gamma} = (\sin\theta, 0, \cos\theta)$. As regards of the projections of an angular velocity then they satisfy the equations:

$$\omega_X = \dot{\Phi}\sin\Theta\sin\Psi + \dot{\Theta}\cos\Psi, \quad \omega_Y = \dot{\Phi}\sin\Theta\cos\Psi - \dot{\Theta}\sin\Psi, \quad \omega_Z = \dot{\Phi}\cos\Theta - \dot{\Psi}. \tag{5}$$

Taking into account (4) and (5) we can find the metric tensor:

$$\gamma^{\alpha\beta} = \begin{pmatrix} \gamma^{11} & 0 & 0 & 0 & 0 & 0 \\ 0 & \gamma^{22} & 0 & 0 & 0 & 0 \\ 0 & 0 & \gamma^{33} & 0 & 0 & 0 \\ 0 & 0 & 0 & \gamma^{44} & \gamma^{45} & \gamma^{46} \\ 0 & 0 & 0 & \gamma^{54} & \gamma^{55} & \gamma^{56} \\ 0 & 0 & 0 & \gamma^{64} & \gamma^{65} & \gamma^{66} \end{pmatrix} \tag{6}$$

where the following notations introduced:

$$\gamma^{11} = \gamma^{22} = 1, \quad \gamma^{33} = R^2, \quad \gamma^{44} = r^2 + R^2 \cos^2 \Psi \cos^2 \theta, \quad \gamma^{55} = r^2 \sin \Theta + R^2(\sin^2 \Theta \times$$

$$\sin^2 \Psi \cos^2 \theta + \cos^2 \Theta \sin^2 \theta - 2^{-1} \sin 2\Theta \sin 2\theta \sin \Psi), \quad \gamma^{66} = R^2 \sin^2 \theta, \quad \gamma^{45} = \gamma^{54} =$$

$$R^2(\sin \Theta \sin 2\Psi \cos^2 \theta - 2 \cos \Theta \cos \Psi \sin 2\theta), \quad \gamma^{46} = \gamma^{64} = R^2 \sin 2\theta \cos \Psi, \quad \gamma^{56} = \gamma^{65} =$$

$$R^2(\sin \Theta \sin \Psi \sin 2\theta - 2 \cos \Theta \sin^2 \theta).$$

Without going into details note that the considered problem has 12 integrals of motion. Using them, the initial 18th order system is reduced to the 8th order system [9].

3. The geodesic equations on the hypersurface of energy of three-body system

As it is easy to see, the classical system of three bodies moving in a $3D$ Euclidean space permanently forms the triangle, and hence Newton's equations describe the dynamical system on the space of such triangles [8]. The latter means that we can formally consider the motion of a system consisting of two parts. The first is the rotational motion of the body-triangle in the $3D$ Euclidian space and the second is the internal motion of bodies on the plane defined by the triangle. Mathematically, the configuration manifold of solid body R^6 can be represented as a direct product of two subspaces [10]:

$$R^6 :\Leftrightarrow R^3 \times S^3,$$

where R^3 is the manifold which is defined as an orthonormal space of relative distances between bodies while S^3 denotes the space of the rotation group $SO(3)$. However in the problem under consideration the connections between the bodies are not holonomic and consequently we must change the representation for the configuration manifold: $\mathbf{M} :\Leftrightarrow R^6$.

Let us consider the region of localization of the dynamical system (further named *the internal space* $\mathcal{M}_t$):

$$x^1 = ||\boldsymbol{r}|| \in [0, \infty), \quad x^2 = ||\boldsymbol{R}|| \in [0, \infty), \quad x^3 = ||\boldsymbol{r} + \boldsymbol{R}|| = \sqrt{(x^1)^2 - 2x^1x^2 \cos \theta + (x^2)^2} \in L, \tag{7}$$

where θ is the angle between the vectors $\boldsymbol{r}$ and $\boldsymbol{R}$, which in the Jacobi coordinates system is the scattering angle, in addition $L = [x^1 - x^2, x^1 + x^2]$. The set of *internal coordinatesis* giving by: $\{\bar{x}\} = (x^1, x^2, x^3) \in \mathcal{M}_t$. The rotation of a plane defined by body-triangle will be described by the set of three *external coordinates* $(x^4, x^5, x^6) \in S_t^3$, where S_t^3 is a space of the rotation group $SO(3)$ in a neighborhood of interior points $M_i\{(x^1, x^2, x^3)_i\} \in \mathcal{M}_t$. The subset of all interior points $\breve{\mathbf{M}} \subset \mathbf{M}$ is represented as:

$$\breve{\mathbf{M}} \cong \mathcal{M}_t \times S_t^3.$$

The set $\mathbf{M} \setminus \breve{\mathbf{M}}$ has zero measure. However, in some cases it can be important for the dynamics of the classical three-body system.

So, we can define a local system of coordinates in which further studies will be carried out:

$$\overline{x^1, x^6} = \{x\} \in \breve{\mathbf{M}}. \tag{8}$$

Taking into account the well-known work of Krylov [11], we will study the motion of three-body system on the hypersurface of potential energy (HPE) of bodies system. It is obvious that HPE is the curved space the metric tensor of which is defined by the relations:

$$g_{\mu\nu}(\{x\}) = g(\{x\})\delta_{\mu\nu}, \quad g^{\mu\nu} = g^{-1}(\{x\})\delta^{\mu\nu}, \quad g(\{x\}) = [E - U(\{x\})]U_0^{-1} > 0, \tag{9}$$

where E and $U(\{x\}) \equiv V(\mathbf{r})$ are the total energy and the total interaction potential of bodies system respectively, $\delta_{\mu\nu}$ is the Kronecker symbol and $U_0 = \max ||U(\{x\})||$ denotes the maximal depth of the potential. In the case when the potential depends on the relative distances between the particles the metric tensor is equal: $g_{\mu\nu}(\{x\}) = g_{\mu\nu}(\{\bar{x}\})$.

Now, using the variational principle of Maupertuis we can derive the geodesic equations [10, 12]:

$$\ddot{x}^{\alpha} + \Gamma^{\alpha}_{\beta\gamma}\dot{x}^{\beta}\dot{x}^{\gamma} = 0, \qquad \alpha, \beta, \gamma = \overline{1,6}, \tag{10}$$

where $\dot{x}^{\alpha} = dx^{\alpha}/ds$ and $\ddot{x}^{\alpha} = d^2x^{\alpha}/ds^2$; in addition, s is a scalar parameter of motion (e.g. the proper time), the Christoffel symbol: $\Gamma^{\alpha}_{\beta\gamma}(\{x\}) = \frac{1}{2}g^{\alpha\mu}(\partial_{\gamma}g_{\mu\beta} + \partial_{\beta}g_{\gamma\mu} - \partial_{\mu}g_{\beta\gamma})$, where $\partial_{\alpha} \equiv \partial_{x^{\alpha}}$.

Taking into account the definition for the metric tensor (9), from (10) we can find the following system of equations describing geodesic flows on the potential energy hypersurface:

$$\begin{aligned}
\ddot{x}^1 &= a_1\Big\{(\dot{x}^1)^2 - \sum_{\mu\neq 1,\, \mu=2}^{6}(\dot{x}^{\mu})^2\Big\} + 2\dot{x}^1\Big\{a_2\dot{x}^2 + a_3\dot{x}^3\Big\}, \quad &\ddot{x}^4 = \dot{x}^4\Big\{a_1\dot{x}^1 + a_2\dot{x}^2 + a_3\dot{x}^3\Big\},\\
\ddot{x}^2 &= a_2\Big\{(\dot{x}^2)^2 - \sum_{\mu=1,\, \mu\neq 2}^{6}(\dot{x}^{\mu})^2\Big\} + 2\dot{x}^2\Big\{a_3\dot{x}^3 + a_1\dot{x}^1\Big\}, \quad &\ddot{x}^5 = \dot{x}^5\Big\{a_1\dot{x}^1 + a_2\dot{x}^2 + a_3\dot{x}^3\Big\},\\
\ddot{x}^3 &= a_3\Big\{(\dot{x}^3)^2 - \sum_{\mu=1,\, \mu\neq 3}^{6}(\dot{x}^{\mu})^2\Big\} + 2\dot{x}^3\Big\{a_1\dot{x}^1 + a_2\dot{x}^2\Big\}, \quad &\ddot{x}^6 = \dot{x}^6\Big\{a_1\dot{x}^1 + a_2\dot{x}^2 + a_3\dot{x}^3\Big\},
\end{aligned} \tag{11}$$

where $g(\{\bar{x}\}) = g_{11}(\{\bar{x}\}) = ... = g_{66}(\{\bar{x}\})$, in addition; $a_i(\{\bar{x}\}) = -(1/2)\partial_{x^i}\ln g(\{\bar{x}\})$. In the system (11), the last three equations are integrated exactly:

$$\dot{x}^{\mu} = J_{\mu-3}/g(\{\bar{x}\}), \qquad J_{\mu-3} = const_{\mu-3}, \qquad \mu = \overline{4,6}. \tag{12}$$

Note that J_1, J_2 and J_3 are integrals of motion. They can be interpreted as projections of the total angular momentum of three-body system $J = \sqrt{J_1^2 + J_2^2 + J_3^2} = const$ on the corresponding axis.

Finally, substituting (12) into equations (11), we obtain the following system of a non-linear second-order ordinary differential equations:

$$\begin{aligned}
\ddot{x}^1 &= a_1(\{\bar{x}\})\Big\{(\dot{x}^1)^2 - (\dot{x}^2)^2 - (\dot{x}^3)^2 - [J/g(\{\bar{x}\})]^2\Big\} + 2\dot{x}^1\Big\{a_2(\{\bar{x}\})\dot{x}^2 + a_3(\{\bar{x}\})\dot{x}^3\Big\},\\
\ddot{x}^2 &= a_2(\{\bar{x}\})\Big\{(\dot{x}^2)^2 - (\dot{x}^3)^2 - (\dot{x}^1)^2 - [J/g(\{\bar{x}\})]^2\Big\} + 2\dot{x}^2\Big\{a_3(\{\bar{x}\})\dot{x}^3 + a_1(\{\bar{x}\})\dot{x}^1\Big\},\\
\ddot{x}^3 &= a_3(\{\bar{x}\})\Big\{(\dot{x}^3)^2 - (\dot{x}^1)^2 - (\dot{x}^2)^2 - [J/g(\{\bar{x}\})]^2\Big\} + 2\dot{x}^3\Big\{a_1(\{\bar{x}\})\dot{x}^1 + a_2(\{\bar{x}\})\dot{x}^2\Big\}.
\end{aligned} \tag{13}$$

Thus, the system of equations (13) describes the dynamics of an imaginary point with the effective mass μ_0 on the Riemannian manifold: $\mathcal{M} = [\{\bar{x}\} \equiv (x^1, x^2, x^3) \in \mathcal{M}_t;\, g_{ij} = (E - U(\{\bar{x}\}))U_0^{-1}\delta_{ij} > 0]$, by taking into account the rotation of body-triangle. The system of equations (13) can be represented as a system of six ODEs of the first order.

By taking into account (9) and (12), we can obtain the reduced Hamiltonian:

$$\mathcal{H}(\{\bar{x}\}; \{\dot{\bar{x}}\}) = g_{\mu\nu}(\{\bar{x}\})p^{\mu}p^{\nu} = \frac{\mu_0}{2}g(\{\bar{x}\})\Big\{\sum_{i=1}^{3}(\dot{x}^i)^2 + [J/g(\bar{x})]^2\Big\}. \tag{14}$$

Substituting (14) into Hamilton equations:

$$\dot{x}^{\mu} = \frac{\partial \mathcal{H}}{\partial p^{\mu}}, \qquad \dot{p}^{\mu} = -\frac{\partial \mathcal{H}}{\partial x^{\mu}}, \tag{15}$$

and by making simple calculations we can get the system of geodesic equations (13).

4. Transformation of $6D$ Euclidean space to the $6D$ conformal-Euclidean space

In deriving the system of equations (13), we have used some physical considerations that, from the mathematical point of view, are insufficiently rigorous, to argue that the dynamical system (13) is equivalent to the Newtonian problem of three-body. For a strict proof of the equivalence of the approaches, we need to prove that there is one-to-one mapping between the two sets of coordinates: $\overline{\rho^1,\rho^6}=\{\rho\}$ and $\overline{x^1,x^6}=\{x\}$, in addition that is very important in result of transformations $6D$ Euclidean space must pass into the conformal-Euclidean form.

Let us consider two spaces $\mathbf{E}^6\cong R^6$ and $\mathbf{M}$ which satisfy the condition of one-to-one mapping; $\mathbf{E}^6 :\Leftrightarrow \mathbf{M}$. We will suppose that the Euclidean space $\mathbf{E}^6$ is defined by the set of coordinates: $\overline{\rho^1,\rho^6}=\{\rho\}$ and by the metric tensor $\gamma_{\mu\nu}(\{\rho\})$, while the configuration space of an imaginary point $\mathbf{M}$ is defined by the local coordinate frame: $\overline{x^1,x^6}=\{x\}$ and by the metric tensor $g_{\mu\nu}(\{x\})$, respectively. The linear infinitesimal element in both coordinate systems can be represented as:

$$ds^2=\gamma^{\alpha\beta}(\{\varrho\})d\varrho_\alpha d\varrho_\beta=g_{\mu\nu}(\{x\})dx^\mu dx^\nu,\qquad \alpha,\beta,\mu,\nu=\overline{1,6},\tag{16}$$

where the metric tensor $g_{\mu\nu}(\{x\})$ is defined as:

$$g_{\mu\nu}(\{x\})=\gamma^{\alpha\beta}(\{\varrho\})\rho_{\alpha;\mu}\rho_{\beta;\nu},\qquad \varrho_{\alpha;\mu}=\partial\varrho_\alpha/\partial x^\mu.\tag{17}$$

Since the tensor $g_{\mu\nu}(\{x\})$ is still defined in a rather arbitrary way, we can impose an additional conditions on it. In particular, we will require that the metric tensor $g_{\mu\nu}(x)$ describes the conformal-Euclidean space: $g_{\mu\nu}(x)=g(\{x\})\delta_{\mu\nu}$ (see (9)). The latter means that the following algebraic equations take place:

$$\gamma^{\alpha\beta}(\{\varrho\})\varrho_{\alpha;\mu}\varrho_{\beta;\nu}=g(\{x\})\delta_{\mu\nu}.\tag{18}$$

As one can see, the system of algebraic equations (18) is underdetermined since it consists of 21 equations while the number of unknown variables is 36. It is obvious that when these equations are compatible, the system (18) has an infinite number of real and complex solutions. These solutions form two different manifolds of the 15th order. For a classical problem real solutions are important ones, therefore below we will investigate properties of the real manifold. In a similar way, we can obtain the system of algebraic equations for inverse transformations:

$$\gamma_{\alpha\beta}(\{\varrho\})g^{-1}(\{\bar{x}\})=x^\mu_{;\alpha}x^\nu_{;\beta}\delta_{\mu\nu},\qquad x^\mu_{;\alpha}=\partial x^\mu/\partial\rho^\alpha.\tag{19}$$

It is obvious that if there are direct transformations then there are inverse transformations too.

Let us introduce new notations:

$$\mathrm{x}_\mu=\rho_{1;\mu},\quad \mathrm{y}_\mu=\rho_{2;\mu},\quad \mathrm{z}_\mu=\rho_{3;\mu},\quad \mathrm{u}_\mu=\rho_{4;\mu},\quad \mathrm{v}_\mu=\rho_{5;\mu},\quad \mathrm{w}_\mu=\rho_{6;\mu}.\tag{20}$$

Taking into account the fact that the tensor: $g_{\mu\nu}(\{\bar{x}\})$ still is an arbitrary one, we can require fulfillment of the following conditions for its elements:

$$\begin{aligned}&\mathrm{x}_4=\mathrm{x}_5=\mathrm{x}_6=0,\qquad \mathrm{y}_4=\mathrm{y}_5=\mathrm{y}_6=0,\qquad \mathrm{z}_4=\mathrm{z}_5=\mathrm{z}_6=0,\\ &\mathrm{u}_1=\mathrm{u}_2=\mathrm{u}_3=0,\qquad \mathrm{v}_1=\mathrm{v}_2=\mathrm{v}_3=0,\qquad \mathrm{w}_1=\mathrm{w}_2=\mathrm{w}_3=0.\end{aligned}\tag{21}$$

Using (6), (20) and conditions (21), from the equation (18) we can obtain two independent systems of algebraic equations:

$$\begin{aligned}&\mathrm{x}_1^2+\mathrm{y}_1^2+\gamma^{33}\mathrm{z}_1^2=g(\{\bar{x}\}),\qquad \mathrm{x}_1\mathrm{x}_2+\mathrm{y}_1\mathrm{y}_2+\gamma^{33}\mathrm{z}_1\mathrm{z}_2=0,\\ &\mathrm{x}_2^2+\mathrm{y}_2^2+\gamma^{33}\mathrm{z}_2^2=g(\{\bar{x}\}),\qquad \mathrm{x}_1\mathrm{x}_3+\mathrm{y}_1\mathrm{y}_3+\gamma^{33}\mathrm{z}_1\mathrm{z}_3=0,\\ &\mathrm{x}_3^2+\mathrm{y}_3^2+\gamma^{33}\mathrm{z}_3^2=g(\{\bar{x}\}),\qquad \mathrm{x}_2\mathrm{x}_3+\mathrm{y}_2\mathrm{y}_3+\gamma^{33}\mathrm{z}_2\mathrm{z}_3=0,\end{aligned}\tag{22}$$

and correspondingly:

$$\begin{array}{ll}\gamma^{44}u_4^2+\gamma^{55}v_4^2+\gamma^{66}w_4^2+2(\gamma^{45}u_4v_4+\gamma^{46}u_4w_4+\gamma^{56}v_4w_4)=g(\{\bar{x}\}), & a_4u_4+a_5v_4+a_6w_4=0,\\ \gamma^{44}u_5^2+\gamma^{55}v_5^2+\gamma^{66}w_5^2+2(\gamma^{45}u_5v_5+\gamma^{46}u_5w_5+\gamma^{56}v_5w_5)=g(\{\bar{x}\}), & b_4u_5+b_5v_5+b_6w_5=0,\\ \gamma^{44}u_6^2+\gamma^{55}v_6^2+\gamma^{66}w_6^2+2(\gamma^{45}u_6v_6+\gamma^{46}u_6w_6+\gamma^{56}v_6w_6)=g(\{\bar{x}\}), & c_4u_6+c_5v_6+c_6w_6=0.\end{array} \tag{23}$$

In (23) $a_i=\gamma^{i4}u_5+\gamma^{i5}v_5+\gamma^{i6}w_5$, $b_j=\gamma^{j4}u_6+\gamma^{j5}v_6+\gamma^{j6}w_6$ and $c_k=\gamma^{k4}u_4+\gamma^{k5}v_4+\gamma^{k6}w_4$, where $i=\overline{4,6}$. Note that solutions of systems (22) and (23) form two different manifolds of 3rd order which are in one-to-one mapping with the hypersurface of the potential energy of the three-body system. Recall that similar systems of algebraic equations can be obtained for inverse transformations. It is easy to see that the coordinates defined by formulas (7) and (8) in general case do not satisfy the conditions of transformations (22) and (23) and correspondingly the system of equations (13), generally speaking, is not equivalent to the Newtonian three-body problem. Nevertheless, starting from general considerations, we can prove that there exists a such system of local coordinates. Apparently, the form of these coordinates can be determined explicitly after study of equations properties (22)-(23). Finally, it is important to note that the dynamical equations (13) conserve their previous form and only are changed dependence of the potential from coordinates.

5. Conclusion

As it is shown, with the help of coordinates transformations the general three-body problem can be formulated as a problem of geodesic flows on Riemannian manifold. In the framework of this representation we derive an additional symmetry that allows to integrate an exactly the three non-linear equations of the system containing six equations. In this way, the problem reduced to the autonomous system of the 6th order. Despite the fact that the explicit form of the local coordinate system in general case is not found, it can be proved that there is a class differentiable functions which allow to realize the corresponding coordinate transformations and satisfy the equations (22)-(23), that brings the metric tensor to the conformally-Euclidean form.

Finally it is important to note that the system of equations (13), with definitions of local coordinates (7)-(8), in spite of its inequivalency to the Newtonian problem of three-body, is an important model to study the different dynamical and statistical properties of classical three-body system, and can be interesting for the construction of its quantum analogue.

References

[1] Marchal C 2006 *The three-body problem* (Amsterdam: Elsevier)
[2] Bruno A D 1994 *The restricted three-body problem: Plane Periodic Orbits* (De Gruyter)
[3] Poincare H 1972 *New methods of selestial mechanics.* In 3 vol., Sel. work trans. from french. (Moscow: Nauka)
[4] Arnold I Kozlov and Neishtadt A I 1997 *Mathematical aspects of classical and celestial mechanics Springer-Verlag*
[5] Gutzwiller M 1998 *Moon-Earth-Sun: The oldest three-body problem* Rev. Mod. Phys. **70** No.2 589
[6] Delves L M 1958/59 *Tertiary and general-order collisions* Nucl. Phys. **9**, 391
[7] Ayryana E A Gevorkyan A S and Sevastyanova L A 2013 *On the Motion of a Three Body System on Hypersurface of Proper Energy* Physics of Particles and Nuclei Letters **10** No.7 669
[8] Fiziev P P and Fizieva Ts Ya 1987 *Modification of Hyperspherical Coordinates in the Classical Three-Particle Problem* Few-Body Systems **2** 71
[9] Whittaker E T 1988 *A Treatise on the Analytical Dynamics of Particles and Rigid Bodies. With an Introduction to the Problem of Three Bodies* (University Press in Cambridge)
[10] Arnold L 2000 *Mathematical Methods of Classical Mechanics* (Moscow: Editorial URSS in Russian)
[11] Krylov N S 1950 *Studies on the Foundation of Statistical Physics* Publ. AN SSSR (Leningrad: in Russian)
[12] Dubrovin B A Fomenko A T and Novikov S P 1984 *Modern Geometry Methods and Applications* Part I (Berlin: Springer- erlag)

2nd International Symposium on the Modern Physics of Compact Stars and Relativistic Gravity IOP Publishing
Journal of Physics: Conference Series **496** (2014) 012031 doi:10.1088/1742-6596/496/1/012031

Quantum Mechanics and physical calculations

H. S. Karayan

Yerevan State University, 1 Alex Manoogian Street, 0025 Yerevan, Armenia

E-mail: hskarayan@ysu.am

Abstract. We suggest to realize the computer simulation and calculation by the algebraic structure built on the basis of the logic inherent to processes in physical systems (called physical computing). We suggest a principle for the construction of quantum algorithms of neuroinformatics of quantum neural networks. The role of academician Sahakyan is emphasized in the development of quantum physics in Armenia.

1. Academician Sahakyan and the development of quantum physics in Armenia

Gurgen Sahakyan, being a well-known physicist in the fields of gravity, astrophysics and microphysics, showed a big interest to quantum physics. In the middle of last century, he founded the Chair of Theoretical Physics at the Physics Department of Yerevan State University (YSU). At that time, the quantum concept entered into all areas of science and technology. Already the first quantum-mechanical calculations revealed a band character of the energy spectrum of multi-particle systems which became the basis for solid-state physics, semiconductors and semiconductor electronics, and then for quantum electronics. That time in Armenia well-known experts in microphysics were developing their directions. Being a highly erudite and multi-sided developed scientist, G.S. Sahakyan felt unacceptable the indirect role of quantum mechanics and bravely took to put it on a fundamental course. Times were hard, there were no textbooks in Armenian language. On the basis of his first lectures he and his student E.V. Chubaryan have written the first, unfortunately so far the only textbook on quantum physics in the Armenian language (it is republished, as well as translated into Russian language [1]). Academician Sahakyan could induce students' great interest in quantum physics and give them his enthusiasm. I still remember with gratitude his pride and pathos in the introductory lecture on quantum mechanics in our group when he presented the essence of the concept of quantum physics: "At present, the quantum theory is the most common, complete, beautiful and experimentally justified among all scientific theories... But more radical properties are inherent to the concept of quantum physics: a new way of thinking, a new level of civilization and a new logic – the logic of Nature."

2. Quantum physics in informatics

The first machines carried out calculations in the classical Boolean logic system by using given algorithms and programs. Algebra of Boolean logic (ABL) is universal, so a mathematical machine realizing it is universal as well. This is reflected in the famous Church-Turing thesis [2]. In the middle of the last century the mathematical theory of information and cybernetics were built. In fact, it was believed that any problem of informatics is principally solvable on such machines and the physics was considered simply and only as a means of realization of mathematical ideas and algorithms. The complete set of these tools composed the element base of informatics, which is responsible for the efficient operation of information systems. For this purpose, high expectations were associated with quantum mechanics. The successes are impressive to the extent that mathematicians-informatics were convinced in its "omnipotence." The classics of physics did not shared this opinion and predicted more "omnipotence" but principally new information systems (Sahakyan was also a supporter of the last statement).

The behavior of a micro-object is described by the amplitude of probability, the set of which forms a Hilbert space $\mathcal{H}$. It was discovered that the quantum probabilities are not described by the classical probability theory [2]-[4]. Further, the nature of the physical and mathematical information is different [5]. L. Brillouin developed the foundations of the theory of physical informatics with the use of the quantum concept of physics and showed that in the nature we meet the physical rather than abstract information and he received a number of restrictions on the mathematical information [5]. Further, infinite amount of information is inherent to a quantum system, a part of which is redundant and can not be put out from the system, but that is involved in all information processes. This property can not be modelled by finite resources on the basis of ABL.

A. Einstein, E. Schrodinger, and others have indicated entangled states of combined quantum systems which have no counterparts in macro-world and contain more information than non-entangled states. This is a manifestation of the properties of logic, in accordance of which processes occur in Nature. G. Birkhoff and J. von Neumann have constructed the algebra of quantum logic, which contains as a subalgebra ABL [2, 6]. It is more precise to say that "the classical theory is an essential part of quantum theory, which also contains all the limiting cases." Moreover, in macro-world ABL is only valid, so Feynman concluded that any physical system (machine) realizing only the ABL, can not model quantum mechanics, but it can only imitate. He also noted that the information performance of ABL-machines depends of their size polynomially, but for quantum machines, this dependence is exponential. R. Feynman formulated these ideas as quantum informatics [2]-[4].

3. Algebras of physical logic and physical mathematics

Mathematical logic (and its algebras) is an axiomatic science, which is not always possible to realize in nature. Being abstract, it may contain an infinitely large, infinitely small or the other kind of elements which do not exist in nature. Only the completeness and consistency are required from its axioms, so it can be physically unrealizable. If, in addition, physical realizability is required, not for all axiomatics in our universe the "Maxwell's Demon", realizing them, exists [2, 3, 5]. But judging from the condition of the realizability and universality of logic, we come to the inverse problem: using the logic of the processes occurring in nature to form the axiomatic of the logic algebra with the properties of completeness, consistency and universality. Then its existence and realizability are automatically provided. We refer to such logical systems as physical and the computation in them physical [7]. A physical logical system contains only physical quantities as carriers of information and processes of their evolution. Mathematical physics is directed to construct mathematical images of physical objects. The representation of mathematical objects by physical phenomena is the essence of the so-called physical mathematics. Both of them have a structure on the logic, abstract mathematical and physical, respectively. We present two examples of algebraic structure of multi-valued physical logic.

Let one has a set $A_p = \{0, 1, 2, ..., (p-1) = E\}$ of equidistant dimensionless p values for some parameter of the state of a physical system. Logical p-level statements can be presented by the numbers $0, 1, \ldots, E$. Moreover, in order to construct an appropriate algebra for p-valued physical logic, we can identify the logical operations, algebraic actions and physical transformations of these parameters. As transformers one can take a multilevel controllable physical system, the input (control) and output parameters of which coincide with A_p (or with $A_{q\leq p}$, in particular, A_2). Such systems exist, for example, polistors [7], and it is proved that the logic of processes in the group $\mathcal{L}_p$ of polistors is sufficient to form a closed, complete, consistent and universal p-valued deterministic physical logical system. $\mathcal{L}_p$ consists of 5 polistors that produce the following transformations:

$$I_2x \equiv 1-x,\ A_2,\ I_px \equiv E-x,\ S(x_1,x_2,\ldots,x_n) = \begin{cases} 1, \text{ if } \sum_{i=1}^n x_i \leq E \\ 0, \text{ if } \sum_{i=1}^n x_i > E \end{cases},$$

$$h(x_1,\ldots,x_n) = \begin{cases} E, \text{ if } \sum_{i=1}^n x_i \leq E \\ 0, \text{ if } \sum_{i=1}^n x_i > E \end{cases},\ K(x_1,\ldots,x_n) = \begin{cases} I_p(\sum_i x_i), \text{ if } \sum_{i=1}^n x_i \leq E \\ 0, \text{ if } \sum_{i=1}^n x_i > E \end{cases}.$$

It is not difficult to find in $\mathcal{L}_p$ physical images (that is, to make physical modelling) of the action of ABL and of mathematical models.

The presentation of indeterministic processes of nature is not possible in any deterministic system of finite resources, whether mathematical or physical. G. Birkhoff and J. von Neumann noted the identity (up to isomorphism) of lattices for a number of constructions. Lattice (similar to the atomic one) is the set $\mathcal{M}$, ordered by the relation of order $a \leq$ and satisfying the conditions (in standard notations): 1.$\forall x, y \in \mathcal{M}$, $\max(x,y) \equiv x \vee y$; $x \leq x \vee y$; $y \leq x \vee y$; $z \leq x$ and $z \leq y \Rightarrow x \vee y \leq zz$; 2.$\forall x, y \in \mathcal{M}, \min(x,y) \equiv x \wedge y$; $x \wedge y \leq x$; $x \wedge y \leq y$; $z \leq x$ and $z \leq y \Rightarrow z \leq x \wedge y$. The element a in the lattice is called an atom if $x \leq a \Rightarrow x = 0$ or $x = a$. In particular, the set of statements with the implication closed with respect to the operations of conjunction and disjunction (that is $Q \leq P \iff P$ implies Q) and the set of subspaces of the space of states for a quantum system are isomorphic lattices. Both these lattices are orthocomplementary (that is $\forall x$ is associated with the orthogonal complement $\overline{x}$ of such that the conditions: $\overline{\overline{x}} = x$; $x \vee \overline{x}$; $x \leq y \iff \overline{y} \leq \overline{x}$ are satisfied) if $\overline{x}$ belongs to the orthogonal complement of the subspace spanned on the state x. In the algebra of logic the measure of the truth 0 and 1 correspond to the "lie" and "truth", and in the Hilbert space it corresponds to eigenvalues of the orthogonal projection operator. The indeterminism of microworld is reflected in projectors, so Gleason's theorem states that the metastates $\mu(M) = Tr(P_M\rho)$, where ρ is the density matrix and P_M is a projector. If an observable A takes a value a in the state ψ, then it takes the same value in all the states of the form $(\psi + \phi)$, where ϕ is an arbitrary state orthogonal to ψ. Because of this the lattice of quantum states becomes non-distributive and is called orthomodular: $x \leq y \Rightarrow (x \vee \overline{y}) \wedge y = x$. In general, its elements are not compatible, but in the lattice there are also many compatible elements for which the distributive law takes place: $(x \wedge y) \vee (x \wedge \overline{y}) = x$. The ortho-complementary distributive lattice is called classical logic, and the ortho-complementary orthomodular lattice is called quantum logic. That the classical logic is completely incorporated into the quantum logic is the law of Nature. The quantum logic is isomorphic to the lattice of projective space, which contains all his hyperplanes as points and which generates the geometric Clifford algebra. It became clear that quantum logic, being the logic of nature and able to implement the algebras $S(U_n)$ and $S(O_n)$ of matrices, can be used to build quantum informatics. In this case, the carriers of the information are multi-level quantum systems, in particular, qubits.

4. Quantum computers and quantum computing

For modelling and exponentially fast computing only machines are suitable based on quantum-correlated systems, because the dependence of their productivity on the number of qubits is exponential. In this way, two classes of problems arise: (i) how to build a quantum algorithm for computing, that is to present it by quantum processes, and (ii) what kind of machine is the best for the computing?

Journal of Physics: Conference Series **496** (2014) 012031 doi:10.1088/1742-6596/496/1/012031

Feynman has constructed an example of a quantum computer working by the algebra of quantum logic in the "language of second quantization" [4]. To this end, he chose a redundant set of operators, represented by $n \times n$ matrices and realized by the application of the phase change for a state by the angle φ. The main operations are: annihilation and creation operators, a and a^+, the operator of the particle number n, the negation operator N, Walsh-Hadamard operator h, the operator of the phase shift s_φ, the operator of the "element $\pi/8$" and Pauli matrices $\sigma_i, i = 0, 1, 2, 3$, rotation operators $R_{e_i}(\varphi) = \exp(i\varphi e_i/2)$ around the axis OX (similarly OY) and OZ:

$$R_x(\varphi) = \begin{pmatrix} \cos\varphi/2 & -i\sin\varphi/2 \\ -i\sin\varphi/2 & \cos\varphi/2 \end{pmatrix}, \ R_z(\varphi) = \begin{pmatrix} e^{-i\varphi/2} & 0 \\ 0 & e^{i\varphi/2} \end{pmatrix}.$$

Multiplace operations are realized by the conditional dynamics of a closed quantum combined system. We give just a matrix representations of the basic double-place operations: controllable phase element S, controllable element C_{σ_z}, element C_N (controllable N) and the element of exchange Sw (see [2, 4] and references cited therein):

$$S \equiv \begin{pmatrix} \sigma_0 & 0 \\ 0 & s_\varphi \end{pmatrix}, \ C_{\sigma_z} \equiv \begin{pmatrix} \sigma_0 & 0 \\ 0 & \sigma_0 \end{pmatrix}, \ C_N \equiv \begin{pmatrix} \sigma_0 & 0 \\ 0 & N \end{pmatrix}, \ Sw \equiv \begin{pmatrix} n & a^+ \\ a & 1-n \end{pmatrix}.$$

In essence, a quantum algorithm of this problem is reduced to the construction of such a time-dependent Hamiltonian, which contains these operators and their compositions, with the exception of the input and output processes (measurements), which require projectors as well. Currently, a number of quantum algorithms are found for specific tasks, such as Deutsch - Jos, Grover, Shor and others algorithms [2].

Here we should distinguish two classes of problems. The first one is reduced to the organization of the mathematical calculation process on a quantum computer in order to speed up the calculations in comparison with a classical computer. This direction is most likely a quantum cybernetics. The second class of problems does not contain mathematical calculations, as such, an example of which is Shor's algorithm on the factorization of an integer N. The expansion of the number N in prime factors is realized on the base of quantum Fourier transform and by measuring the period of some N-dependent function by the interference pattern. Shor's algorithm is a physical computing and the first example of computing exponentially faster than classical one.

5. Artifical Intelligence and Quantum Neuroinformatics

Classically correlated combined physical systems are widely used in the associative and parallel processing of information, in performing calculations and in physical modelling. In particular, currently the only way to create artificial intelligence is based so far on the use of these systems, also known as neural networks [8]. However, classically correlated combined physical systems are valid to perform polynomial algorithms only. At the same time for the construction of strong artificial intelligence, exponentially high efficiency of information processing is required. Now it becomes actual another new and more complicated problem: informal simulation of the mind and thinking. But in order the physical system could think of (that is, it is able to make an algorithm of its action, not only to realize the algorithms imposed in the system from outside), it is necessary the realization by it of the non-distributive algebra of quantum logic.

At least one of the ways to solve the above problems is associated with the unification of capabilities of classical neuroinformatics and quantum informatics, that is it is associated with the creation of quantum neuroinformatics. However, this unification is not trivial because of the incompatibility of the methods classical neuroinformatis and quantum dynamics of a closed system.

Consider a quantum implementation of a cyclic process of conditionally controlled iteration for the formation of knowledge. For the formation of conditions of repetition of an each cycle of iteration, it is necessary to compare the state of a physical system to its state in the previous cycle or to some given state (expert state), the formation, by the results of comparison, of the conditions of the system dynamics variation with the help of feedback, the formation of the operator of conditionally controllable iteration and composition of the projection operator in the subspace of "codes of knowledge" of the state vector.

Since the state of a classical physical system can be known at any time, the realization of these operations is reduced merely to the realization of program operators induced by the algorithm [8]. The iteration is continued as long as the specified state is not found or some condition of closeness of the two output states is not satisfied. The output state of a macroscopic system is described by the values of characteristic quantities, the numerical values of which determine their equality or closeness. The physical system is only responsible for the efficiency of realizing the mathematical operations by the program. Such a method of knowledge formation and getting artificial intelligence does not apply when using quantum-correlated systems. First of all, the difficulties are associated with the fact that in principle we can not know the state of a quantum system without destroying the state itself or its coherence. This fact prevents the use of classical methods of comparison between different states by proximity and consequently, the formation of conditions for the variation of dynamics [9]. Moreover, the comparison must be made only by unitary operations to ensure the physical reversibility. Moreover, the comparison must be made only by unitary operations to ensure the physical reversibility. Finally, in the quantum case the proximity of states is not a trivial notion by itself. The quantum state is presented by a function of a complex unitary space and it is impossible to establish the equality or closeness of two states by one or a finite number of numerical values without a significant loss of the information.

The basis in the space of states of a coherently correlated combined quantum system of several qubits A_i is formed by the tensor product of the bases of all the qubits. In the space $\mathcal{H}_k$ of states of combined system, constructed from this basis, if the coefficients of two superpositions of the states coincide in one basis, then these superpositions, up to a global phase, present the same state of the system. This known fact allows to realize the comparison of quantum states in a given basis unitary by components, i.e. by the numerical values of the components of amplitude. $\mathcal{H}_k$ also includes the space $\mathcal{H}_e$ of all entangled states, which are absent in $\mathcal{H}_{\otimes A_i}$, they have no classical analog and contain information on the states of two or more qubits, that is: $\mathcal{H}_k = \otimes_i \mathcal{H}_i = \mathcal{H}_{\otimes A_i} \oplus \mathcal{H}_e$. The latter property is the basis for the unitary implementation of logical scheme "if ... then ..., if ... then ...", to which the comparison operation of quantum states is reduced with the subsequent formation of the condition of the iteration cycle of the shift operator $\widehat{T}_n(U)$ for the cycle number:

$$\widehat{T}_n(U) \begin{pmatrix} \psi_n(0) \\ \psi_{n-1}(\tau) > \end{pmatrix} = \begin{pmatrix} \psi_{n+1}(0) \\ \psi_n(\tau) > \end{pmatrix}.$$

Let $\Delta_{ab} = |\,|a\rangle - |b\rangle\,|$ be the symmetric difference of the basis states $|a\rangle$ and $|b\rangle$. Then, based on the identity

$$\widehat{C}_N \begin{pmatrix} |a\rangle \\ |b\rangle \end{pmatrix} \equiv \begin{pmatrix} |a\rangle \\ \Delta_{ab} \end{pmatrix}. \tag{1}$$

, the condition of the repeating of iteration cycle can be formed by using the difference $\Delta_n \equiv |\,|\psi_n(\tau)\rangle - |\psi_{n-1}(\tau)\rangle\,|$ (or $\Delta_n \equiv |\,|\psi_n(\tau)\rangle - |\psi_e\rangle\,|$, in case of the presence of expert state). Indeed, if the qubit with the state Δ_n is used as controlling, then the similarity transformation $\widehat{C}_N \Delta_n \widehat{C}_N$ does not change the state of two involved qubits (because the operator $\widehat{C}_N$ is self-reversing) . But the qubit with Δ_n can act as an intermediate supervisory by combining with the third qubit to perform conditional operations. In particular, with the help of Δ_n we form the operator $\widehat{U}_\varphi$ that shifts the phase of controlled qubit by a specific angle φ, if the qubit is in the state $|1\rangle$, otherwise, it produces no changes in the state of the third qubit.

If for some number n_{np} the condition $\Delta_n = 0$ is obeyed, then it is obeyed for all $n \geq n_{np}$ and the corresponding state is a limiting state $|\psi_{np}\rangle$. It is obvious that in this case $\widehat{T}_n$ becomes and remains as identity operator. Then, for $n \geq n_{np}$ the whole iteration process is presented by the operator:

$$\widehat{T}_{it} = \widehat{1}\widehat{1}\ldots\widehat{T}_{n_{np}}\widehat{T}_{n_{np}-1}\ldots\widehat{T}_2\widehat{T}_1 \equiv (\widehat{T})^{n\geq n_{np}}, \quad (2)$$

because the structure of the operators $\widehat{T}_n$ is the same for all cycles, it is denoted by $\widehat{T}$:

$$\widehat{T} \equiv \widehat{I}(2)\widehat{S}_w(2,3)\widehat{C}_N(3,2)\widehat{U}_\varphi(2,1)\widehat{C}_N(3,2)\widehat{U}_\tau(2)\widehat{C}_N(1,2). \quad (3)$$

Here, in (2) and (3), in order to avoid cumbersome matrix form of the presentation, near each operator the number of the qubit is written in the parentheses on which it acts and for binary operators the first number corresponds to the controlling, and the second one to the controlled. However, the algorithm presented above does not contain the indication to establish the existence of the limiting state, and also to find the value of n_{np}. This narrows the range of the algorithm application and reduces the efficiency of the search of limiting state. The probability of finding the limiting state can be increased by random variation of the phase in the operator $\widehat{U}_\varphi$. But this does not solve the problem of finding the values of n_{np}.

It seems that the extension to invariance of the search algorithm for a limiting state is the most consistent. In other words, it is necessary to find an algorithm that for a given N steps would lead to the state closest to limiting one regardless of the existence of the limiting state and the nature of the convergence to it. For this purpose, in turn, it is necessary to "extend" the measure of closeness of quantum states, based on the identity (1). In this relation $\widehat{C}_N$ and Δ_{ab} can be generalized. For example, we replace the operator $\widehat{C}_N$ by the operator $\widehat{K}_N$ which reverses the phase of the state $|b\rangle$ in the case when the eigenvalues of the states $|a\rangle$ and $|1\rangle$ differ by some nonzero quantity δ. Physically, this means that instead of a "naked" entangled states of vectors $|a\rangle$ and $|b\rangle$ in the subspace $\mathcal{H}_e$, we use a kind of "dressed" entanglement of "δ-dressed" state $|a\rangle$ with the state $|b\rangle$. It is mathematically equivalent to replace $\mathcal{H}_e$ by some of its non-orthogonal "δ-dressed" extension.

6. Conclusion

Academician Sahakyan managed to put on a good track the birth and development of a powerful school of physicists of the microworld. And this school has participated in the implementation of predictions by Sahakyan. Fundamental researches were conducted on quantum field theory, on the creation of quantum neuroinformatics. He and his followers, E. Chubaryan, D. Sedrakian, Yu. Vardanyan, R. Avakian, G. Harutyunyan, A. Saharian and many others transferred and will transfer the ideas and methods of quantum physics to astrophysics and gravitation.

References

[1] Sahakyan G S and Chubaryan E V 2006 *Quantum Mechanics* (Yerevan: Yerevan State University Press)
[2] Nielsen N A and Chuang I L 2002 *Quantum Computation and Quantum Information* (Cambridge: Cambridge University Press)
[3] Klyshko D N 1998 UFN **168** 975
[4] Feynman R 1982 Inter. J. Theor. Phys. **21** 467
[5] Brillouin L 1960 *Science and Information Theory* (New York: Academic Press)
[6] Birkhoff G and von Neumann J 1936 The Logic of Quantum Mechanics **37** 823
[7] Karayan H S at al 2004 Proceedings of 2nd IEEE International Conference on Circuits and Systems for Communications (Moscow)
[8] Kruglov V V and Borisov V V 2002 *Artificial neuron circuits* (Moscow)
[9] Karayan H S 2007 Izv. NAS RA: Series Tech. S. **LX** 608

2nd International Symposium on the Modern Physics of Compact Stars and Relativistic Gravity IOP Publishing
Journal of Physics: Conference Series **496** (2014) 012032 doi:10.1088/1742-6596/496/1/012032

Kaluza-Klein theory revisited: projective structures and differential operators on algebra of densities

H. M. Khudaverdian

School of Mathematics, University of Manchester, Oxford Road, Manchester M13 9PL, UK

E-mail: khudian@manchester.ac.uk

Abstract. We consider differential operators acting on densities of arbitrary weights on manifold M identifying pencils of such operators with operators on algebra of densities of all weights. This algebra can be identified with the special subalgebra of functions on extended manifold $\widehat{M}$. On one hand there is a canonical lift of projective structures on M to affine structures on extended manifold $\widehat{M}$. On the other hand the restriction of algebra of all functions on extended manifold to this special subalgebra of functions implies the canonical scalar product. This leads in particular to classification of second order operators with use of Kaluza-Klein-like mechanisms.

1. Algebra of densities

In mathematical physics it is very useful to consider differential operators acting on densities of various weights on a manifold M (see [1] and citations there). We say that $\mathbf{s} = s(x)|Dx|^{\lambda}$ is a density of weight λ on M if under change of local coordinates

$$\mathbf{s} = s(x)|Dx|^{\lambda} = s\left(x\left(x'\right)\right)\left|\det\left(\frac{\partial x}{\partial x'}\right)\right|^{\lambda}|Dx'|^{\lambda},$$

(λ is an arbitrary real number). We denote by $\mathcal{F}_{\lambda}(M)$ the space of densities of weight λ on manifold M. (The space of functions on M is $\mathcal{F}_0(M)$, densities of weight $\lambda = 0$.)

Densities can be multiplied. If $\mathbf{s}_1 = s_1(x)|Dx|^{\lambda_1}$ and $\mathbf{s}_2 = s_2(x)|Dx|^{\lambda_2}$ are densities of weights λ_1, λ_2 respectively then $\mathbf{s} = \mathbf{s}_1 \cdot \mathbf{s}_2 = s_1(x)s_2(x)|Dx|^{\lambda_1+\lambda_2}$ is a density of weight $\lambda_1 + \lambda_2$. We come to the algebra $\mathcal{F}(M) = \oplus_{\lambda}\mathcal{F}_{\lambda}(M)$ of finite linear combinations of densities of arbitrary weights. Use a formal variable t instead volume form $|Dx|$. Thus an arbitrary density $\mathbf{s} = s_1(x)|Dx|^{\lambda_1} + \cdots + s_k|Dx|^{\lambda_k}$ can be written as a function on x, t which is *quasipolynomial* on t, $\mathbf{s}(x,t) = s_1(x)t^{\lambda_1} + \cdots + s_k(x)t^{\lambda_k}$. An arbitrary density $\mathbf{s} \in \mathcal{F}(M)$ can be identified with function $\sum s_r(x)t^{\lambda_r}$ on the extended manifold $\widehat{M}$, which is quasipolynomial on 'vertical' variable t. There is a natural fibre bundle structure $\widehat{M} \to M$. Extended manifold $\widehat{M}$ is the frame bundle of the determinant bundle of M, (x^i, t) are local coordinates on $\widehat{M}$. Changing of local coordinates is:

$$(x^{i'}, t'):\quad x^{i'} = x^{i'}(x^i),\quad t' = t'(x^i, t) = \det\left(\frac{\partial x^{i'}}{\partial x^i}\right)t\,.$$

The fibre bundle $M \to M$ can be used for studying projective geometry on M since there is a canonical construction which assigns to an arbitrary projective connection on manifold M an usual affine connection on $\widehat{M}$ (see the last section). This affine connection on $\widehat{M}$ can be used for describing the 'projective geometry' on the base manifold M. Such investigations can be traced to H. Weyl, Weblen and Thomas. On the other hand we will come to additional geometrical structures on fibre bundle $\widehat{M} \to M$ if instead algebra of all smooth functions on extended manifold $\widehat{M}$ we consider only the subalgebra of functions on $\widehat{M}$, which are quasipolynomial on vertical variable t, i.e. algebra $\mathcal{F}(M)$ of densities on M. This algebra can be equipped with the canonical scalar product: If $\mathbf{s}_1 = s_1(x)|Dx|^{\lambda_1}$ and $\mathbf{s}_2 = s_2(x)|Dx|^{\lambda_2}$ are two densities with a compact support then

$$\langle \mathbf{s}_1, \mathbf{s}_2 \rangle = \begin{cases} \int_M s_1(x) s_2(x) |Dx| \,, & \text{if} \quad \lambda_1 + \lambda_2 = 1 \,, \\ \\ 0 & \text{if} \quad \lambda_1 + \lambda_2 \neq 1 \,. \end{cases} \tag{1}$$

This construction turns out to be very important tool to study geometry of differential operators on M [2], [3]. We give short exposition of these results. Here we come to these results and formulate new ones based on alternative approach of Kaluza-Klein-like reduction.

2. Differential operators on algebra of densities

Consider linear operator $\widehat{w}$ such that $\widehat{w}(\mathbf{s}) = \lambda \mathbf{s}$ in the case if $\mathbf{s}$ is a density of weight λ, ($\mathbf{s} \in \mathcal{F}_\lambda(M)$). If $\mathbf{s}_1$ is a density of weight λ_1 and $\mathbf{s}_2$ is a density of weight λ_2 then

$$\widehat{w}(\mathbf{s}_1 \cdot \mathbf{s}_2) = (\lambda_1 + \lambda_2) \mathbf{s}_1 \cdot \mathbf{s}_2 = (\widehat{w}\mathbf{s}_1) \cdot \mathbf{s}_2 + \mathbf{s}_1 \cdot (\widehat{w}\mathbf{s}_2) \,.$$

Leibnitz rule is obeyed, $\widehat{w}$ is first order differential operator on the algebra of densities. In local coordinates (x^i, t) on $\widehat{M}$, $\widehat{w} = t\frac{\partial}{\partial t}$. A differential operator $\widehat{\Delta}$ on algebra of densities has appearance $\widehat{\Delta} = \widehat{\Delta}\left(x, \frac{\partial}{\partial x}, t, \widehat{w}\right)$ in local coordinates. An arbitrary operator $\widehat{\Delta}$ on algebra of densities defines the pencil of operators:

$$\widehat{\Delta} \mapsto \{\Delta_\lambda\} \colon \Delta_\lambda = \widehat{\Delta}\big|_{\widehat{w}=\lambda} \,.$$

E.g. the operator $\widehat{\Delta} = A(\widehat{w}) S^{ik} \partial_i \partial_k + B(\widehat{w}) T^i \partial_i + C(\widehat{w}) R$ on algebra $\mathcal{F}(M)$ defines the pencil of operators $\{\Delta_\lambda\}$: $\Delta_\lambda = A(\lambda) S^{ik} \partial_i \partial_k + B(\lambda) T^i \partial_i + C(\lambda) R$. Here A, B, C are polynomials on $\widehat{w}$. If for example $A = 1 + \widehat{w}$, $B = \widehat{w}^2$ and $C = 1$, then $\widehat{\Delta}$ is third order operator on the algebra of densities, which defines the pencil of second order operators. Operators on algebra of densities can be identified with operator pencils which depend polynomially on pencil parameter λ.

Remark 1 *Here we consider only operators which do not change the weight of densities:* $\widehat{\Delta} = \widehat{\Delta}(x, \frac{\partial}{\partial x}, \widehat{w})$, *i.e. for corresponding pencil* $\{\Delta_\lambda\}$, $\Delta_\lambda \colon \mathcal{F}_\lambda(M) \to \mathcal{F}_\lambda(M)$.

Canonical scalar product (1) defines adjointness of linear operators. Linear operator $\widehat{\Delta}$ acting on the algebra of densities has an adjoint $\widehat{\Delta}^*$: $\langle \widehat{\Delta}\mathbf{s}_1, \mathbf{s}_2 \rangle = \langle \mathbf{s}_1, \widehat{\Delta}^* \mathbf{s}_2 \rangle$. One can see that $(x^i)^* = x^i$, $\partial_i^* = -\partial_i$ and $\widehat{w}^* = 1 - \widehat{w}$.

To consider self-adjoint and anti-self-adjoint operators on extended manifold $\widehat{M}$ is very illuminating for studying geometry of operators on base manifold M. (See for details Refs. [2, 3, 4, 5].)

3. First order operators and Kaluza-Klein mechanism

Consider an arbitrary first order operator $\widehat{K}$ such that it does not change the weight of densities (see Remark 1) and obeys normalisation condition $\widehat{K}(1) = 0$. In local coordinates it has the following appearance

$$\widehat{K} = K^i(x) \partial_i + K^0(x) \widehat{w} \,.$$

Journal of Physics: Conference Series **496** (2014) 012032 doi:10.1088/1742-6596/496/1/012032

One can see that its adjoint is equal to $\widehat{K}^* = -K^i(x)\partial_i - \partial_i K^i(x) + K^0(x)(1-\widehat{w})$. $\widehat{K}$ is a vector field on extended manifold $\widehat{M}$. One can define divergence of this vector field:

$$\operatorname{div}\widehat{K} = -(\mathbf{K}+\mathbf{K}^*) = \partial_i K^i(x) - K^0(x)\,. \tag{2}$$

Notice that the divergence is defined in spite of the absence of well-defined volume form on $\widehat{M}$ (see for details Refs. [2] and [5]).

We see that vector field $\widehat{K}$ is divergenceless iff it is anti-self-adjoint:

$$\widehat{K} = -\widehat{K}^* \Leftrightarrow \operatorname{div}\widehat{K} = 0 \Leftrightarrow \widehat{K} = K^i(x)\partial_i + \widehat{w}\partial_i K^i(x)\,.$$

One can see that divergenceless vector field $\widehat{K}$ is a Lie derivative of densities along its projection, vector field $\mathbf{K}$ on M. An arbitrary vector field $\mathbf{X} = X^i(x)\partial_i$ can be lifted to anti-self-adjoint (i.e. divergenceless) vector field on extended manifold $\widehat{M}$, which is nothing but Lie derivative of densities: $\mathbf{X} \mapsto \widehat{\mathcal{L}}_{\mathbf{X}}$ such that for arbitrary $\mathbf{s} = s(x)|Dx|^\lambda$,

$$\widehat{\mathcal{L}}_{\mathbf{X}}(\mathbf{s}) = \widehat{X}(\mathbf{s}) = \left(X^i(x)\partial_i + \widehat{w}\partial_i X^i(x)\right)s(x)|Dx|^\lambda = \left(X^i\partial_i s(x) + \lambda\partial_i X^i s(x)\right)|Dx|^\lambda\,.$$

It is useful to consider a connection on a bundle $\widehat{M} \to M$. It assigns to every vector field $\mathbf{X} = X^i(x)\partial_i$ on M its lifting, the horizontal vector field $\widehat{X}_{\mathrm{hor}} = X^i(x)\partial_i + \gamma_i(x)X^i(x)\widehat{w}$ on $\widehat{M}$. Connection defines derivation $\nabla_{\mathbf{X}}$ on algebra of densities: for $\mathbf{s} = s(x)|Dx|^\lambda$

$$\nabla_{\mathbf{X}}(\mathbf{s}) = \widehat{X}_{\mathrm{hor}}(\mathbf{s}) = \left(X^i(x)\partial_i s(x) + \lambda\gamma_i(x)X^i(x)s(x)\right)|Dx|^\lambda\,. \tag{3}$$

Under changing of local coordinates $x^i \to x^{i'} = x^{i'}(x^i)$ components γ_i of connection are transforming in the following way:

$$\gamma_{i'} = \frac{\partial x^i}{\partial x^{i'}}\left(\gamma_i + \frac{\partial}{\partial x^i}\log\left(\det\frac{\partial x^{j'}}{\partial x^j}\right)\right)\,, \quad (\gamma_i(x)|Dx| = \nabla_i|Dx|)\,.$$

Connection $\gamma_i(x)$ defines divergence $\operatorname{div}_\gamma$ of vector fields on M, which is equal to divergence (2) of horizontal lifting of this vector field: $\operatorname{div}_\gamma\mathbf{X} = \operatorname{div}\widehat{X}_{hor} = \partial_i X^i(x) - \gamma_i(x)X^i(x)$.

Remark 2 *Let $\mathbf{X} = X^i\partial_i$ be a projection on M of a vector field $\widehat{X} = X^i\partial_i + \widehat{w}X^0$ on M, and $\widehat{X}_{\mathrm{hor}}$ be a horizontal lifting of $\mathbf{X}$. Then $\widehat{X} - \widehat{X}_{\mathrm{hor}} = \widehat{w}(X^0 - \gamma_i X^i)$ is a vertical vector field and $X^0 - \gamma_i X^i$ is a scalar field.*

Remark 3 *A volume form $\boldsymbol{\rho} = \rho(x)|Dx|$ on M naturally defines a connection $\gamma_i = -\partial_i\log\rho(x)$. A Riemannian metric $G = g_{ik}dx^i dx^k$ on M naturally defines a volume form $\boldsymbol{\rho} = \sqrt{\det g}|Dx|$. The corresponding connection $\gamma_i = -\Gamma^k_{ik}$, where Γ^i_{km} are Cristoffel symbols of Levi-Civita connection of the metric. In this case $\operatorname{div}_\gamma$ is a standard divergence operator (with respect to a volume form).*

4. Second order operators and Kaluza-Klein reduction

Let $\widehat{\Delta}$ be an arbitrary second order operator on algebra of densities $\mathcal{F}(M)$:

$$\widehat{\Delta} = \underbrace{S^{ik}(x)\partial_i\partial_k + 2\widehat{w}B^i(x)\partial_i + \widehat{w}^2C(x)}_{\text{second order derivatives}} + \underbrace{D^i(x)\partial_i + \widehat{w}E(x)}_{\text{first order derivatives}} + F(x)\,. \tag{4}$$

(As always we consider only operators which do not change weight of densities (see remark 1).)

Principal symbol of this operator is

$$\widehat{S} = \begin{pmatrix} S^{ik} & B^i \\ B^k & C \end{pmatrix}, \qquad \text{(in coordinates } x^i, x^0 = \log t).$$

$\widehat{S}$ is a contravariant symmetric tensor field on the extended manifold $\widehat{M}$. 'Space components' S^{ik} of the tensor field $\widehat{S}$ are components of symmetric contravariant tensor field on M. Operator $\widehat{\Delta}$ defines a pencil of second order operators $\{\Delta_\lambda\}$, $\Delta_\lambda = S^{ik}\partial_i\partial_k + \dots$, and all these operators have the same principal symbol S^{ik}.

Put normalisation condition $F = \widehat{\Delta}(1) = 0$ and consider the operator which is adjoint to operator (4):

$$\widehat{\Delta}^* = \partial_k\partial_i \left(S^{ik}\dots\right) - \widehat{w}^*\partial_i \left(2B^i + (\dots)\right) + (\widehat{w}^*)^2 (C\dots) - \partial_i \left(D^i\dots\right) + \widehat{w}^* E\,, \; (\widehat{w}^* = 1 - \widehat{w})\,.$$

The condition that operator $\widehat{\Delta}$ is self-adjoint, $\widehat{\Delta}^* = \widehat{\Delta}$ implies that

$$\widehat{\Delta} = S^{ik}\partial_i\partial_k + \partial_k S^{ki}\partial_i + (2\widehat{w} - 1)\, B^i\partial_i + \hat{w}\partial_i B^i + \widehat{w}\,(\widehat{w} - 1)\, C\,. \tag{5}$$

Thus self-adjoint second order operator on algebra of densities, which obeys normalisation condition $\widehat{\Delta}(1) = 0$ is uniquely defined by its symbol.

The geometry of operator (5) was studied in detail in articles [2, 3, 4]. Here we present and analyze these results, using Kaluza-Klein-like mechanism.

Kaluza-Klein mechanism defines a connection (gauge field) and Riemannian metric on a base manifold through Riemannian metric on a total space of fibre bundle $\widehat{M} \to M$. Connection, i.e. the distribution of horizontal hyperplanes (subspaces which are transversal to the fibres) is defined by the condition that these hyperplanes are orthogonal to the fibres with respect to Riemannian metric in the bundle space.

One can slightly alter this mechanism. Contravariant tensor field $\widehat{S}$, principal symbol of operator (5) maps 1-forms (covectors) to vectors on $\widehat{M}$. Consider the following Kaluza-Klein-like mechanism: take an arbitrary 1-form Ω on $\widehat{M}$ such that $\Omega(\widehat{w}) \neq 0$, i.e. Ω is proportional to form $dx^0 + \dots$ ($x^0 = \log t$), and the following condition is obeyed: vector field $\widehat{S}\Omega$ is proportional to vertical vector field $\widehat{w}$ ($\widehat{w} = t\frac{\partial}{\partial t} = \frac{\partial}{\partial x^0}$) . This means that for 1-form $\Omega = a(x)(dx^0 - \gamma_k(x)dx^k)$ the following condition holds:

$$\begin{pmatrix} S^{ik} & B^i \\ B^k & C \end{pmatrix} \begin{pmatrix} -\gamma_k \\ 1 \end{pmatrix} \text{ is proportional to vector } \begin{pmatrix} 0 \\ 1 \end{pmatrix}. \tag{6}$$

This condition canonically defines distribution of horizontal hyperplanes in $\widehat{M}$, which are sets of vectors which annihilate the form Ω. (One can take $\widehat{S}\Omega = 0$ in the case if $\widehat{S}$ is degenerate.) Every vector field $\mathbf{X} = X^i(x)(x)\partial_i$ on the base M can be lifted to horizontal vector field which annihilates the connection form Ω (see also equation (3)). This construction works if condition (6) is obeyed, i.e. in the case if the equation

$$S^{ik}(x)\gamma_k(x) = B^i(x) \tag{7}$$

has a solution. In this case second order operator (4) via its principal symbol $\widehat{S}$ defines a connection γ_k. In the case where S^{ik} is non-degenerate, Eq. (7) has a unique solution. In this case operator defines uniquely canonical connection and Riemannian metric on the base.

The field $B^i(x) = S^{ik}(x)\gamma_k(x)$ can be considered as an *upper connection*. It follows from (6) that in this case $C - B^i\gamma_i$ is a scalar and C is related with *Brans-Dicke function*.

In general case if the condition (7) is not obeyed then more detailed analysis shows that $B^i - S^{ik}\Gamma_k$ is a vector field and $C - 2B^i\Gamma_i + S^{ik}\Gamma_i\Gamma_k$ is a scalar, where Γ_i is an arbitrary connection.

The importance of operator (5) is defined by the following uniqueness Theorem:

Journal of Physics: Conference Series **496** (2014) 012032 doi:10.1088/1742-6596/496/1/012032

Theorem 1 *Let Δ be second order operator acting on densities of weight λ_0, where $\lambda_0 \neq 0, 1, 1/2$. Then there exists a unique self-adjoint operator $\widehat{\Delta}$ ($\widehat{\Delta}^* = \widehat{\Delta}$) which obeys the following conditions*

- $\widehat{\Delta}\big|_{\widehat{w}=\lambda_0} = \Delta$,
- *normalisation condition* $\widehat{\Delta}(1) = 0$.

In other words there exists unique self-adjoint normalised pencil of second order operators which passes through a given operator.

This Theorem was formulated and proved in [2] (see also [3]).

Example 4.1 *Consider second order operator $\Delta_{\lambda_0} = \mathcal{L}_\mathbf{X} \circ \mathcal{L}_\mathbf{Y}$, where $\mathbf{X}, \mathbf{Y}$ are arbitrary vector fields, $\mathcal{L}_\mathbf{X}, \mathcal{L}_\mathbf{Y}$ are Lie derivatives of densities of weight λ_0 along vector fields $\mathbf{X}, \mathbf{Y}$ respectively. Construct the following operator on algebra of densities:*

$$\widehat{\Delta} = \frac{1}{2}\left(\widehat{\mathcal{L}}_\mathbf{X}\widehat{\mathcal{L}}_\mathbf{Y} + \widehat{\mathcal{L}}_\mathbf{Y}\widehat{\mathcal{L}}_\mathbf{X}\right) + \frac{1}{2}\left(\frac{2\widehat{w}-1}{2\lambda_0-1}\right)\left(\widehat{\mathcal{L}}_\mathbf{X}\widehat{\mathcal{L}}_\mathbf{Y} - \widehat{\mathcal{L}}_\mathbf{Y}\widehat{\mathcal{L}}_\mathbf{X}\right).$$

This operator is obviously self-adjoint operator and it passes through operator Δ_{λ_0}. One can see that it is equal to

$$\widehat{\Delta} = \widehat{\mathcal{L}}_\mathbf{X}\widehat{\mathcal{L}}_\mathbf{Y} + \left(\frac{\widehat{w}-\lambda_0}{2\lambda_0-1}\right)\widehat{\mathcal{L}}_{[\mathbf{X},\mathbf{Y}]}.$$

Let $\Delta_{\lambda_0} = A^{ij}\partial_i\partial_j + A^i\partial_i + A$ be an arbitrary second order operator acting on space of densities of a given weight λ_0, ($\lambda_0 \neq 0, 1/2, 1$). Then we see that the self-adjoint operator (5) passes via the operator Δ_{λ_0} if for upper connection B^i and Brans-Dicke function C

$$B^i = \frac{A^i - \partial_k A^{ki}}{2\lambda_0 - 1}, \qquad C = \frac{A}{\lambda_0(\lambda_0-1)} - \frac{\partial_i A^i - \partial_i\partial_k A^{ki}}{(\lambda_0-1)(2\lambda_0-1)}.$$

The upper connection B^i is induced by a connection γ_i ($B^i = A^{ik}\gamma_k$) iff equation (7) has a solution (for $S^{ik} = A^{ik}$). The condition (7) defines this special property of second order operators on densities. It is interesting to analyze its geometrical meaning.

5. Thomas bundle and projective geometry

The canonical constructions studied in the previous sections were successfully performed since we consider not the algebra of all functions on extended manifold $\widehat{M}$, but only functions which are quasipolynomial on vertical variable t, since scalar product (1) is not well-defined on algebra of all (smooth) functions on x, t. Nevertheless in general case for fibre bundle $\widehat{M} \to M$ there exists the remarkable construction which assigns to projective class of connections on M the affine connection on $\widehat{M}$. This construction is due to T. Y. Thomas [6] (see also Refs. [7, 8]). The bundle $\widehat{M} \to M$ sometimes is called Thomas bundle. Now we sketch this construction.

We say that two symmetric affine connections ∇ and $\tilde{\nabla}$ on manifold M belong to the same projective class $[\nabla] = [\tilde{\nabla}]$ if

$$\tilde{\nabla} - \nabla = \tilde{\Gamma}^i_{km} - \Gamma^i_{km} = t_k\delta^i_m + t_m\delta^i_k, \qquad (t_i \text{ is covector}),$$

where $\tilde{\Gamma}^i_{km}$ and Γ^i_{km} are Christoffel symbols of connections $\tilde{\nabla}$ and ∇ respectively. Equivalence class of symmetric connections is a projective connection. (A projective connection in particular defines non-parametrised geodesics: two symmetric affine connection belong to the same class iff they have the same non-parameterised geodesics.)

For affine connection ∇ on n-dimensional manifold M with Christoffel symbols Γ^i_{km} one can consider symbols

$$\Pi^i_{km}(\nabla) = \Pi^i_{km} = \Gamma^i_{km} + \frac{1}{n+1}\left(\gamma_k\delta^i_m + \gamma_m\delta^i_k\right), \tag{8}$$

where $\gamma_i = -\Gamma^k_{ik}$ define connection on densities on M (see also remark 3). Two symmetric connections $\nabla, \tilde{\nabla}$ belong to the same projective class iff $\Pi^i_{km}(\nabla) = \Pi^i_{km}(\tilde{\nabla})$.

Let $[\nabla]$ be a projective class of symmetric connections on n-dimensional manifold M. Then Thomas construction assigns to this projective class $[\nabla]$ the symmetric affine connection $\widehat{\nabla}$ on the extended manifold $\widehat{M}$. with following Christoffel symbols

$$\widehat{\Gamma}^i_{km} = \Pi^i_{km}, \widehat{\Gamma}^0_{km} = \frac{1}{n+1}\left(\partial_r\Pi^r_{km} - \Pi^r_{sk}\Pi^s_{rm}\right), \widehat{\Gamma}^i_{k0} = -\frac{\delta^i_k}{n+1}, \widehat{\Gamma}^i_{00} = \widehat{\Gamma}^0_{i0} = 0, \widehat{\Gamma}^0_{00} = -\frac{1}{n+1}.$$

Here Π^i_{km} are symbols (8) corresponding to Christoffel symbols of a connection in the class $[\nabla]$. (We use local coordinates $(x^i, x^0) = (x^i, \log t))$ in the extended space.)

Acknowledgments

This article is based on my talk at the conference "The Modern Physics of Compact Stars and Relativistic Gravity" (Yerevan, Septemebr 2013). I am grateful to organisers of this conference for financial support, and also to Engineering and Physical Sciences Research Council (EPSRC) for financial support via the Mathematics Platform Grant (MAPLE) EP/IO1912X/1. I am grateful to T. Voronov for fruitful discussions.

References

[1] Ovsienko V, Tabachnikov S 2005 *Projective Differential Geometry Old and New. From Schwarzian Derivative to the Cohomology of Diffeomorphism Groups.* (Cambridge University Press).

[2] Khudaverdian H M, Voronov T 2004 On Odd Laplace operators II *In Amer.Math.Soc.Transl.(2)* **212** 179

[3] Khudaverdian H M, Voronov T 2013 Geometry of differential operators of second order, the algebra of densities, and groupoids *J. Geom. Phys.* **64**, 31

[4] Biggs A, Khudaverdian H M 2013 Operator pencil passing through a given operator *J.Math.Phys.* **54** 123503

[5] Khudaverdian H M, Voronov T 2013 Geometric constructions on algebra of densities math-archiv arXiv:1310.0784.

[6] Thomas T Y 1926 A projective theory of affinely connected manifolds *Math. Zeit.* **25**, 723-733

[7] George J 2008 Projective Connections and the Algebra of Densities math-archiv arXiv:0808.2831.

[8] George J 2010 it Differential operators and brackets on projectively connected (super)manifolds. PhD Thesis, University of Manchester.

AUTHOR INDEX